Buildings and Structures under Extreme Loads II

Buildings and Structures under Extreme Loads II

Editors

Chiara Bedon
Flavio Stochino
Daniel Honfi

MDPI • Basel • Beijing • Wuhan • Barcelona • Belgrade • Manchester • Tokyo • Cluj • Tianjin

Editors

Chiara Bedon
Department of Engineering and
Architecture
University of Trieste
Trieste
Italy

Flavio Stochino
Department of Civil
Environmental Engineering and
Architecture
University of Cagliari
Cagliari
Italy

Daniel Honfi
Transport Department
City of Stockholm
Stockholm
Sweden

Editorial Office
MDPI
St. Alban-Anlage 66
4052 Basel, Switzerland

This is a reprint of articles from the Special Issue published online in the open access journal *Applied Sciences* (ISSN 2076-3417) (available at: www.mdpi.com/journal/applsci/special_issues/ Buildings_Structures_Loads).

For citation purposes, cite each article independently as indicated on the article page online and as indicated below:

LastName, A.A.; LastName, B.B.; LastName, C.C. Article Title. *Journal Name* **Year**, *Volume Number*, Page Range.

ISBN 978-3-0365-3872-3 (Hbk)
ISBN 978-3-0365-3871-6 (PDF)

Cover image courtesy of Kimon Maritz

Contents

About the Editors

Chiara Bedon

Chiara is Professor at University of Trieste, Department of Engineering and Architecture, Italy, where she is instructor for the course of 'Structural Analysis'. She received a PhD degree in 2012, and since 2009 she has been involved in international research projects and networks, as an expert member, national delegate, task leader or advisor (JRC-ERNCIP, NATO-SPS, EU-COST, etc.). Editorial Board member for several ISI journals. Top Scientist for research impact in 2019 and 2020 (Stanford std cit indicators), and in the list of Top Italian Scientists (TIS) since 2021 (Engineering >Civil).

Flavio Stochino

Flavio is Assistant Professor of Theory and Design of Structures at University of Cagliari (Italy), Department of Civil Environmental Engineering and Architecture. He obtained a Ph.D. in Structural Engineering at University of Cagliari in 2013. Then, he moved to the University of Sassari (Italy), working on computational mechanics problems as a Post Doc researcher until 2016. In that year, he became Dresden Junior Fellow at TU Dresden (Germany) where he worked on some advanced computational techniques for fracture mechanics. In 2017, he won the PhD ITalents call, funded by CRUI and Confindustria.

Daniel Honfi

Daniel is a structural engineer dealing with infrastructure maintenance planning at the Transport Department of the City of Stockholm. His research activities cover the topics of structural engineering, structural safety, structural serviceability, structural robustness and infrastructure resilience.

Preface to "Buildings and Structures under Extreme Loads II"

Exceptional loads on buildings and structures are known to take origin and manifest from different causes, like natural hazards and possible high-strain dynamic effects, human-made attacks and impact issues for load-bearing components, possible accidents, and even unfavorable/extreme operational conditions. All these aspects can be critical for specific structural typologies and/or materials that are particularly sensitive to external conditions. In this regard, dedicated analysis methods and performance indicators are required for the design and maintenance under the expected lifetime. Typical issues and challenges can find huge efforts and clarification in research studies, which are able to address with experiments and/or numerical analyses the expected performance and capacity of a given structural system, with respect to demands. Accordingly, especially for existing structures or strategic buildings, the need for retrofit or mitigation of adverse effects suggests the definition of optimal and safe use of innovative materials, techniques, and procedures. This Special Issue follows the first successful edition and confirms the need of continuous research efforts in support of building design under extreme loads, with a list of original research papers focused on various key aspects of structural performance assessment for buildings and systems under exceptional design actions and operational conditions.

Chiara Bedon, Flavio Stochino, and Daniel Honfi
Editors

Editorial

Special Issue on "Buildings and Structures under Extreme Loads II"

Chiara Bedon [1,*], Flavio Stochino [2] and Daniel Honfi [3]

1 Department of Engineering and Architecture, University of Trieste, 34127 Trieste, Italy
2 Department of Civil Environmental Engineering and Architecture, University of Cagliari, 09123 Cagliari, Italy; fstochino@unica.it
3 Monitoring and Analyses of Existing Structures, Ramboll, 2300 Copenhagen, Denmark; dhon@ramboll.dk
* Correspondence: chiara.bedon@dia.units.it; Tel.: +39-040-558-3837

Citation: Bedon, C.; Stochino, F.; Honfi, D. Special Issue on "Buildings and Structures under Extreme Loads II". *Appl. Sci.* **2022**, *12*, 2660. https://doi.org/10.3390/app12052660

Received: 23 December 2021
Accepted: 3 March 2022
Published: 4 March 2022

Publisher's Note: MDPI stays neutral with regard to jurisdictional claims in published maps and institutional affiliations.

1. Introduction

Exceptional loads on buildings and structures are known to take origin and manifest from different causes, like natural hazards and possible high-strain dynamic effects, human-made attacks and impact issues for load-bearing components, possible accidents, and even unfavorable/extreme operational conditions. All these aspects can be critical for specific structural typologies and/or materials that are particularly sensitive to external conditions. In this regard, dedicated analysis methods and performance indicators are required for the design and maintenance under the expected lifetime. Typical issues and challenges can find huge efforts and clarification in research studies, which are able to address with experiments and/or numerical analyses the expected performance and capacity of a given structural system, with respect to demands. Accordingly, especially for existing structures or strategic buildings, the need for retrofit or mitigation of adverse effects suggests the definition of optimal and safe use of innovative materials, techniques, and procedures. This Special Issue follows the first successful edition [1] and confirms the need of continuous research efforts in support of building design under extreme loads, with 13 original research papers focused on various key aspects of structural performance assessment for buildings and systems under exceptional design actions and operational conditions.

2. Contents

Generally speaking, it is known that joints and connections represent a first critical component in buildings and systems. Depending on their actual mechanical features and capacities, as well as sensitivity to external conditions, the structural performance assessment of the building as a whole can be strongly affected. In this regard, a first set of research papers is dedicated to the experimental, numerical or hybrid analysis of special joints for constructional systems, under a variety of mechanical properties and loading conditions [2–6]. For timber structures, for example, it is known that both material and geometrical parameters for connections arrangement can have severe effects on mechanical performances, in the same way of moisture, loading protocol, etc. Experimental studies are thus reported in [2] for Bonded-in-Rod (BIR) connections under various operational conditions for bonding adhesives, while the study in [3] presents a critical analysis of performance indicators for timber-to-timber screwed connections with various configurations. As far as connections parameters are taken into account for steel structures are considered, several calculation approaches can be notoriously taken into account in support of design [4]. The study presented in [5] and validated to literature experiments for bolted connections proves that artificial intelligence can offer useful feedback and support for component design optimization. The structural performance of slab-to-column connections in concrete frames can be especially critical when subjected to blast loads. The investigation summarized in [6], in this regard, gives recommendations with the support of efficient and realistic finite element numerical models. A prediction equation for the blast-resistant performance of slab-column joints under blast is also presented in [6], as obtained based on additional numerical analyses.

Blast loads, impact events and earthquakes certainly need a special attention for design and verification of critical infrastructures. Besides, critical situations for buildings and components may derive from a multitude of scenarios. First, in any case, it is required to properly describe the structural dynamic problem with reliable mathematical models [7]. Successively, the use of more complex and advanced numerical tools can provide additional important indications about the actual performance and capacity.

The study in [8] focuses on cable-stayed bridges under blast accidents resulting from explosions produced by vehicle collisions or terrorist attacks. An efficient procedure to assess structural blast-resistance performance of key structural components in cable-stayed bridges is suggested, based on a numerical analysis approach. Impact test methods are presented in [9] to assess the structural capacity of bridge piers, given that the stability of bridge substructures is closely related to safety problems. Often, however, safety inspections are mostly focused on materials and structural issues problems, and there are no recommendations for quantitative analysis of substructures. When existing bridge structures are subjected to earthquakes, finally, special retrofit interventions should be needed for piers and connection details. An experimental study is presented in [10], based on six small scale specimens reconstructed to describe existing bridge piers under seismic (laboratory) conditions.

For bridges, but also for structures in general and building components, there is a strong correlation of mechanical capacities and natural characteristics, such as wind phenomena, soil effects, and ambient. The study in [11] proves that uncertainty due to the randomness of wind tunnel wind flows is a critical aspect for the design of high-rise buildings under wind. This uncertainty should be taken into account during experiments and output processing by examining the cumulative probability trends and assuming a reliable level of confidence for the estimated crucial magnitudes, such as the modal shapes, frequencies and damping ratio parameters.

The investigation reported in [12] presents a field analysis for damage issues incurred in a historical cathedral subjected to foundation strengthening works. Considering the historic data related to the cathedral, the on-site registered damages, the results of geodetic measurements and the experiences gained from rehabilitation works, the authors conclude that soil-structure interactions for historical buildings represent a very complex, multidisciplinary problem. Damage propagation and patterns in the cathedral facade are monitored in support of intervention design and optimization.

The study in [13] starts from the analysis of the failure mechanism for a roof, with steel truss construction, of a factory building in the northwestern part of Turkey. The failure occurred under hefty weather conditions including lightning strikes, heavy rain, and fierce winds. In order to interpret the reason for the failure, the effects of different combinations of factors on the design and dimensioning of the roof are studied by authors, with the support of finite element analysis and on-site investigations.

Finally, the investigation reported in [14] proves that creep properties of Balau wood timber cross-arms reinforced with additional braced arms can be significantly reduced, compared to existing design wooden cross-arms. Accordingly, the implementation of bracing system in cross-arm structures displays a more stable stress independent material exponent in members. This results in improved dimensional structural stability in service life.

Funding: This research received no external funding.

Institutional Review Board Statement: Not applicable.

Informed Consent Statement: Not applicable.

Acknowledgments: This second edition of the Special Issue would not be possible without the contributions of various talented authors, hardworking and professional reviewers, and dedicated editorial team members of the *Applied Sciences* journal. The guest editors would like to express their sincere gratefulness to all the involved scientists, both authors and reviewers, for the valuable contribution to this collection. Finally, a special thanks goes to the editorial team of *Applied Sciences*.

Conflicts of Interest: The authors declare no conflict of interest.

References

1. Bedon, C.; Stochino, F.; Honfi, D. Special Issue on Buildings and Structures under Extreme Loads. *Appl. Sci.* **2020**, *10*, 5676. [CrossRef]
2. Barbalić, J.; Rajčić, V.; Bedon, C.; Budzik, M.K. Short-Term Analysis of Adhesive Types and Bonding Mistakes on Bonded-in-Rod (BiR) Connections for Timber Structures. *Appl. Sci.* **2021**, *11*, 2665. [CrossRef]
3. Bedon, C.; Sciomenta, M.; Fragiacomo, M. Mechanical Characterization of Timber-to-Timber Composite (TTC) Joints with Self-Tapping Screws in a Standard Push-Out Setup. *Appl. Sci.* **2020**, *10*, 6534. [CrossRef]
4. Faridmehr, I.; Hajmohammadian Baghban, M. An Overview of Progressive Collapse Behavior of Steel Beam-to-Column Connections. *Appl. Sci.* **2020**, *10*, 6003. [CrossRef]
5. Faridmehr, I.; Nikoo, M.; Pucinotti, R.; Bedon, C. Application of Component-Based Mechanical Models and Artificial Intelligence to Bolted Beam-to-Column Connections. *Appl. Sci.* **2021**, *11*, 2297. [CrossRef]
6. Lim, K.M.; Yoo, D.G.; Lee, B.Y.; Lee, J.H. Prediction of Damage Level of Slab-Column Joints under Blast Load. *Appl. Sci.* **2020**, *10*, 5837. [CrossRef]
7. Momeni, M.; Beni, M.R.; Bedon, C.; Najafgholipour, M.A.; Dehghan, S.M.; JavidSharifi, B.; Hadianfard, M.A. Dynamic Response Analysis of Structures Using Legendre–Galerkin Matrix Method. *Appl. Sci.* **2021**, *11*, 9307. [CrossRef]
8. Lee, J.; Choi, K.; Chung, C. Numerical Analysis-Based Blast Resistance Performance Assessment of Cable-Stayed Bridge Components Subjected to Blast Loads. *Appl. Sci.* **2020**, *10*, 8511. [CrossRef]
9. Lee, M.; Yoo, M.; Jung, H.-S.; Kim, K.H.; Lee, I.-W. Study on Dynamic Behavior of Bridge Pier by Impact Load Test Considering Scour. *Appl. Sci.* **2020**, *10*, 6741. [CrossRef]
10. Kim, J.H.; Kim, I.-H.; Lee, J.H. Experimental Study on the Behavior of Existing Reinforced Concrete Multi-Column Piers under Earthquake Loading. *Appl. Sci.* **2021**, *11*, 2652. [CrossRef]
11. Rizzo, F. Investigation of the Time Dependence of Wind-Induced Aeroelastic Response on a Scale Model of a High-Rise Building. *Appl. Sci.* **2021**, *11*, 3315. [CrossRef]
12. Santrač, P.; Grković, S.; Kukaras, D.; Đuric, N.; Svilar, M. Case Study—An Extreme Example of Soil–Structure Interaction and the Damage Caused by Works on Foundation Strengthening. *Appl. Sci.* **2021**, *11*, 5201. [CrossRef]
13. Tüfekci, M.; Tüfekci, E.; Dikicioğlu, A. Numerical Investigation of the Collapse of a Steel Truss Roof and a Probable Reason of Failure. *Appl. Sci.* **2020**, *10*, 7769. [CrossRef]
14. Asyraf, M.R.M.; Ishak, M.R.; Sapuan, S.M.; Yidris, N. Influence of Additional Bracing Arms as Reinforcement Members in Wooden Timber Cross-Arms on Their Long-Term Creep Responses and Properties. *Appl. Sci.* **2021**, *11*, 2061. [CrossRef]

applied sciences

Article

Dynamic Response Analysis of Structures Using Legendre–Galerkin Matrix Method

Mohammad Momeni [1,*], Mohsen Riahi Beni [2], Chiara Bedon [3,*], Mohammad Amir Najafgholipour [1], Seyed Mehdi Dehghan [1], Behtash JavidSharifi [1] and Mohammad Ali Hadianfard [1]

[1] Department of Civil and Environmental Engineering, Shiraz University of Technology, Shiraz 7155713876, Iran; najafgholipour@sutech.ac.ir (M.A.N.); smdehghan@sutech.ac.ir (S.M.D.); b.javidsharifi@sutech.ac.ir (B.J.); hadianfard@sutech.ac.ir (M.A.H.)

[2] Department of Mathematics, Higher Education Complex of Saravan, Saravan 9951634145, Iran; m.riahi@saravan.ac.ir

[3] Department of Engineering and Architecture, University of Trieste, 34127 Trieste, Italy

* Correspondence: m.momeni@sutech.ac.ir (M.M.); chiara.bedon@dia.units.it (C.B.)

Abstract: The solution of the motion equation for a structural system under prescribed loading and the prediction of the induced accelerations, velocities, and displacements is of special importance in structural engineering applications. In most cases, however, it is impossible to propose an exact analytical solution, as in the case of systems subjected to stochastic input motions or forces. This is also the case of non-linear systems, where numerical approaches shall be taken into account to handle the governing differential equations. The Legendre–Galerkin matrix (LGM) method, in this regard, is one of the basic approaches to solving systems of differential equations. As a spectral method, it estimates the system response as a set of polynomials. Using Legendre's orthogonal basis and considering Galerkin's method, this approach transforms the governing differential equation of a system into algebraic polynomials and then solves the acquired equations which eventually yield the problem solution. In this paper, the LGM method is used to solve the motion equations of single-degree (SDOF) and multi-degree-of-freedom (MDOF) structural systems. The obtained outputs are compared with methods of exact solution (when available), or with the numerical step-by-step linear Newmark-β method. The presented results show that the LGM method offers outstanding accuracy.

Keywords: differential equation of motion; Legendre–Galerkin matrix (LGM) method; algebraic polynomials; single degree of freedom (SDOF); multi degree of freedom (MDOF)

Citation: Momeni, M.; Riahi Beni, M.; Bedon, C.; Najafgholipour, M.A.; Dehghan, S.M.; JavidSharifi, B.; Hadianfard, M.A. Dynamic Response Analysis of Structures Using Legendre–Galerkin Matrix Method. *Appl. Sci.* **2021**, *11*, 9307. https://doi.org/10.3390/app11199307

Academic Editor: José A. F. O. Correia

Received: 1 August 2021
Accepted: 29 September 2021
Published: 7 October 2021

Publisher's Note: MDPI stays neutral with regard to jurisdictional claims in published maps and institutional affiliations.

1. Introduction and State-of-Art

Most structural systems in civil engineering applications, as known, are either discrete or can be estimated as discrete equivalents. The dynamic equation of motion of a continuous system can be hence handled as a discrete system. The advantage of this assumption is that the solution of fundamental equations of motion of discrete systems—which is of utmost importance for engineering applications—can be efficiently calculated to predict the responses of structural systems. Among others, let us consider the following well-known second-order differential equation with given initial conditions:

$$\begin{cases} \ddot{y}(t) = f(t, y(t), \dot{y}(t)), & t_0 < t \leq T, \\ y(t_0) = y_0, & \dot{y}(t_0) = \dot{y}_0. \end{cases} \tag{1}$$

Equation (1) is commonly used to express the governing equation of mechanical vibrations, quantum dynamic calculations, dynamic equilibrium of structures, etc. Employing and developing new numerical methods for approaching the exact solution is of great importance.

In structural dynamics, algorithms of direct time integration are usually used to obtain the solution of discrete temporal equations of motion at selected time steps [1]. In the past, to this aim, different integration algorithms in the time domain have been developed using various methods. The same algorithms are widely used to solve the equation of systems under dynamic loading. Several well-known algorithms have been presented by researchers, among which the Newmark-β [2], Wilson-Theta [3], Runge-Kutta [4], etc., can be pointed out. Detailed explanations can be found in structural dynamics textbooks [1,5,6].

Since 1994, the Chebyshev polynomials [7], Legendre polynomials [8], Bessel polynomials [9], Hermite polynomials [10], Laguerre polynomials [11], and matrix methods have been used in many research studies to solve linear and nonlinear equations with high orders including partial differential equations, hyperbolic partial differential equations, delay equations, integral and integro-differential equations, SDOF and MDOF systems, etc. One of the approaches to find the solution to an initial value problem is taking semi-analytical procedures such as the differential transform method (DTM) [12–23]. The nature of dynamic equations of motion of SDOF systems makes them differential equations with initial values; hence, DTM has been also applied to solve non-linear SDOF problems [24–33].

Alternatively, Equation (1) can also be solved through spectral methods of discretization [34–36], which are strategic for the numerical solution of differential equations. The main advantage of these methods lies in their accuracy for a given number of unknowns. For smooth problems with simple geometries, these methods offer exponential rates of convergence/spectral accuracy. In contrast, methods such as finite difference and finite element yield only algebraic convergence rates. Three spectral methods, namely the Galerkin [37], Collocation [38], and Tau [39] are extensively used in the literature. Spectral methods are preferable in numerical solutions of partial differential equations due to their high-order accuracy [40,41]. The Standard Spectral and Galerkin methods have been extensively investigated to handle different types of problems [42–45]. Several numerical methods have been also developed to solve different types of differential equations. Some of these methods can be used to solve more practical equations. These methods include sparse multiscale representation of the Galerkin method [46], spectral collocation method [47], Jacob spectral method [48], and many others. Recently, Erfanian et al. [49] developed a new method for solving two-dimensional nonlinear Volterra integral equations, based on the use of rationalized Haar functions in the complex plane. Bernoulli Galerkin matrix method has been also proposed by Hesameddini and Riahi [50] for solving the system of Volterra-Fredholm integro-differential equations. A new hybrid orthonormal Bernstein and improved block-pulse functions method has been studied by Ramadan and Osheba [51] for solving mathematical physics and engineering problems.

In recent years, researchers have developed a number of numerical methods to find the solution for such equations. In this regard, matrix methods including the Euler method [52], Bernoulli collocation method [53], hybrid Legendre block-pulse function method [54], least squares method [55], Bessel colocation method [56], etc., have received much attention since 2010. Given that the governing equations of natural phenomena are usually nonlinear and complex, so the chosen method of solving must be commensurate with the problem's complexity and its dimensions. A nonlinear differential equation system consists of several nonlinear equations that solving such a system requires numerical methods. The basis of matrix methods is the expression of each of the functions in the problem based on selected polynomials. After selecting the type of polynomials and expressing each of the functions, the differential equation system becomes a system of linear equations with several unknown coefficients, which can be solved by Galerkin and collocation methods.

In this paper, the Legendre–Galerkin matrix (LGM) method is used to solve the equations of motion of different SDOF and multi-degree-of-freedom (MDOF) structural systems. The results are compared with those from the exact solution (when available), or from the numerical step-by-step Newmark-β method with linear acceleration. The accuracy of the LGM formulation is thus highlighted in the discussion of comparative calculations.

2. Basics for SDOF and MDOF Systems

In the dynamic analysis of structures, from a practical point of view, a real structure is defined as a system with infinite degrees of freedom that can be modeled as a discrete system with SDOF or MDOF in a finite element approach. In many cases, these simple models include complex information of the real structure and are able to simulate the behavior of real structure with a good level of accuracy and high calculation efficiency. As an example, in blast-resistant design of structures, the basis of current books and codes and also simplified engineering tools have been established based on equivalent SDOF models of real structures [57–59].

2.1. Governing Equation for SDOF Systems

A structure with one degree of freedom with mass m, stiffness k, and damping c is assumed to be under dynamic load $P(t) = P_0 \sin(\Omega t)$ with excitation frequency ς (Figure 1).

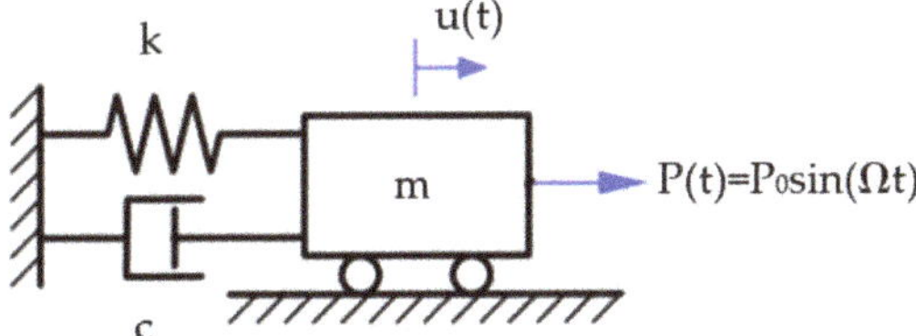

Figure 1. SDOF structure subjected to dynamic load P(t).

The governing equation of this system is [1]:

$$m\ddot{u}(t) + c\dot{u}(t) + ku(t) = P_0 \sin(\Omega t) \tag{2}$$

where Equation (2) is written with similar connotations to Equation (1). The exact solution to the motion equation of a SDOF system under harmonic load $P_0 \sin(\Omega t)$ with initial conditions $u(0) = u_0$ and $\dot{u}(0) = \dot{u}_0$ is the sum of the displacements of the transient state of the system ($u_{transient}$) and the steady state ($u_{steady\,state}$), and is expressed as [5]:

$$u_{total} = u_{transient} + u_{steady\,state}, \tag{3}$$

in which $u_{transient}$ and $u_{steady\,state}$ are defined as:

$$u_{transient} = e^{-\xi\omega t}\left(u_0 \cos\omega_d t + \frac{\dot{u}_0 + u_0\omega\xi}{\omega_d}\sin\omega_d t\right) + \frac{P_0}{k}\frac{e^{-\xi\omega t}}{(1-\beta^2)^2+(2\xi\beta)^2}\left[2\xi\beta\cos\omega_d t + \frac{\omega}{\omega_d}\{2\beta\xi^2 - \beta(1-\beta^2)\}\sin\omega_d t\right], \tag{4}$$

and

$$u_{steady\,state} = \frac{P_0}{k}\frac{1}{(1-\beta^2)^2+(2\xi\beta)^2}\left\{(1-\beta^2)\sin\Omega t - 2\xi\beta\cos\Omega t\right\} \tag{5}$$

where ξ and $\omega = \sqrt{\frac{k}{m}}$ are the damping ratio and natural frequency of the system, respectively, and $\omega_d = \omega\sqrt{1-\xi^2}$ indicates the damped frequency. Finally, $\beta = \frac{\Omega}{\omega}$ is the ratio of the excitation frequency to the system natural frequency.

2.2. Linear Newmark-β Method

One of the most efficient methods of handling equations of motion of SDOF and MDOF systems has been proposed by Newmark [2]. In 1959, Newmark also presented a set of step-by-step numerical equations that are entirely known as the Newmark-β method [5]. In the present study, the linear Newmark-β method is taken into account from [5] to verify the results of the newly developed LGM method.

In order to solve the equation of motion of a MDOF system using the linear Newmark-β method, it is important to remind that the following steps should be generally followed:

1. Determination of the initial conditions, $\{u_0\}$, $\{\dot{u}_0\}$ and $\{\ddot{u}_0\}$, where the value of $\{\ddot{u}_0\}$ is given by:

$$\ddot{u}_0 = \frac{p_0 - C\dot{u}_0 - Ku_0}{M} \tag{6}$$

2. Determination of Δt (assumed time step, e.g., 0.005)
3. Determination of the effective stiffness matrix:

$$\hat{k} = k + \frac{\gamma}{\beta\Delta t}C + \frac{1}{\beta(\Delta t)^2}M \tag{7}$$

where $\beta = 1/6$ and $\gamma = 1/2$ for linear acceleration method.

4. Calculation of a and b factors:

$$\begin{cases} a = \frac{1}{\beta\Delta t}M + \frac{\gamma}{\beta}C \\ b = \frac{1}{2\beta}M + \Delta t\left(\frac{\gamma}{2\beta} - 1\right)C \end{cases} \tag{8}$$

5. Calculation of the effective load:

$$\Delta\hat{p}_i = \Delta p_i + a\dot{u}_i + b\ddot{u}_i \tag{9}$$

6. Calculation of displacement, velocity, and acceleration:

$$\begin{cases} u_{i+1} = u_i + \Delta u_i, \\ \dot{u}_{i+1} = \dot{u}_i + \Delta\dot{u}_i, \\ \ddot{u}_{i+1} = \ddot{u}_i + \Delta\ddot{u}_i, \end{cases} \tag{10}$$

where,

$$\Delta u_i = \frac{\Delta\hat{p}_i}{\hat{K}} \tag{11}$$

$$\Delta\dot{u}_i = \frac{\gamma}{\beta\Delta t}\Delta u_i - \frac{\gamma}{\beta}\dot{u}_i + \Delta t\left(1 - \frac{\gamma}{2\beta}\right)\ddot{u}_i \tag{12}$$

$$\Delta\ddot{u}_i = \frac{\gamma}{\beta(\Delta t)^2}\Delta u_i - \frac{1}{\beta\Delta t}\dot{u}_i - \frac{1}{2\beta}\ddot{u}_i. \tag{13}$$

3. Legendre–Galerkin Matrix Method

Legendre polynomials are among the most important functions in the function approximation theory. The LGM method is highly successful in approximating different functions which mainly stems from the orthogonality of Legendre basis components. This orthogonality makes it easy to find the unknown coefficients of the problem. Another reason for using these basis components is their weight. The weight function will not be a problem for the calculation of the integrals of the Galerkin method and hence the operational matrices of differentiations and other existing functions in the problem can be easily found.

The Galerkin method is an efficient and easy-to-implement approach for solution of the equations of motion of MDOF systems:

$$\sum_{j=1}^{f}\left\{m_{ij}\ddot{y}_j(t) + c_{ij}\dot{y}_j(t) + k_{ij}y_j(t)\right\} = p_i(t), \qquad i = 1, 2, \ldots f \qquad t \in [0, t_1], \tag{14}$$

under the following initial conditions:

$$\begin{cases} y_i(0) = \lambda_i, \\ y'_i(0) = \gamma_i, \end{cases} \quad for \quad i = 1, 2, \ldots, f. \tag{15}$$

In Equation (14), m_{ij}, c_{ij} and k_{ij} are, respectively, components of mass, damping and stiffness matrices. $p_i(t)$ is the load prescribed at the i-th degree of freedom and λ_i and γ_i are, respectively, the initial displacement and velocity applied to the i-th degree of freedom. y_i is the displacement of the i-th degree of freedom which is the unknown of the problem, and t represents time.

3.1. Approximation of the Function Using Shifted Legendre Polynomials

- Legendre polynomials: introduced by $L_m(t)$, the Legendre polynomials are defined in the interval $[-1, 1]$ and can be obtained using the following recursive equations:

$$\begin{aligned} &L_0(t) = 1, \\ &L_1(t) = t, \\ &L_2(t) = \tfrac{3}{2}t^2 - \tfrac{1}{2}, \\ &\vdots \\ &L_{m+1}(t) = \tfrac{2m+1}{m+1}t\, L_m(t) - \tfrac{m}{m+1} L_{m-1}(t). \end{aligned} \tag{16}$$

For further application of these polynomials, they are shifted to the interval $[0, L]$ with the change of variable $\frac{2}{L}t - 1$. Therefore, these shifted polynomials which are shown by $L_m^*(t)$ are represented as the following:

$$L_m^*(t) = L_m\!\left(\frac{2}{L}t - 1\right) \quad t \in [0, L]. \tag{17}$$

- Inner product of two functions: the inner product of two functions $f(t)$ and $g(t)$ is represented by $\langle f(t), g(t) \rangle$. If the functions are known and continuous in the interval $[0, b]$, then:

$$\langle f(t), g(t) \rangle = \int_a^b f(t)g(t)w(t)dt. \tag{18}$$

- Orthogonality of two functions: two functions $f(t)$ and $g(t)$ are orthogonal with respect to the weight function $w(t)$ on $[a, b]$ if:

$$\langle f(t), g(t) \rangle = 0. \tag{19}$$

The shifted Legendre polynomials are orthogonal with respect to the weight function $w(t) = 1$ in the interval $[0, L]$. This means that:

$$\int_a^b L_m^*(t) L_n^*(t) dt = \begin{cases} 0 & m \neq n, \\ \frac{L}{2m+1} & m = n. \end{cases} \tag{20}$$

Orthogonality of the functions has wide application in the function approximation theory. Any continuous function such as $f(x)$ can be approximated in the interval $[0, L]$ using these polynomials as below:

$$f(t) = \sum_{m=0}^{\infty} f_m L_m^*(t), \tag{21}$$

where f_m can be obtained from:

$$f_m = \frac{\langle f(t), L_m^*(t) \rangle}{\langle L_m^*(t), L_m^*(t) \rangle} = \frac{2m+1}{L} \int_0^L f(t) L_m^*(t) dt. \tag{22}$$

Only the first $N+1$ terms of Equation (21) are used in practice. Therefore:

$$f(t) = \sum_{m=0}^{N} f_m L_m^*(t) = \mathbf{F}^T \boldsymbol{\phi}_N(t), \tag{23}$$

where $\mathbf{F}$ and $\boldsymbol{\phi}_N(t)$ are factors of the shifted Legendre vectors expressed as:

$$\mathbf{F} = [f_0\ f_1\ \dots\ f_N]^T, \tag{24}$$

$$\boldsymbol{\phi}_N(t) = [L_0^*(t)\ L_1^*(t)\ \dots\ L_N^*(t)]^T. \tag{25}$$

3.2. Expression of the LGM Method

Assuming that Equation (14) has a unique solution under the initial conditions in Equation (15), the goal is to find an approximate analytical solution to Equation (14) using the discretized Legendre series as below:

$$y_{i,N}(t) = \sum_{m=0}^{N} a_{i,m} L_m^*(t) = \mathbf{A}_{i,N}^T \boldsymbol{\phi}_N(t) , \tag{26}$$

where,

$$\begin{aligned}\mathbf{A}_{i,N} &= [a_{i,0}\ a_{i,1}\ \dots\ a_{i,N}]^T, \\ \boldsymbol{\phi}_N(t) &= [L_0^*(t)\ L_1^*(t)\ \dots\ L_N^*(t)].\end{aligned} \tag{27}$$

The matrix form of the derivative vector will be:

$$\frac{d}{dt} \boldsymbol{\phi}_N(t) = \mathbf{D} \boldsymbol{\phi}_N(t), \tag{28}$$

where $\mathbf{D}$ is a matrix of dimensions $(N+1) \times (N+1)$ and is obtained as below:

$$\mathbf{D} = [d_{ij}] = \begin{cases} \frac{2(2j+1)}{L}, & for \quad j = i - k, \quad \begin{cases} k = 1,3,\dots,N & \textit{if N is odd number,} \\ k = 1,3,\dots,N-1 & \textit{if N is even number,} \end{cases} \\ 0, & \textit{otherwise.} \end{cases} \tag{29}$$

For example, for different values of N, the following equations are obtained:

$$If \quad N = 1 \quad \rightarrow \quad \mathbf{D} = \begin{bmatrix} 0 & 0 \\ \frac{2}{L} & 0 \end{bmatrix}$$

$$If \quad N = 2 \quad \rightarrow \quad \mathbf{D} = \begin{bmatrix} 0 & 0 & 0 \\ \frac{2}{L} & 0 & 0 \\ 0 & \frac{6}{L} & 0 \end{bmatrix}$$

$$If \quad N = 3 \quad \rightarrow \quad \mathbf{D} = \begin{bmatrix} 0 & 0 & 0 & 0 \\ \frac{2}{L} & 0 & 0 & 0 \\ 0 & \frac{6}{L} & 0 & 0 \\ \frac{2}{L} & 0 & \frac{10}{L} & 0 \end{bmatrix}$$

Using Equation (27), the k-th derivative of $\boldsymbol{\phi}_N(t)$ vector is:

$$\frac{d^k}{dt^k} \boldsymbol{\phi}_N(t) = \mathbf{D}^k \boldsymbol{\phi}_N(t), \tag{30}$$

Therefore, by replacing $k = 1$ and 2 in Equation (29), and then combining the results with Equation (25), one can write:

$$\dot{y}_{i,N}(t) = \sum_{m=0}^{N} a_{i,m} L_m^*(t) = \mathbf{A}_{i,N}^T \mathbf{D} \boldsymbol{\phi}_N(t), \tag{31}$$

and

$$\ddot{y}_{i,N}(t) = \sum_{m=0}^{N} a_{i,m} L_m^*(t) = \mathbf{A}_{i,N}^T \mathbf{D}^2 \boldsymbol{\phi}_N(t). \tag{32}$$

Moreover, the functions $p_i(t)$ can be shown in matrix form as follows:

$$p_i(t) = \sum_{m=0}^{N} p_{i,m} L_m^*(t) = \mathbf{P}_{i,N}^T \boldsymbol{\phi}_N(t), \tag{33}$$

where $\mathbf{P}_{i,N}^T$ is a (N+1) dimensional vector with the following expression:

$$\mathbf{P}_{i,N} = [p_{i,0} \; p_{i,1} \; \cdots \; p_{i,N}]^T, \tag{34}$$

where,

$$p_{i,m} = \frac{2m+1}{L} \int_0^L p_i(t) L_m^*(t) dt. \tag{35}$$

By introducing Equations (25) and (30)–(32) in Equation (14), it is thus possible to obtain:

$$\sum_{j=1}^{f} \left\{ m_{ij} \mathbf{A}_{j,N}^T \mathbf{D}^2 + c_{ij} \mathbf{A}_{j,N}^T \mathbf{D} + k_{ij} \mathbf{A}_{j,N}^T \right\} \boldsymbol{\phi}_N(t) = \mathbf{P}_{i,N}^T \boldsymbol{\phi}_N(t) \tag{36}$$

Now, let us assume that:

$$\mathbf{Y}_N = \mathbf{A}\boldsymbol{\phi}_N(t) = \begin{bmatrix} y_{1,N} \\ y_{2,N} \\ \vdots \\ y_{f,N} \end{bmatrix}, \quad \mathbf{A} = \begin{bmatrix} \mathbf{A}_{1,N}^T \\ \mathbf{A}_{2,N}^T \\ \vdots \\ \mathbf{A}_{f,N}^T \end{bmatrix}, \quad \mathbf{P} = \begin{bmatrix} \mathbf{P}_{1,N}^T \\ \mathbf{P}_{2,N}^T \\ \vdots \\ \mathbf{P}_{f,N}^T \end{bmatrix}, \tag{37}$$

$$\mathbf{M} = [m_{ij}], \quad \mathbf{C} = [c_{ij}], \quad \mathbf{K} = [k_{ij}], \quad i,j = 1,2,\ldots,f,$$

where $\mathbf{Y}_N$ is an f dimensional vector, $\mathbf{A}$ and $\mathbf{P}$ are $f \times (N+1)$ dimensional and $\mathbf{M}, \mathbf{C}$ and $\mathbf{K}$ are the known $f \times f$ dimensional matrices, respectively.

Using Equations (35) and (36), the matrix form for approximation of Equation (14) can be obtained as:

$$(\mathbf{MAD}^2 + \mathbf{CAD} + \mathbf{KA})\boldsymbol{\phi}_N(t) = \mathbf{P}\boldsymbol{\phi}_N(t), \tag{38}$$

or

$$\mathbf{U}\boldsymbol{\phi}_N(t) = \mathbf{P}\boldsymbol{\phi}_N(t), \tag{39}$$

where,

$$\mathbf{U} = \mathbf{MAD}^2 + \mathbf{CAD} + \mathbf{KA}.$$

Equation (38) can be solved using different methods, namely the direct method (i.e., omitting $\boldsymbol{\phi}_N(t)$ from both sides of the equation and assuming it to vanish), the collocation method, or the Galerkin method.

In the present paper, the Galerkin method is selected to solve the equation. This method tries to minimize the error toward zero with the inner product of the equations in Legendre basis $L_m^*(t)$. Consequently:

$$\langle \mathbf{U}\boldsymbol{\phi}_N(t), L_m^*(t) \rangle = \langle \mathbf{P}\boldsymbol{\phi}_N(t), L_m^*(t) \rangle, \quad m = 0,1,\ldots,N. \tag{40}$$

The above equation is an algebraic system of linear equations with $N+1$ equations and $N+1$ unknowns that can be easily solved. It is needed, however, to apply the boundary conditions to the problem, as also in accordance with Equation (15). Following Equations (25) and (30), it is:

$$\begin{aligned} \mathbf{A}\boldsymbol{\phi}_N(0) &= \boldsymbol{\Lambda}, \\ \mathbf{AD}\boldsymbol{\phi}_N(0) &= \boldsymbol{\Pi}, \end{aligned} \tag{41}$$

where,

$$\mathbf{\Lambda} = [\lambda_1 \ \lambda_2 \ \dots \ \lambda_f]^T, \quad \mathbf{\Pi} = [\gamma_1 \ \gamma_2 \ \dots \ \gamma_f]^T.$$

Finally, in order to find the response of the system in Equation (14) under the initial conditions from Equation (15), by replacing $2f$ of the rows of Equation (40) with Equation (38), a system of algebraic equations with $(N + 1 - 2f)$ equations (Equation (38)) and $2f$ equations (Equation (40)) is created. Their solution yields the analytic approximation of the original problem. The last rows of Equation (38) are usually replaced with those equations from Equation (40); however, this is not always obligatory and the rows which will cause the system of equations to be singular can be alternatively replaced.

3.3. Solution of a Calculation Example

Consider the following equation as a SDOF structure without damping under a prescribed load:

$$\begin{cases} 0.1y''(t) + 4y(t) = 4(e^{-t} - e^{-15t}), \\ y(0) = 1, \quad y'(0) = -1. \end{cases} \tag{42}$$

The exact solution of this system is:

$$\begin{aligned} y(t) = \ & -0.3618\sin(6.3245t) + 0.1753\cos(6.3245t) \\ & + 0.9756e^{-t} - 0.1509e^{-15t}. \end{aligned} \tag{43}$$

Choosing $N = 5$, the LGM method is used to solve the system. The matrix system can be obtained as below:

$$(\mathbf{MAD}^2 + \mathbf{KA})\boldsymbol{\phi}_N(t) = \mathbf{P}\boldsymbol{\phi}_N(t), \tag{44}$$

where $\mathbf{M} = 0.1$ and $\mathbf{K} = 4$. Furthermore:

$$\mathbf{D}^2 = \begin{bmatrix} 0 & 0 & 0 & 0 & 0 & 0 \\ 0 & 0 & 0 & 0 & 0 & 0 \\ 12 & 0 & 0 & 0 & 0 & 0 \\ 0 & 60 & 0 & 0 & 0 & 0 \\ 40 & 0 & 140 & 0 & 0 & 0 \\ 0 & 168 & 0 & 252 & 0 & 0 \end{bmatrix}, \quad \boldsymbol{\phi}_N(t) = \begin{bmatrix} 1 \\ 2t - 1 \\ 6t^2 - 6t + 1 \\ 20t^3 - 30t^2 + 12t - 1 \\ 70t^4 - 140t^3 + 90t^2 - 20t + 1 \\ 252t^5 - 630t^4 + 560t^3 - 210t^2 + 30t - 1 \end{bmatrix}, \tag{45}$$

$$\mathbf{A} = \begin{bmatrix} a_0 & a_1 & \dots & a_5 \end{bmatrix}, \quad \mathbf{P} = \begin{bmatrix} 2.2618 & -0.5503 & -0.6653 & 0.7842 & -0.6008 \end{bmatrix},$$

where $\boldsymbol{\phi}_N(t)$ is the shifted Legendre vector on the interval $[0,1]$.

Using the Galerkin matrix method, the following algebraic system of equations is obtained:

$$\begin{bmatrix} 4 & 0 & 1.2 & 0 & 4 & 0 \\ 0 & 1.33 & 0 & 2 & 0 & 5.6 \\ 0 & 0 & 0.8 & 0 & 2.8 & 0 \\ 0 & 0 & 0 & 0.5714 & 0 & 3.6 \\ 1 & -1 & 1 & -1 & 1 & -1 \\ 0 & 2 & -6 & 12 & -20 & 30 \end{bmatrix} \begin{bmatrix} a_0 \\ a_1 \\ a_2 \\ a_3 \\ a_4 \\ a_5 \end{bmatrix} = \begin{bmatrix} 2.2618 \\ -0.1834 \\ -0.1331 \\ 0.1120 \\ 1 \\ -1 \end{bmatrix} \tag{46}$$

Solving the above algebraic equations, the unknown Legendre coefficients are finally obtained as follows:

$$a_0 = 0.60901, a_1 = 0.07098, a_2 = 0.27735, a_3 = -0.40712, a_4 = -0.12676, a_5 = 0.09574.$$

By replacing the obtained values in Equation (17), the approximate solution for $y(t)$ is:

$$y(t) \cong 24.1271\,t^5 - 69.1914\,t^4 + 63.2204\,t^3 - 17.6372\,t^2 - t + 1. \tag{47}$$

Figure 2 illustrates the results of the exact solution as a function of time, along with its approximation. The accuracy of the LGM method compared to the exact solution can be noticed in the whole time interval.

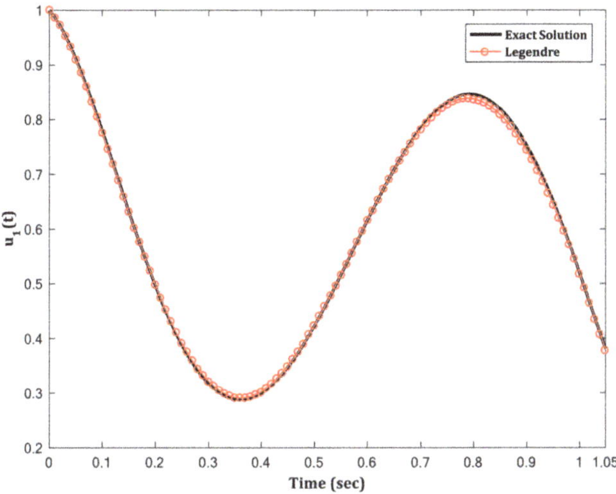

Figure 2. Comparison of LGM and exact solutions for an SDOF system.

4. Worked Examples and Discussion of Results

For a more exhaustive discussion of the developed LGM method and for a reliable assessment of its potential in structural analysis, the motion equation of an MDOF system is solved in accordance with Equation (14), and some numerical examples are presented. These examples are representative of simple models of real structures (by introducing mass, stiffness, and damping matrices) with and without damping. Accordingly, the current study and the presented formulation of the LGM method represents a first step towards its further extension and use for other applications, such as reliability analysis, optimization, and modal analysis of structures.

4.1. Two-Degree-of-Freedom (2DOF) Structure

A structure with two degrees of freedom (2DOF) is analyzed in different dynamic conditions. The structure is first analyzed under free vibration with no effective damping, and then with additional damping. Successively, the analysis is further performed with the assumption of forced vibration, with and without damping. Basic input data and parameters are summarized as follows:

- Free vibration, without damping: it is $\mathbf{M} = \begin{bmatrix} 1.5 & 0 \\ 0 & 2 \end{bmatrix}$, $\mathbf{C} = 0$ and $\mathbf{K} = \begin{bmatrix} 300 & -300 \\ -300 & 800 \end{bmatrix}$ for mass, damping and stiffness parameters.

  The initial conditions are $u(0) = \begin{bmatrix} 1 \\ 1/2 \end{bmatrix}$ and $\dot{u}(0) = \begin{bmatrix} 0 \\ 0 \end{bmatrix}$.

- Free vibration, with damping: it is $\mathbf{C} = \begin{bmatrix} 1.628 & -0.256 \\ -0.256 & 2.512 \end{bmatrix}$, while all the other parameters are equal to the undamped case.

The solution of the motion equation using Legendre's method for $u_1(t)$ and $u_2(t)$ are respectively shown in Figures 3 and 4. Moreover, the LGM results are compared with the exact method, giving evidence of acceptable consistency. It is worth noting that the exact solution is obtained using the Solver of Systems of Equation in the Maple software. Comparisons are then made with the approximate solution.

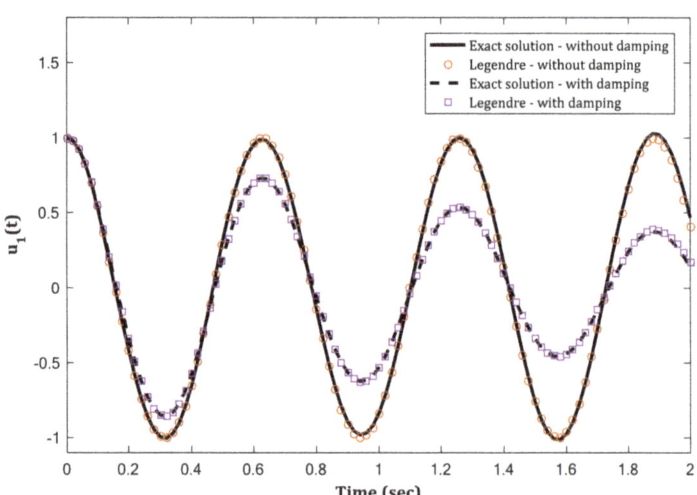

Figure 3. Free vibration analysis for a 2DOF system. Comparison of $u_1(t)$ as a function of time, as obtained from the LGM method or the exact solution.

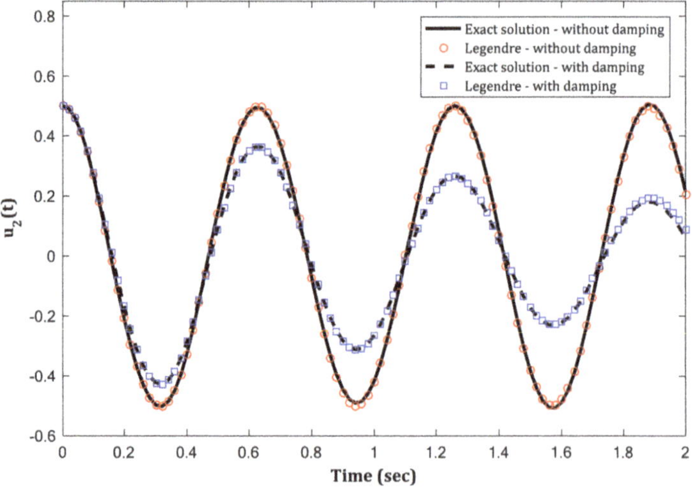

Figure 4. Free vibration analysis for a 2DOF system. Comparison of $u_2(t)$ as a function of time, as obtained from the LGM method or the exact solution.

Successively, the examined structure is analyzed under forced vibration. In this case, it is:

$$p(t) = \begin{bmatrix} \sin 3t \\ 0 \end{bmatrix}, \, u(0) = \begin{bmatrix} 0 \\ 0 \end{bmatrix} \text{ and } \dot{u}(0) = \begin{bmatrix} 0 \\ 0 \end{bmatrix}$$

Once again, the structure is analyzed with and without damping, using the developed LGM method. The analytical results obtained from the LGM method for $u_1(t)$ and $u_2(t)$ are shown in Figures 5 and 6, respectively, and compared with the exact solution, proving again a rather acceptable consistency in the examined time interval.

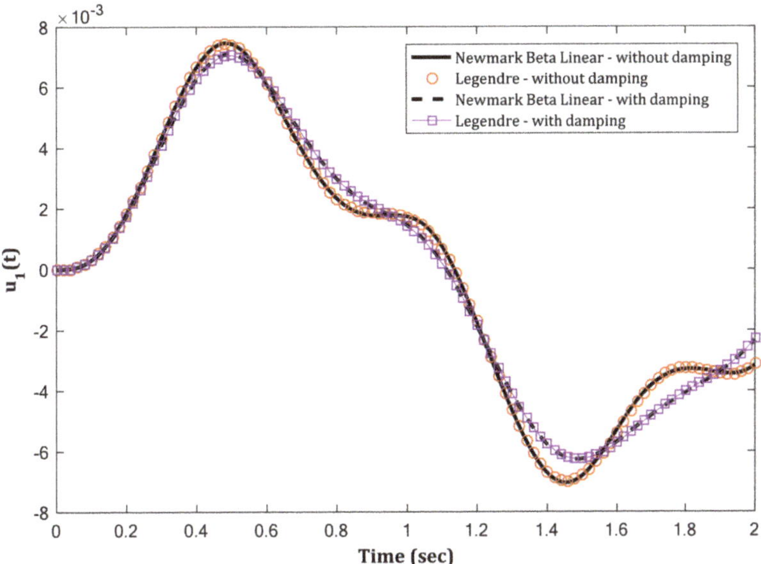

Figure 5. Forced vibration analysis of a 2DOF system. Comparison of $u_1(t)$ as a function of time, as obtained from the LGM method or the linear Newmark-β method.

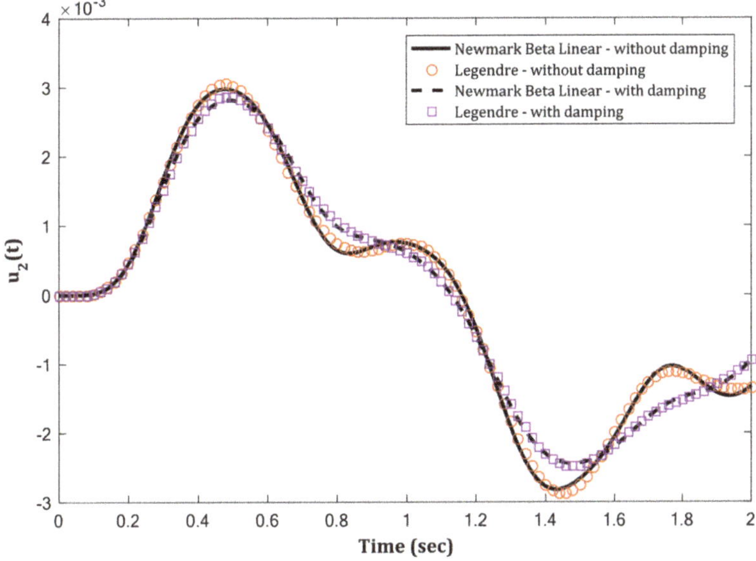

Figure 6. Forced vibration analysis of a 2DOF system. Comparison of $u_2(t)$ as a function of time, as obtained from the LGM method or the linear Newmark-β method.

4.2. Three-Degree-of-Freedom Structure

As a further validation example, a structure with three degrees of freedom (3DOF) is analyzed in forced and free vibration conditions, with or without damping.

- Free vibration, without damping: it is $\mathbf{M} = \begin{bmatrix} 2 & 0 & 0 \\ 0 & 2 & 0 \\ 0 & 0 & 2 \end{bmatrix}$, $\mathbf{C} = 0$ and $\mathbf{K} = \begin{bmatrix} 2 & -2 & 0 \\ -2 & 3 & -1 \\ 0 & -1 & 1 \end{bmatrix}$ for mass, damping and stiffness, respectively. The initial conditions are $u(0) = \begin{bmatrix} 1 \\ 2/3 \\ 1/3 \end{bmatrix}$ and $\dot{u}(0) = \begin{bmatrix} 0 \\ 0 \\ 0 \end{bmatrix}$.

- Free vibration, with damping: it is $\mathbf{C} = \begin{bmatrix} 0.8 & 0 & 0 \\ 0 & 0.7 & 0 \\ 0 & 0 & 0.6 \end{bmatrix}$, while the other parameters are the same as without damping.

The results for $u_1(t)$, $u_2(t)$ and $u_3(t)$ are shown in Figures 7–9, respectively, where it is possible to see the comparison of the LGM method and linear Newmark-β method, with rather good consistency.

The equation of motion of the structure under forced vibration (with and without damping) for $p(t) = \begin{bmatrix} \sin t \\ 0 \\ 0 \end{bmatrix}$ with initial conditions $u(0) = \begin{bmatrix} 0 \\ 0 \\ 0 \end{bmatrix}$ and $\dot{u}(0) = \begin{bmatrix} 0 \\ 0 \\ 0 \end{bmatrix}$ is solved further by using Legendre's method.

The results obtained in this case for $u_1(t)$, $u_2(t)$ and $u_3(t)$ are proposed in Figures 10–12, respectively, and compared with the results by the linear Newmark-β method. Moreover, in this case, the comparative data show the consistency of the LGM with the linear Newmark-β method.

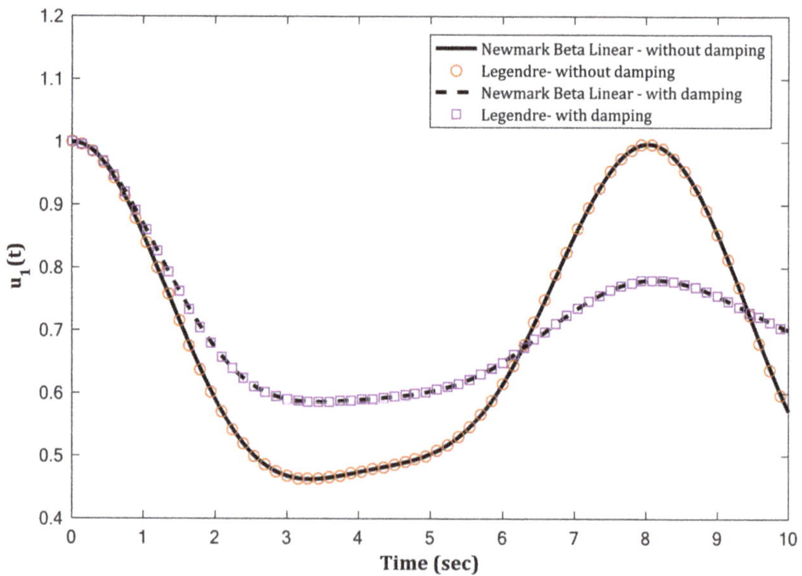

Figure 7. Free vibration analysis of a 3DOF system. Comparison of $u_1(t)$ as a function of time, as obtained using the LGM method or the linear Newmark-β method.

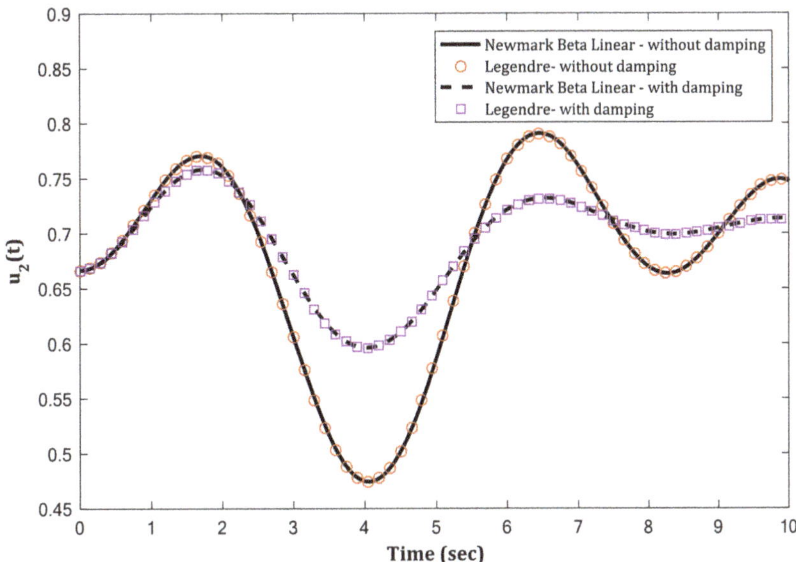

Figure 8. Free vibration analysis of a 3DOF system. Comparison of $u_2(t)$ as a function of time, as obtained using the LGM method or the linear Newmark-β method.

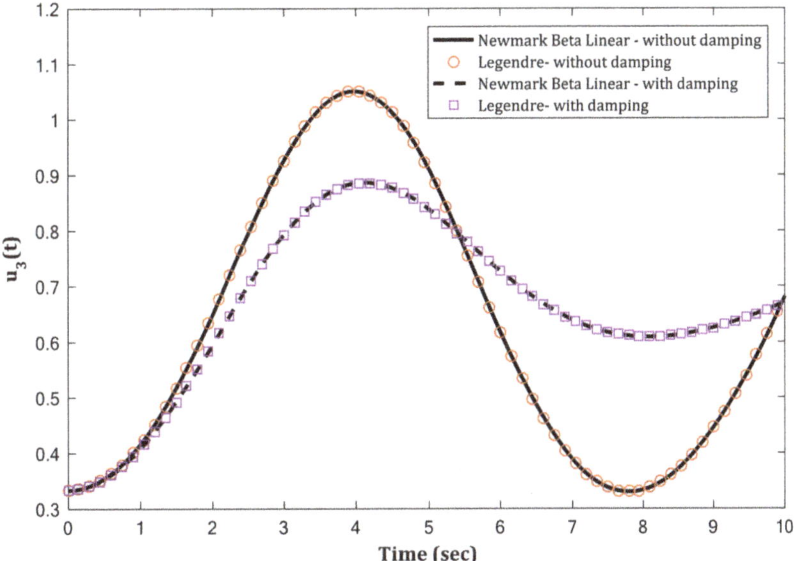

Figure 9. Free vibration analysis of a 3DOF system. Comparison of $u_3(t)$ as a function of time, as obtained using the LGM method or the linear Newmark-β method.

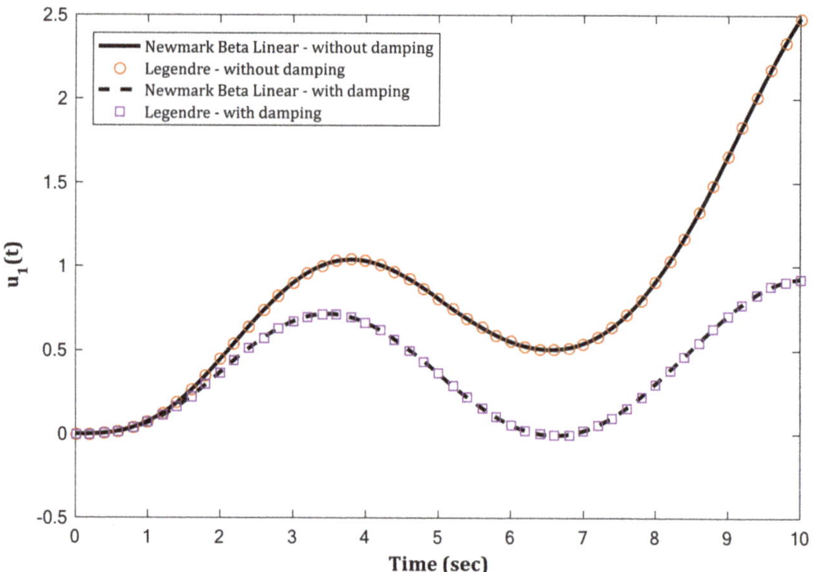

Figure 10. Forced vibration analysis of a 3DOF system. Comparison of $u_1(t)$ as a function of time, as obtained using the LGM method or the linear Newmark-β method.

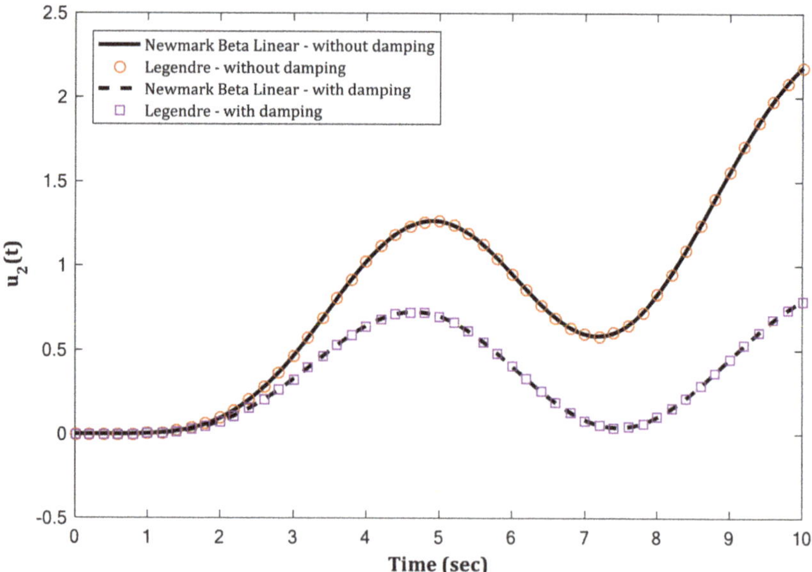

Figure 11. Forced vibration analysis of a 3DOF system. Comparison of $u_2(t)$ as a function of time, as obtained using the LGM method or the linear Newmark-β method.

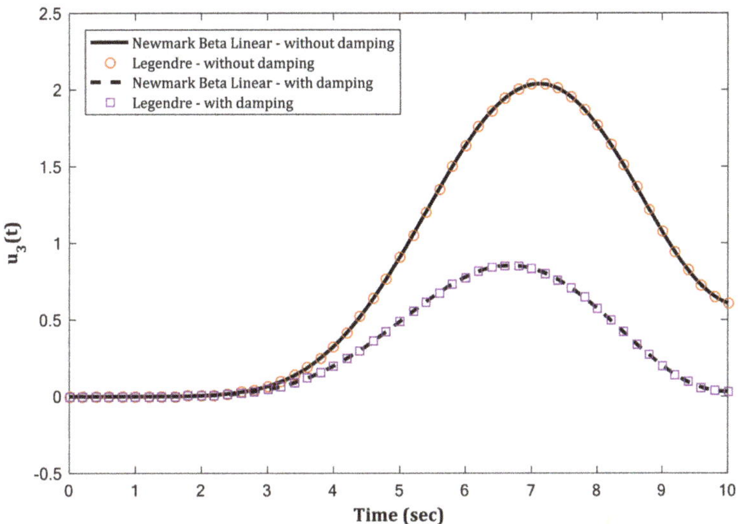

Figure 12. Forced vibration analysis of a 3DOF system. Comparison of $u_3(t)$ as function of time, as obtained using the LGM method or the linear Newmark-β method.

4.3. Five-Degree-of-Freedom Structure

In conclusion, a structure with five degrees of freedom (5DOF) is analyzed with the LGM method, both under free vibration with and without damping.

- Free vibration, without damping: it is assumed that $\mathbf{M} = \begin{bmatrix} 1 & 0 & 0 & 0 & 0 \\ 0 & 2 & 0 & 0 & 0 \\ 0 & 0 & 2 & 0 & 0 \\ 0 & 0 & 0 & 2 & 0 \\ 0 & 0 & 0 & 0 & 2 \end{bmatrix}, \mathbf{C} = 0$

and $\mathbf{K} = \begin{bmatrix} 4 & -4 & 0 & 0 & 0 \\ -4 & 12 & -8 & 0 & 0 \\ 0 & -8 & 20 & -12 & 0 \\ 0 & 0 & -12 & 24 & -12 \\ 0 & 0 & 0 & 12 & 28 \end{bmatrix}$, respectively.

The initial conditions are $u(0) = \begin{bmatrix} 1 \\ 4/5 \\ 3/5 \\ 2/5 \\ 1/5 \end{bmatrix}$ and $\dot{u}(0) = \begin{bmatrix} 0 \\ 0 \\ 0 \\ 0 \\ 0 \end{bmatrix}$.

- Free vibration, with damping: it is assumed that $\mathbf{C} = \begin{bmatrix} 0.5 & 0 & 0 & 0 & 0 \\ 0 & 0.5 & 0 & 0 & 0 \\ 0 & 0 & 0.5 & 0 & 0 \\ 0 & 0 & 0 & 0.5 & 0 \\ 0 & 0 & 0 & 0 & 0.5 \end{bmatrix}$,

while the other parameters are the same as in the undamped case.

The solutions of the equation of motion of the structure using the LGM for damped and undamped states, in terms of $u_1(t)$ to $u_5(t)$, are shown in Figures 13–17 and compared with those by the linear Newmark-β method. As for the previously discussed calculation examples, the comparative plots prove an appropriate consistency and a high level of accuracy of the LGM procedure..

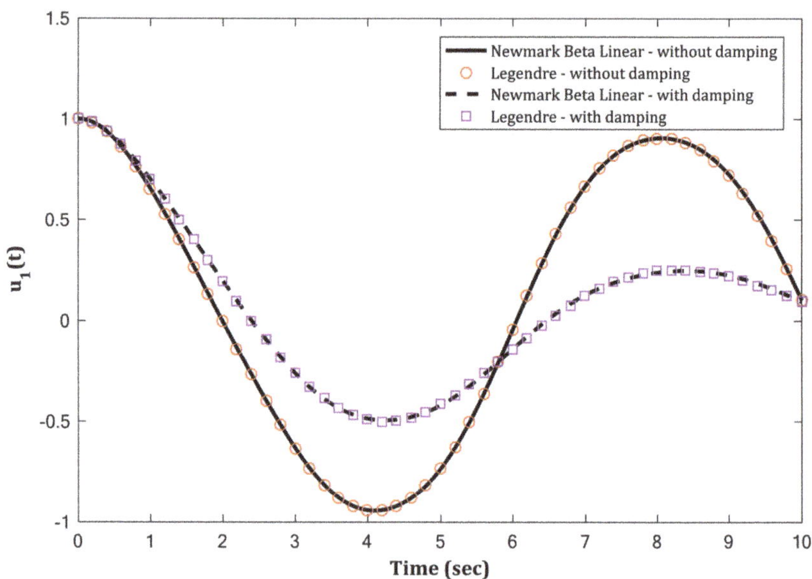

Figure 13. Free vibration analysis of a 5DOF system. Comparison of $u_1(t)$ as a function of time, as obtained using the LGM method or the linear Newmark-β method.

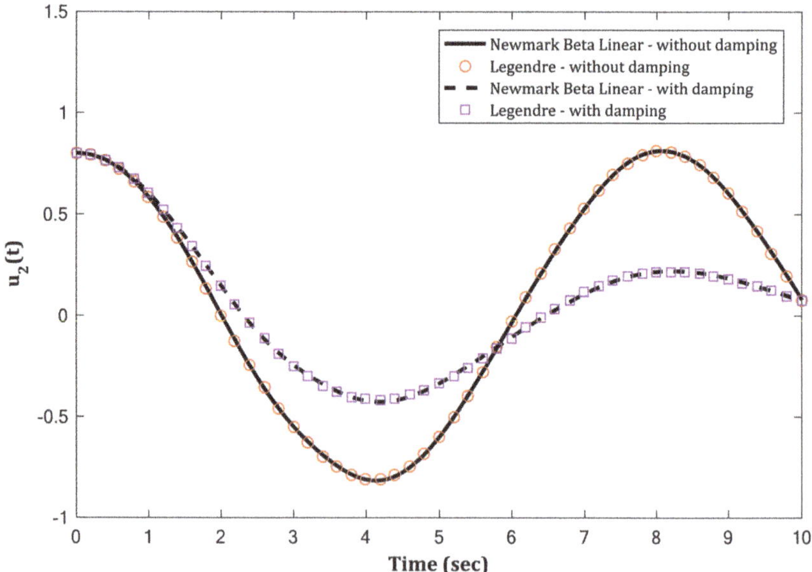

Figure 14. Free vibration analysis of a 5DOF system. Comparison of $u_2(t)$ as a function of time, as obtained using the LGM method or the linear Newmark-β method.

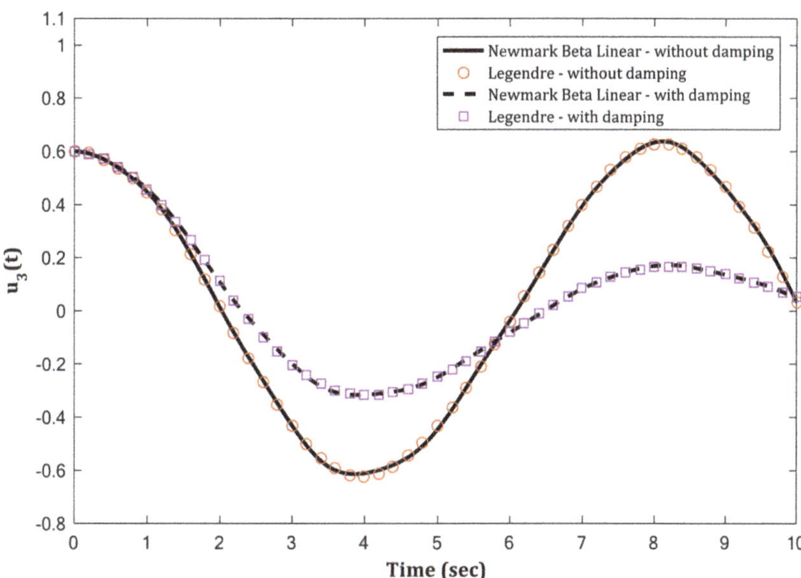

Figure 15. Free vibration analysis of a 5DOF system. Comparison of $u_3(t)$ as a function of time, as obtained using the LGM method or the linear Newmark-β method.

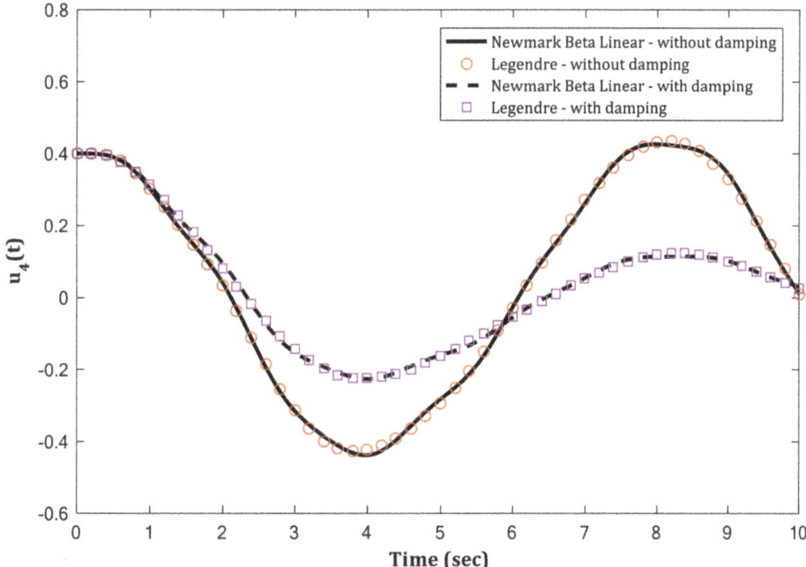

Figure 16. Free vibration analysis of a 5DOF system. Comparison of $u_4(t)$ as a function of time, as obtained using the LGM method or the linear Newmark-β method.

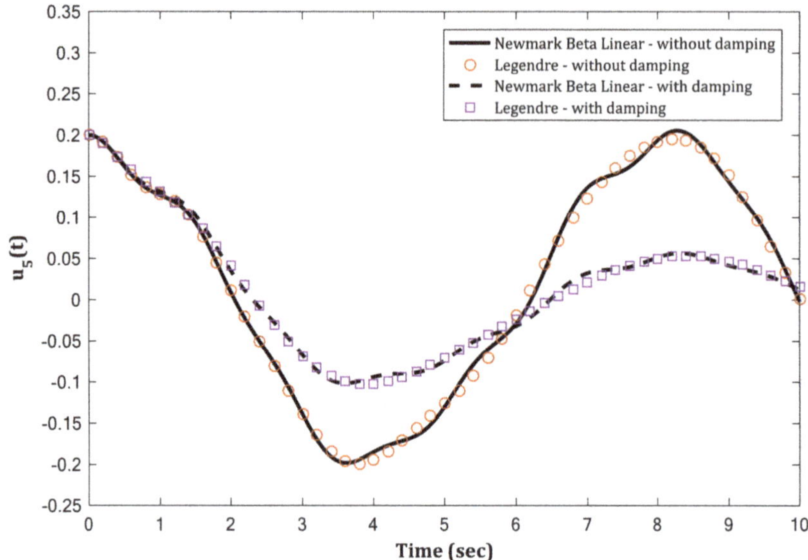

Figure 17. Free vibration analysis of a 5DOF system. Comparison of $u_5(t)$ as a function of time, as obtained using the LGM method or the linear Newmark-β method.

It should be noted that, in the case of stochastic analysis that is interpreted classically as a repetition of solution for different input values with defined probability density functions (taking into account uncertainties associated with input parameters), the methods such as Monte Carlo simulation (MCS) [60] can be easily used along with the results of LGM method to investigate the problem in a probabilistic manner. This is due to the fact that, in the LGM method, the solution is approximated by discretized Legendre series that can be used as a state function in reliability analysis where no mathematical closed-form state function can be found (e.g., through finite element method). Furthermore, such a solution obtained from the LGM method can also be utilized with analytical reliability methods (e.g., jointly distributed random variables method [61]) or approximate ones (e.g., first and second-order reliability methods (FORM and SORM), point estimate method (PEM), etc. [62]), with less computational efforts in comparison to the MCS method.

5. Conclusions

The Legendre–Galerkin matrix (LGM) method was developed in this study to solve systems of differential equations of motion. As shown, the advantage of this spectral method is that it converts the governing differential equation of a given problem to a system of algebraic equations, based on a set of orthogonal Legendre polynomials. The final solution leads to a good estimate of the solution for a system of differential equations. In the present research study, the selected differential equations were typical of single degree (SDOF) and multi-degree-of-freedom (MDOF) structural systems.

In order to prove the accuracy of the proposed method in the response calculation of SDOF and MDOF structures, a number of numerical examples for damped and undamped structural systems under free or forced vibrations were developed and discussed. When available, exact solutions were taken into account for the comparative analysis of LGM predictions, otherwise, the results of the numerical linear Newmark-β method were used to verify the estimates from the developed LGM method. The overall comparative data showed that the LGM method is of high accuracy in estimating the response of SDOF and MDOF systems and can be thus effectively employed in the solution of fundamental motion equations of structures.

Author Contributions: Conceptualization, M.M., M.R.B., B.J., C.B. and M.A.N.; methodology, M.M., M.R.B.; software, M.M., M.R.B. and C.B.; validation, M.M., M.R.B. and M.A.N.; formal analysis, C.B., M.A.N., M.A.H. and S.M.D.; investigation, C.B., M.A.H., M.A.N. and S.M.D.; writing—original draft preparation, M.M. and B.J.; writing—review and editing, C.B., M.A.N., B.J., S.M.D. and M.A.H.; visualization, M.M., M.R.B., B.J. and C.B.; supervision, C.B., M.A.N., S.M.D. and M.A.H. All authors have read and agreed to the published version of the manuscript.

Funding: This research received no external funding.

Institutional Review Board Statement: Not applicable.

Informed Consent Statement: Not applicable.

Data Availability Statement: Supporting data will be made available upon request.

Conflicts of Interest: The authors declare no conflict of interest.

References

1. Chopra, A.K. *Dynamics of Structures: Theory and Applications to Earthquake Engineering*; Prentice Hall: Hoboken, NJ, USA, 1995; Volume 3.
2. Newmark, N.M. A method of computation for structural dynamics. *J. Eng. Mech. Div.* **1959**, *85*, 67–94. [CrossRef]
3. Wilson, E.; Farhoomand, I.; Bathe, K. Nonlinear dynamic analysis of complex structures. *Earthq. Eng. Struct. Dyn.* **1972**, *1*, 241–252. [CrossRef]
4. Butcher, J.C. A history of Runge-Kutta methods. *Appl. Numer. Math.* **1996**, *20*, 247–260. [CrossRef]
5. Humar, J. *Dynamics of Structures*; CRC Press: Boca Raton, FL, USA, 2012.
6. Clough Ray, W.; Penzien, J. *Dynamics of Structures*; Computers & Structures Inc.: Berkeley, CA, USA, 1995.
7. Mason, J.C.; Handscomb, D.C. *Chebyshev Polynomials*; CRC Press: Boca Raton, FL, USA, 2002.
8. Dattoli, G.; Ricci, P.E.; Cesarano, C. A note on Legendre polynomials. *Int. J. Nonlinear Sci. Numer. Simul.* **2001**, *2*, 365–370. [CrossRef]
9. Grosswald, E. *Bessel Polynomials*; Springer: Berlin/Heidelberg, Germany, 2006; Volume 698.
10. Dattoli, G.; Lorenzutta, S.; Maino, G.; Torre, A.; Cesarano, C. Generalized Hermite polynomials and supergaussian forms. *J. Math. Anal. Appl.* **1996**, *203*, 597–609. [CrossRef]
11. Koekoek, R.; Meijer, H. A generalization of Laguerre polynomials. *Siam J. Math. Anal.* **1993**, *24*, 768–782. [CrossRef]
12. Zhou, J. *Differential Transformation and Its Applications for Electronic Circuits*; Huazhong Science & Technology University Press: Wuhan, China, 1986.
13. Odibat, Z.; Momani, S. A generalized differential transform method for linear partial differential equations of fractional order. *Appl. Math. Lett.* **2008**, *21*, 194–199. [CrossRef]
14. Ertürk, V.S.; Momani, S. Solving systems of fractional differential equations using differential transform method. *J. Comput. Appl. Math.* **2008**, *215*, 142–151. [CrossRef]
15. Nazari, D.; Shahmorad, S. Application of the fractional differential transform method to fractional-order integro-differential equations with nonlocal boundary conditions. *J. Comput. Appl. Math.* **2010**, *234*, 883–891. [CrossRef]
16. Alquran, M.T. Applying differential transform method to nonlinear partial differential equations: A modified approach. *Appl. Appl. Math. Int. J.* **2012**, *7*, 155.
17. Fatoorehchi, H.; Abolghasemi, H. Improving the differential transform method: A novel technique to obtain the differential transforms of nonlinearities by the Adomian polynomials. *Appl. Math. Model.* **2013**, *37*, 6008–6017. [CrossRef]
18. Patil, N.; Khambayat, A. Differential transform method for ordinary differential equations. *J. Comput. Math. Sci.* **2012**, *3*, 248–421.
19. Kharrat, B.N.; Toma, G. Differential transform method for solving boundary value problems represented by system of ordinary differential equations of 4 order. *World Appl. Sci. J.* **2020**, *38*, 25–31.
20. Kassem, M.A.; Hemeda, A.; Abdeen, M. Solution of the tumor-immune system by differential transform method. *J. Nonli. Sci. Appl.* **2020**, *13*, 9–21. [CrossRef]
21. Jena, S.K.; Chakraverty, S. Free vibration analysis of Euler–Bernoulli nanobeam using differential transform method. *Int. J. Comput. Mater. Sci. Eng.* **2018**, *7*, 1850020. [CrossRef]
22. Rysak, A.; Magdalena, G. Differential Transform Method as an Effective Tool for Investigating Fractional Dynamical Systems. *Appl. Sci.* **2021**, *11*, 6955. [CrossRef]
23. Najafgholipour, M.; Soodbakhsh, N. Modified differential transform method for solving vibration equations of MDOF systems. *Civ. Eng. J.* **2016**, *2*, 123–139. [CrossRef]
24. El-Shahed, M. Application of differential transform method to non-linear oscillatory systems. *Commun. Nonlinear Sci. Numer. Simul.* **2008**, *13*, 1714–1720. [CrossRef]
25. Momani, S.; Ertürk, V.S. Solutions of non-linear oscillators by the modified differential transform method. *Comput. Math. Appl.* **2008**, *55*, 833–842. [CrossRef]

26. Nourazar, S.; Mirzabeigy, A. Approximate solution for nonlinear Duffing oscillator with damping effect using the modified differential transform method. *Sci. Iran.* **2013**, *20*, 364–368.

27. Chen, C.K.; Ho, S.H. Transverse vibration of a rotating twisted Timoshenko beams under axial loading using differential transform. *Int. J. Mech. Sci.* **1999**, *41*, 1339–1356. [CrossRef]

28. Ho, S.H.; Chen, C.K. Analysis of general elastically end restrained non-uniform beams using differential transform. *Appl. Math. Model.* **1998**, *22*, 219–234. [CrossRef]

29. Özdemir, Ö.; Kaya, M. Flapwise bending vibration analysis of a rotating tapered cantilever Bernoulli–Euler beam by differential transform method. *J. Sound Vib.* **2006**, *289*, 413–420. [CrossRef]

30. Catal, S. Solution of free vibration equations of beam on elastic soil by using differential transform method. *Appl. Math. Model.* **2008**, *32*, 1744–1757. [CrossRef]

31. Arikoglu, A.; Ozkol, I. Vibration analysis of composite sandwich beams with viscoelastic core by using differential transform method. *Compos. Struct.* **2010**, *92*, 3031–3039. [CrossRef]

32. Das, D. The generalized differential transform method for solution of a free vibration linear differential equation with fractional derivative damping. *J. Appl. Math. Comput. Mech.* **2019**, *18*, 19–29. [CrossRef]

33. Bozyigit, B.; Yesilce, Y.; Catal, S. Free vibrations of axial-loaded beams resting on viscoelastic foundation using Adomian decomposition method and differential transformation. *Eng. Sci. Technol. Int. J.* **2018**, *21*, 1181–1193. [CrossRef]

34. Bhrawy, A. An efficient Jacobi pseudospectral approximation for nonlinear complex generalized Zakharov system. *Appl. Math. Comput.* **2014**, *247*, 30–46. [CrossRef]

35. Canuto, C.; Hussaini, M.Y.; Quarteroni, A.; Thomas, A., Jr. *Spectral Methods in Fluid Dynamics*; Springer Science & Business Media: Heidelberg, Germany, 2012.

36. Peyret, R. *Spectral Methods for Incompressible Viscous Flow*; Springer Science & Business Media: Heidelberg, Germany, 2013; Volume 148.

37. Shen, J.; Tang, T. *Spectral and High-Oder Methods with Applications*; Science Press: Beijing, China, 2006.

38. Lucas, T.R.; Reddien, J.; George, W. Some collocation methods for nonlinear boundary value problems. *Siam J. Numer. Anal.* **1972**, *9*, 341–356. [CrossRef]

39. Ortiz, E.L.; Samara, H. An operational approach to the Tau method for the numerical solution of non-linear differential equations. *Computing* **1981**, *27*, 15–25. [CrossRef]

40. Ben-Yu, G. *Spectral Methods and Their Applications*; World Scientific: Singapore, 1998.

41. Bernardi, C. Maday Spectral Methods. In *The Handbook of Numerical Analysis*; Ciarlet, V.P.G., Lions, J.-L., Eds.; Elsevier B.V.: Amsterdam, The Netherlands, 1997.

42. Bialecki, B.; Karageorghis, A. A Legendre spectral quadrature Galerkin method for the Cauchy-Navier equations of elasticity with variable coefficients. *Numer. Algorithms* **2017**, *77*, 491–516. [CrossRef]

43. Guo, S.; Mei, L.; Li, Y. An efficient Galerkin spectral method for two-dimensional fractional nonlinear reaction–diffusion-wave equation. *Comput. Math. Appl.* **2017**, *74*, 2449–2465. [CrossRef]

44. Mao, Z.; Shen, J. Efficient spectral–Galerkin methods for fractional partial differential equations with variable coefficients. *J. Comput. Phys.* **2016**, *307*, 243–261. [CrossRef]

45. Ren, Q.; Tian, H. Legendre–Galerkin spectral approximation for a nonlocal elliptic Kirchhof-type problem. *Int. J. Comput. Math.* **2017**, *94*, 1747–1758. [CrossRef]

46. Nemati Saray, B. Sparse multiscale representation of Galerkin method for solving linear-mixed Volterra-Fredholm integral equations. *Math. Methods Appl. Sci.* **2020**, *43*, 2601–2614. [CrossRef]

47. Amiri, S.; Hajipour, M.; Baleanu, D. A spectral collocation method with piecewise trigonometric basis functions for nonlinear Volterra–Fredholm integral equations. *Appl. Math. Comput.* **2020**, *370*, 124915. [CrossRef]

48. Zaky, M.A.; Ameen, I.G. A novel Jacob spectral method for multi-dimensional weakly singular nonlinear Volterra integral equations with nonsmooth solutions. *Eng. Comput.* **2020**, *37*, 2623–2631. [CrossRef]

49. Erfanian, M.; Zeidabadi, H.; Mehri, R. Solving Two-Dimensional Nonlinear Volterra Integral Equations Using Rationalized Haar Functions in the Complex Plane. *Adv. Sci. Eng. Med.* **2020**, *12*, 409–415. [CrossRef]

50. Hesameddini, E.; Riahi, M. Bernoulli Galerkin Matrix Method and Its Convergence Analysis for Solving System of Volterra–Fredholm Integro-Differential Equations. *Iran. J. Sci. Technol. Trans. A Sci.* **2019**, *43*, 1203–1214. [CrossRef]

51. Ramadan, M.; Osheba, H.S. A new hybrid orthonormal Bernstein and improved block-pulse functions method for solving mathematical physics and engineering problems. *Alex. Eng. J.* **2020**, *59*, 3643–3652. [CrossRef]

52. Mirzaee, F.; Bimesl, S. A new Euler matrix method for solving systems of linear Volterra integral equations with variable coefficients. *J. Egypt. Math. Soc.* **2014**, *22*, 238–248. [CrossRef]

53. Mirzaee, F. Bernoulli collocation method with residual correction for solving integral-algebraic equations. *J. Linear Topol. Algebra* **2015**, *4*, 193–208.

54. Sahu, P.K.; Ray, S.S. Hybrid Legendre Block-Pulse functions for the numerical solutions of system of nonlinear Fredholm–Hammerstein integral equations. *Appl. Math. Comput.* **2015**, *270*, 871–878. [CrossRef]

55. Pourabd, M. Moving least square for systems of integral equations. *Appl. Math. Comput.* **2015**, *270*, 879–889.

56. Yüzbaşı, Ş. Numerical solutions of system of linear Fredholm–Volterra integro-differential equations by the Bessel collocation method and error estimation. *Appl. Math. Comput.* **2015**, *250*, 320–338. [CrossRef]

57. US DoD. *UFC 3-340-02 Structures to Resist the Effects of Accidental Explosions*; US DoD: Washington, DC, USA, 2008.
58. Bounds, W.L. *Design of Blast-Resistant Buildings in Petrochemical Facilities*; ASCE Publications: Reston, VA, USA, 2010.
59. Momeni, M.; Hadianfard, M.A.; Bedon, C.; Baghlani, A. Damage evaluation of H-section steel columns under impulsive blast loads via gene expression programming. *Eng. Struct.* **2020**, *219*, 110909. [CrossRef]
60. Hadianfard, M.A.; Malekpour, S.; Momeni, M. Reliability analysis of H-section steel columns under blast loading. *Struct. Saf.* **2018**, *75*, 45–56. [CrossRef]
61. Johari, A.; Momeni, M.; Javadi, A. An analytical solution for reliability assessment of pseudo-static stability of rock slopes using jointly distributed random variables method. *Iran. J. Sci. Technol. Trans. Civ. Eng.* **2015**, *39C2*, 351–363.
62. Nowak, A.S.; Collins, K.R. *Reliability of Structures*; CRC Press: Boca Raton, FL, USA, 2012.

 *applied sciences*

Article

Case Study—An Extreme Example of Soil–Structure Interaction and the Damage Caused by Works on Foundation Strengthening

Petar Santrač [1], Slobodan Grković [2], Danijel Kukaras [1,*], Neđo Đuric [1] and Mila Svilar [1]

[1] Faculty of Civil Engineering, University of Novi Sad, Kozaracka 2a, 24000 Subotica, Serbia; santrac@gf.uns.ac.rs (P.S.); djuric@gf.uns.ac.rs (N.Đ.); svilarm@gf.uns.ac.rs (M.S.)

[2] Invest Pro Centar Ltd. & Geoexpert Ltd., Adolfa Singera 11, 24000 Subotica, Serbia; grkovic@geoexpert.rs

* Correspondence: dkukaras@gf.uns.ac.rs

Abstract: This paper describes the works on foundation strengthening of the towers of the Cathedral of St. Theresa of Avila in Subotica and the damages caused by these works. Strengthening was performed by means of jacked-in piles and deep soil injection. The construction of the Cathedral began in 1773 and it lasted for several decades with frequent interruptions and changes to the project. The present appearance of the facade was created in 1912. According to historic data, several years after construction, the cracks appeared on the front facade. With time, these cracks became more pronounced, and in 2015, when the remediation project started, the total width of major cracks reached about 15 cm. The first contemporary attempt to repair the towers was made in 2017 by inserting piles beneath the foundations. These works were interrupted due to increased settlements and the appearance of new cracks. In the second attempt, the strengthening was performed by deep injection of soil with expansive resins. During these works, settlements and damages intensified even more, causing the works to be halted in 2018. Analysis of the whole structure and revaluation of all the results, obtained from continuous monitoring of settlements and crack widths from the previous period, led to the new remediation proposal. The imperative was to retain the original appearance of the Cathedral facades while performing the total reconstruction of the upper sections of the front facade. This implies that the overall weight of the reconstructed parts is to be decreased, while the strength is to be increased. Strong structural connections are planned, both among the two towers, and between the towers and the nave. These clear structural solutions will lead to reduced stresses within the existing brick walls, reduced contact soil pressures and ceasing of increased settlements and tilting of the Cathedral towers.

Keywords: cathedral; foundation rehabilitation; jacked-in piles; soil injection; cracks; masonry

Citation: Santrač, P.; Grković, S.; Kukaras, D.; Đuric, N.; Svilar, M. Case Study—An Extreme Example of Soil–Structure Interaction and the Damage Caused by Works on Foundation Strengthening. *Appl. Sci.* **2021**, *11*, 5201. https://doi.org/10.3390/app11115201

Academic Editors: Chiara Bedon, Flavio Stochino and Daniel Honfi

Received: 5 April 2021
Accepted: 27 May 2021
Published: 3 June 2021

1. Introduction

This paper presents an overview of the activities and construction works that were undertaken from 2015 to 2020, aimed at repair and foundation strengthening of the towers of Cathedral of St. Theresa of Avila in Subotica. During this time, two major attempts were made in order to stop the slow, but persistent, differential settlements of the towers. These settlements were observed since the erection of the towers and continue to this day. The first repair attempt consisted of jacked-in-piles beneath the towers, and the second attempt consisted of soil injections of the polyurethane, expansive, resin (PU resin) beneath the towers and the surrounding area. Unfortunately, both attempts failed to stop the settlements and damages to the structure are still present, leading to the new repair proposal.

The Cathedral is east-west oriented, Figure 1. The total length is 61 m, the width of the nave is 21 m, the ridge is 28 m high, and the height of the tower is 54 m. In terms of constructive and decorative elements, regarding the shape of arches and plastic treatment of the facades, the Cathedral has properties of a transitional Baroque-Classicist style. Two tall monumental towers face east and lavishly decorated portal between them is crowned by a tympanum with the sculpture of the Virgin Mary on top of it. The verticality of the building is emphasized by multi-profiled pilasters, which are doubled at the edge of the bell tower and simple on the ground floor; on the first floor, there are ionic capitals, below which the garland descends, and on the second floor there are Corinthian capitals. The windows in the fields are elongated with characteristic, unpretentious, baroque plastic. The façade is horizontally divided into three segments. Above the entrance portal, there are two vertical, arched, openings. The bell towers on the ground floor have rectangular openings, and above them there are two semi-circular arched windows. The rest of the building is less imposing. Simply profiled pilasters are rhythmically arranged along the wall canvas, between which there are arched windows. Above the elevated part of the altar is a crypt, and on the side, there are semi-circular sacristies and a chapel.

b) Interior view towards the altar

a) Cathedral, north-eastern view prior to 2017.   c) Interior view towards the entrance

Figure 1. View of the north-eastern part of the Cathedral prior to 2017. (**a**) Cathedral, north-eastern view prior to 2017, (**b**) Interior view towards the altar, (**c**) Interior view towards the entrance.

According to historical data, the Cathedral of St. Theresa of Avila in Subotica was built from 1773 to 1798. The works began according to the project of the Pest master mason Ferenc Kaufman, to be completed according to the project of Adam Heisler in 1783. It was built on compressible clay soil, on a site bordered by swampy troughs. From the east side, in the north-south direction, a stream flowed until the end of the 19th century, through which water flowed from the nearby shallow lake "Jasi-bara". Construction proceeded slowly with frequent changes to the project, only to be suspended in 1779, and resumed in 1787 to 1789. The interior decoration was completed in 1803, and the copper roof covering of the towers, which replaced the dilapidated wooden shingles, in 1839. Historic archive data show that the construction of the building was constantly accompanied by settlements, cracks and damage to the arches and vaults, which were repaired several times. A newspaper article from 1879 describes the cracks that spread from the foundation to the roof of the facade, as well as cracks in the vaults of the nave. Partial renovation of the altar was carried out in 1888 by replacing the old sacristies with larger semi-circular sacristies with vestibule, staircase, and a small chapel, stiffening and stabilizing the nave arches in the altar region.

The problem of cracks in the frontage remained relevant and attracted the attention of eminent experts of the time, from Subotica, Pest, and Vienna. Professor Csaki Bella from Budapest wrote in 2009 in his "Study on the assessment of the condition of the Basilica of St. Theresa of Avila" [1] about the restoration by the engineer Toth Mihaly in 1876, who used iron ties. As he further states, due to very large horizontal forces, the ties stretched significantly and were unable to prevent the tilting of the towers and the crack growth. A similar idea, with ties, was given in 1883 by the famous Hungarian architect of that time, Geza Ziegler. Traces of attempts to solve the problem of the towers tilting and the crack growth in the frontage wall are still visible today. Out of the nine ties, placed about 150 years ago, seven are broken and the remaining two are overly stretched and dysfunctional. According to the writings of engineer Tot Mihaly, during 1886 to 1877 observations, cracks were measured but appropriate conclusions about the causes could not be formed in such a short time. It was only six years later that a group of prominent experts at the time—Pal Karvayi, Titus Macskovic, Istvan Grundbok and Geza Kocka—came to the correct conclusion that the crack growth in the frontage was the result of very slow and uneven settlements and tilting of towers on soft ground.

The last major intervention in the Cathedral lasted from 1909 to 1912 and included the restoration of the outer facade with "Terranova" mortar. However, after a few years, cracks reappeared, so the problem remained relevant throughout the 20th century, without any measures being taken to repair them. The restoration of the facade was completed in 1912, and since there were no major interventions since, it can be concluded that the current total width of cracks in the crown of the frontage wall of about 15 cm is the result of uneven settlement of the towers in the past 105 years. In 2015, a decision was made to start planning a rehabilitation of the Cathedral. Based on calculations and previous experiences, the initial plan to install new ties [2–5] was rejected. In 2018, an attempt was made to reduce the soil angular pressure by jacked-in piles [6]. However, due to the implementation technology, small, residual, successive, settlements appeared during hydraulic jack transfer from one pile to another, which, in turn, increased the crack widths. The works were halted, resulting in only 11 out of 16 piles being installed—six below the north tower and five below the south tower. The settlements partially subsided after six months. After that, according to the proposal of experts from Budapest, the soil was strengthened by injecting expansive resins [7,8]. These works were completed but, unfortunately, instead of stabilizing, the settlements intensified even more. Consequently, in 2019, all works were temporary halted again, while maintaining continuous monitoring of the settlements, crack widths, and in-depth analysis of other possible remediation strategies. Based on the structural monitoring and additional analyses, a new solution was formed that includes partial dismantling and reconstruction of the upper sections of the front facade wall and of the towers. Within reconstruction, a strong structural connection is planned, among the two towers, and between the towers and the nave. The emphasis is on retaining the original appearance of the Cathedral facades while creating reconstructed elements that are lighter and structurally stronger, thereby producing smaller weight and pressure on the ground, and obtaining more favorable structural behavior overall.

2. Investigative Works That Preceded the Rehabilitation

For the purpose of the rehabilitation project, in 2015, for the first time in the long history of the structure, detailed tests were conducted, which included soil investigations, excavations near existing foundations, testing of bricks and mortar within the foundation and the structure, and mapping of all cracks and damages [5], Figures 2–4.

Figure 2. (**Left**) Cracks in the front facade wall; (**Right**) Crack details on the upper window.

Figure 3. Cracks in the longitudinal wall—north facade.

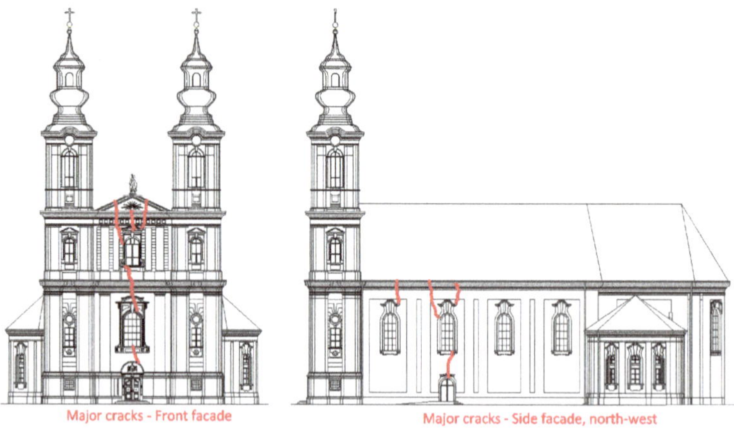

Figure 4. Crack monitoring gauges within the Cathedral—inner side of the front facade wall.

Figure 2 shows the large cracks that developed on the frontal facade. At present, these cracks have a maximum width of approximately 19 cm, and they spread over the full thickness of the wall, over the whole crack lengths.

Installation of crack meters, inclinometers, geodetic markers and measurements, and complete recording of the geometry and condition of the structure and materials of the building were made. As part of soil-mechanic tests, six static penetrations 16 to 21 m deep, and four boreholes of 15 m were made, from which samples were taken. Laboratory tests have shown that the soil profile, up to a depth of 9 m, consists of low/medium plastic soft clay (CL-CI), below which is a layer of sand (SF) 1.8 to 2 m thick (Figure 5).

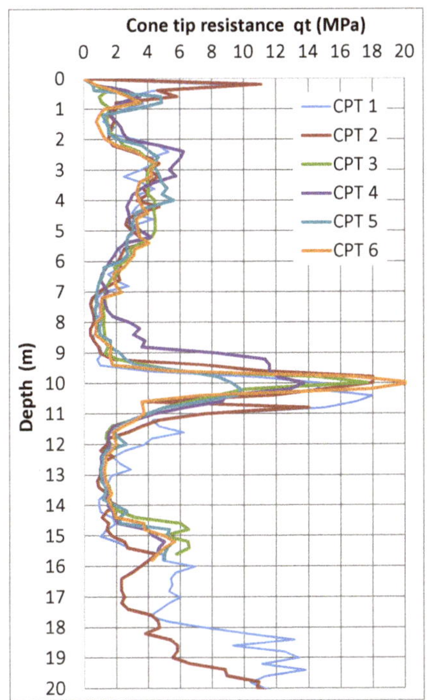
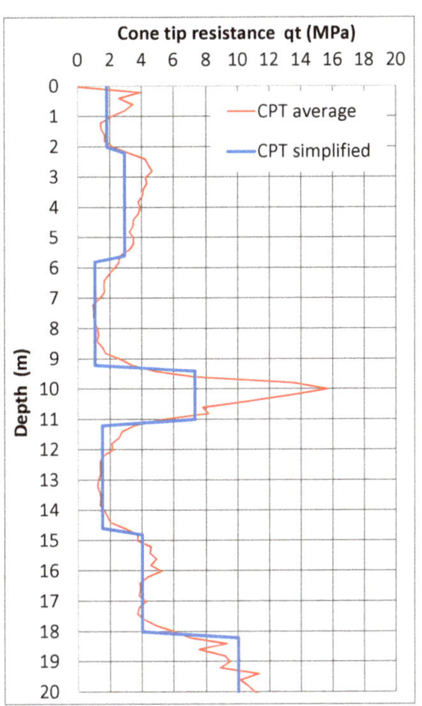

Figure 5. All penetration tests next to the towers (**left**) and average of all tests (**right**) [5].

After that, clays reappear, ending above a layer of compacted sand at a depth of 18 to 20 m. According to historic data, this location in the past was bordered by swampy troughs and canals, which is indicated by the presence of muddy and marshy impurities in the upper parts of the terrain. According to limited piezometric measurements (Figure 6), the groundwater depth is at 3.2 to 4.2 m from the ground level around the Cathedral (112.6 m MASL).

Based on the excavations, the depth of the foundation of the northern tower is approximately 2.5 m, while at the southern tower, it is at approximately 3.0 m. Testing of the bricks from the foundation wall showed that their compressive strength ranges from 1.3 to 2.0 MPa. The bottom parts, the footings, of the tower foundations have a surface of approximately 8 × 8 m, and a thickness of 2.0 m, and were made from an uneven mixture of lime concrete with sand and irregular pieces of larger and smaller bricks and sandstone rocks, Figure 7.

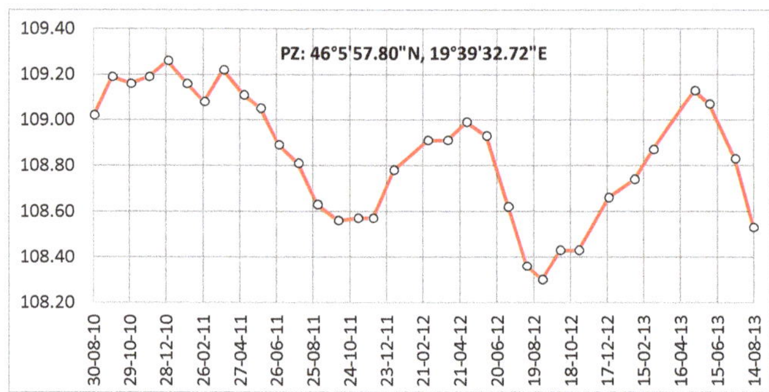

Figure 6. Results of groundwater level monitoring around the Cathedral [5].

Figure 7. Appearance of the cleaned side surface of the lime concrete foundation.

Lime concrete was used in construction approximately 5000 years before cement; it is characterized by significantly lower compressive strength and greater plasticity than cement concrete. These properties make it able to withstand large and long-term deformations before cracks appear. Lime concrete is also very hygroscopic, which is why it turns into a soft mass when cut or drilled with water cooling. The strength of lime concrete used in this structure could not be tested, but based on its resistance to impact and breakage of parts, it could be concluded that it is slightly higher than the compressive strength of bricks within the structure.

It is interesting to note that, despite the significant damage to the Cathedral (Figure 8) caused by long-lasting and uneven settlements, there are no written traces that geodetic measurements were ever performed in the previous 100 to 150 years. Except for the state markers in the front wall of the south tower, there was no other geodetic markers on the building. Therefore, for the purpose settlements monitoring of the Cathedral, about 30 new geodetic markers were installed in the outer walls and two adjacent ones on old buildings nearby. These geodetic markers form a stable and highly accurate levelling network. In order to analyse the inclination of the towers, in 2016, a geodetic measurement of the verticality of the south wall edges of the south tower and the north wall edges of the north tower was performed (Figure 9), between elevation $H = 131$ to 136 m. The $+ X$ axis is in the south direction, and the $+ Y$ is in the west direction. I1 and I4 are the edges of the south tower, while I2 and I3 are of the north tower [9,10].

b) Tower, north-east corner – cladding falling;

a) Large cracks on the front facade;   c) Longitudinal cracks on the first nave vault.

Figure 8. Characteristic damages: facade wall, tower wall cladding, cracks in the nave vaults.

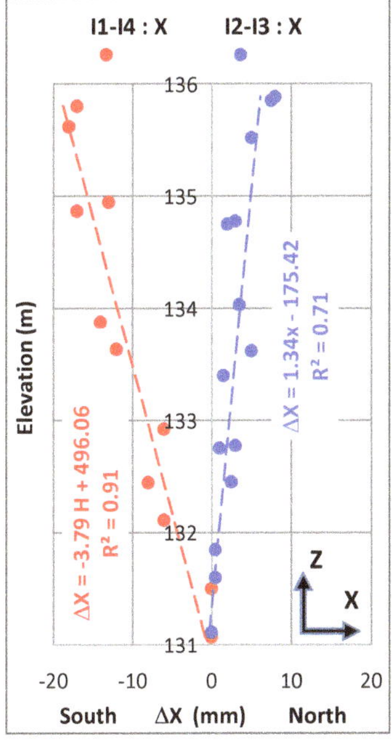

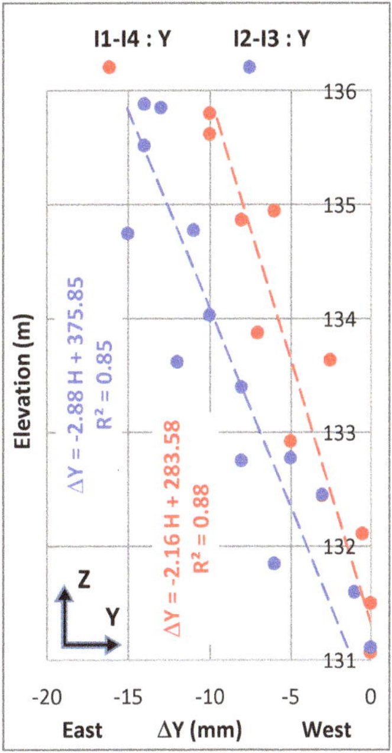

Figure 9. Linear interpolation of observed points on tower lining [9].

Linear interpolation of observed points on the tower edges show the inclination, namely the lining I1–I4 of the south tower to the south by $\theta_X = 3.79$ mm/m, and the lining I2–I3 of the north tower to the north by $\theta_X = 1.33$ mm/m, or divergence of 5.12 mm/m. At the same time, both towers lean to the east about $\theta_Y = 2.16$–2.88 mm/m. At the height of the crown of the frontage wall at 27.0 m (elevation 137.5 m), the spacing of the towers is about $\Delta X = 27 \times 0.512 = 13.8$ cm, which is slightly less than 15 cm, which is approximately the sum of the measured cracks width at the crown of the frontage wall.

3. Analysis and Interpretation of Measurement Results

3.1. Soil Properties

Based on soil investigation, the ground at the location of the Cathedral consists of low-plastic clay which extends to a depth of about 18 to 20 m and is cut at a depth of about 9 to 9.5 m with a 1.8 to 2 m thick layer of sand. Based on laboratory tests, the compressibility of clay, at a depth between 2.5 to 5.0 m, because of the effect of pre-consolidation, reaches the highest values, of a range between 7.0 and 9.0 MPa. At other depths, the modulus is significantly smaller, between 3.1 and 4.8 MPa.

According to the analysis, the weight of the individual tower is up to $V = 32.0$ MN, and the height of the centre of gravity is about $Z_t = 15.0$ m. The average pressure below the quadratic foundation footing $B = 8.0$ m is $q = 500$ kPa. When the pressure on the ground is known, the settlement of the tower can be determined through the method of one-dimensional deformation, and since the foundation is rigid, the settlement is determined by the equivalent point of the foundation.

The soil compressibility modules were adopted by the correlation with one penetration resistance, using a simplified diagram of mean values on Figure 10. The calculated settlement is about $s = 25$ cm.

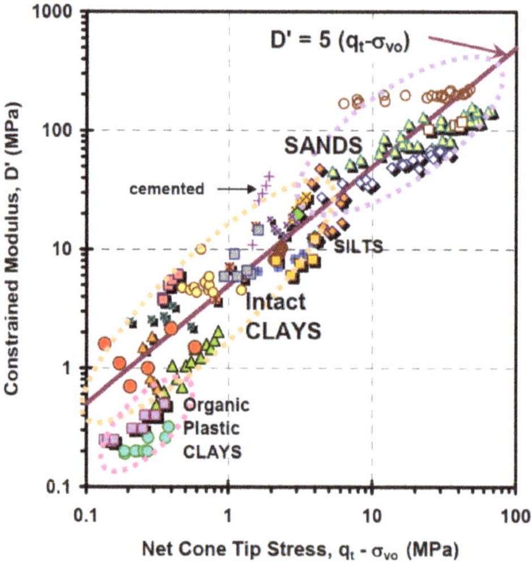

Figure 10. Correlation of soil compressibility modulus and cone resistance [11].

For further analysis, using the calculated settlement of the tower, the average modulus of elasticity was determined by the following analytical expression [12]:

$$s = qB\frac{1-v'^2}{E'}\left(I_3 + \frac{1-2v'}{1-v'}I_2\right)I_1 \, , v' = 0.3 \tag{1}$$

$$\frac{H}{B} = 1.63, \quad \frac{D}{B} = 0.25 \tag{2}$$

In the above expression, $I_1 = 0.8$ is the embedment depth ratio, $I_2 = 0.24$ is the shape ratio, and $I_3 = 0.07$ is the layer thickness ratio. The thickness of the compressible layer is the depth at which the vertical stress from the influence of the foundation is less than 10% of the effective stress due to the self-weight of the soil. The average modulus of elasticity of a deformable layer of thickness H, based on the above expression, is $E' = 3.2$ MPa. The corresponding modulus of compressibility of the layer of $D' = 4.3$ MPa was determined by the following well-known expression:

$$D' = E'\left(1 - \frac{2v'^2}{(1 - v')}\right) \tag{3}$$

An important data for the analysis is the eccentricity of the towers center of gravity (e_x, e_y) in relation to the foundation joint, which is determined based on the height of the centre of gravity Z_t and geodetically measured inclination of the towers θ (Table 1).

Table 1. Measured eccentricities of the towers e_x and e_y.

South Tower of the Cathedral				North Tower of the Cathedral			
θ_x (mm/m)	θ_y (mm/m)	e_x (mm)	e_y (mm)	θ_x (mm/m)	θ_y (mm/m)	e_x (mm)	e_y (mm)
−3.79	−2.16	−57	−32	1.33	−2.88	20	−43

In the next step, based on the inversely calculated modulus of elasticity $E' = 3.2$ MPa of the soil layer and measured eccentricities of the towers (e_x, e_y), the inclinations (θ_x, θ_y) of the towers can be calculated by analytical expression [13], as shown in Table 2.

$$\theta_{x,y} = \frac{1 - v'^2}{E'} \frac{V \cdot e_{y,x}}{B^2 L} I_\theta, \quad I_\theta = 4.17 \tag{4}$$

Table 2. Calculated inclination of the towers θ_x and θ_y.

South Tower of the Cathedral				North Tower of the Cathedral			
θ_x (mm/m)	θ_y (mm/m)	e_x (mm)	e_y (mm)	θ_x (mm/m)	θ_y (mm/m)	e_x (mm)	e_y (mm)
−4.22	−2.41	−57	−32	1.48	−3.21	20	−43

When the divergence of the towers at the height of the frontage is calculated, by applying data from Table 2, the value $\Delta X = 27(2.41/1000 + 3.21/1000) = 0.154$ m is obtained, which is a very good approach to the geodetically determined size $\Delta X = 13.8$ cm and the approximate sum of measured crack widths at the crown of the frontage of about 15.0 cm.

The previous analysis shows good agreement between measured and calculated displacements. Ultimately, this indicates that, with the parameters of soil modulus determined in the laboratory and from field tests, the interaction of the object and the foundation soil can be reliably modelled.

It is also interesting to analyse the critical height of the centre of gravity of the tower, which can be estimated based on the following expression [14]:

$$Z_{t,cr} = \frac{E'}{\left(1 - v'^2\right) I_\theta} \frac{B^2 L}{V}, \quad I_\theta = 4.17 \tag{5}$$

The calculated critical height is about 13.5 m, which is close to $Z_t = 15.0$ m. The critical height of the structure centre of gravity is the height at which even a very small, horizontal force, relative to the structure weight, causes a large increase in stress and deformation in the ground, and thus endangers the stability of the structure. This is corroborated by the measured and visually noticeable damages on the Cathedral, based on cracks in the front facade wall, on the longitudinal facade walls of the Cathedral nave, arches, etc.

3.2. Numerical Modelling

In order to investigate the influence of arch forces, assumed to have caused the initial tilting of the towers, a 3D numerical model of the Cathedral structure was created with discontinuities at the locations of major existing cracks of the facade wall, the joints between the tower and the nave, internal arches, and vaults. Figure 11 depicts the general shape of the model, while Figure 12 depicts the FE mesh of the Cathedral and the surrounding soil that was taken into account for the calculations. The FE analysis was performed with software Ansys® Academic Teaching Mechanical, Release 17.

Figure 11. 3D model of the Cathedral—general shape of the model.

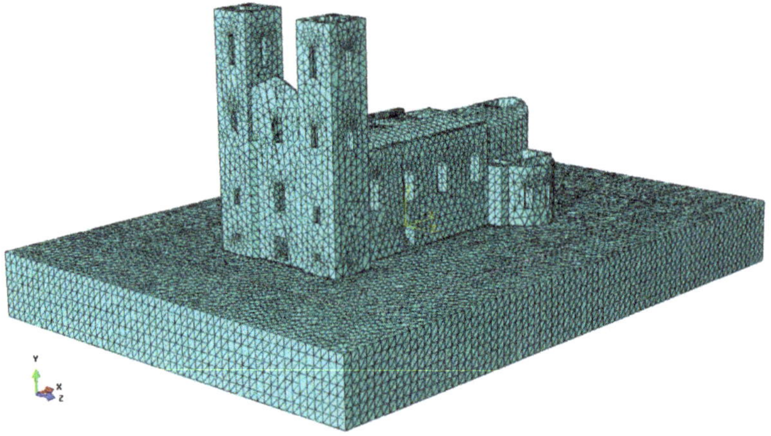

Figure 12. FE mesh of the Cathedral numerical model with the surrounding soil.

The masonry was simulated with the SOLID65, 3D element type from the Ansys Element Library, while for the soil, the SOLID185 element was used. Special attention was devoted to finding an adequate FE mesh, in order to optimize the accuracy of the results, convergence and hardware limitations. As a result, the masonry was modelled with 45,115 elements, and the soil with 192,070 elements, 237,185 elements in total. Timber elements of the Cathedral roof structure were introduced as loads applied on the locations that represent support points for the real structure. Compared to the rest of the structure, these structures have relatively small effects. The bell section of each tower has a weight of 37.5 tons compared to the total weight of each tower, which is 3200 tons. The weight of the roof over the nave is 225 tons, compared to the total weight of the nave of approximately 7000 tons. Since the numerical model is relatively large, material nonlinearities were utilized only for the soil. The masonry was modelled with the linear material model, without cracking. The effect of the existing cracks was introduced in the form of voids that extend throughout the walls. This simplification was adopted after the series of simulations with material nonlinear models both for the masonry and for the soil proved to be too demanding for hardware that was available at the time. On the other hand, the results obtained from the simplified model have a good agreement with the damages observed on the structure itself. The soil was modelled with the Mohr–Coulomb material model. The numerical values used for the definition of this material model were obtained from numerous in situ soil testing, laboratory testing of the soil samples from the site and for the modulus of elasticity, as shown in Section 3.1 of this paper. The boundary conditions on the outer surfaces of the soil, Figure 12, were defined to have no displacements.

Overall calculations related to this structure consisted of both static and dynamic analysis, but for the purpose of this paper, the results showed here are related to the effects of self-weight, which govern the damaging long-term settlements that led to such extreme cracks observed on the Cathedral. Figures 13–17 therefore all show the effects of the permanent gravity loads on the structure.

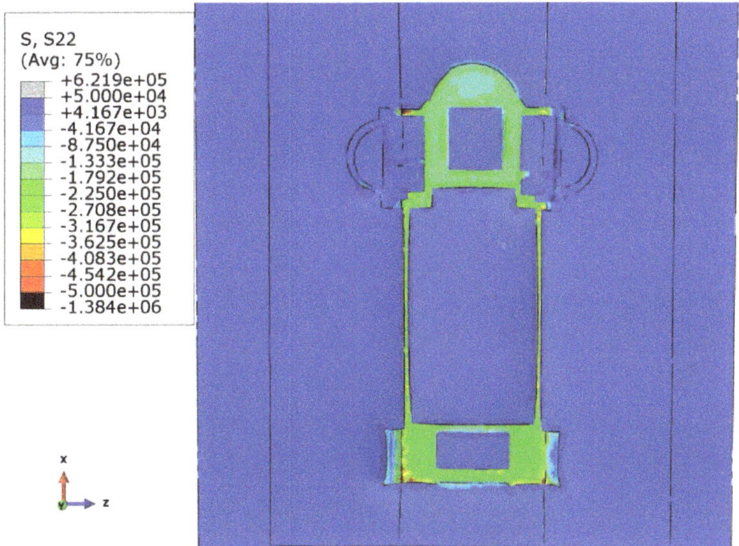

Figure 13. Vertical contact stresses in the soil at the idealized (flat) bottom surface (N/m^2).

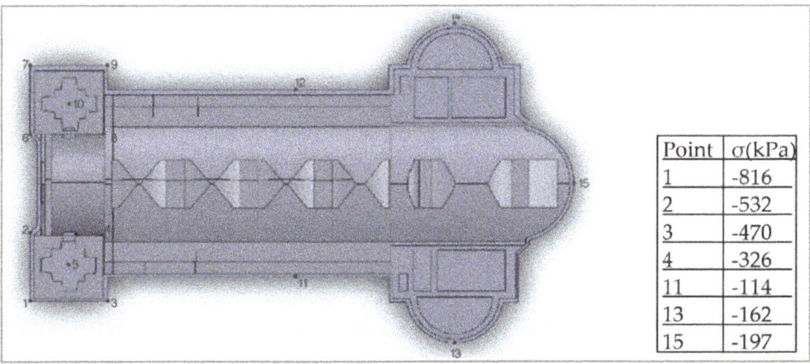

Point	σ(kPa)
1	-816
2	-532
3	-470
4	-326
11	-114
13	-162
15	-197

Figure 14. Characteristic points for contact stresses below the foundations of the Cathedral.

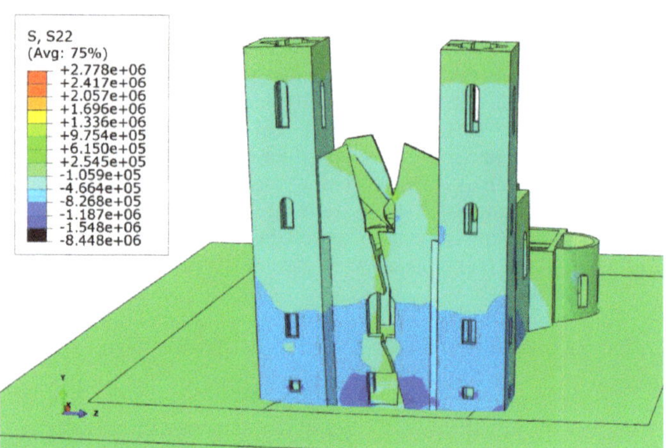

Figure 15. Vertical stresses in the masonry (N/m^2).

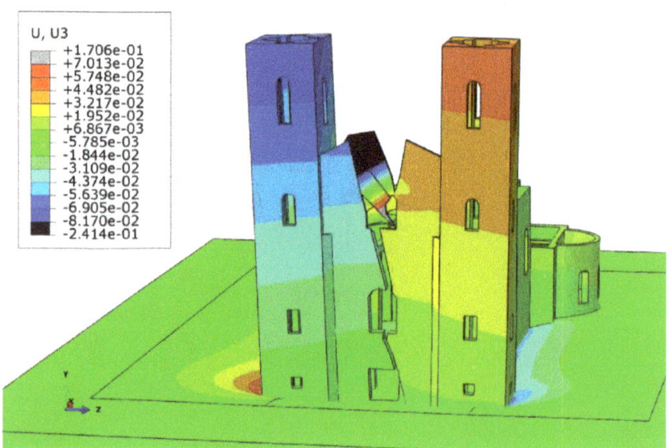

Figure 16. Horizontal displacement of the structure "north-south" direction (m).

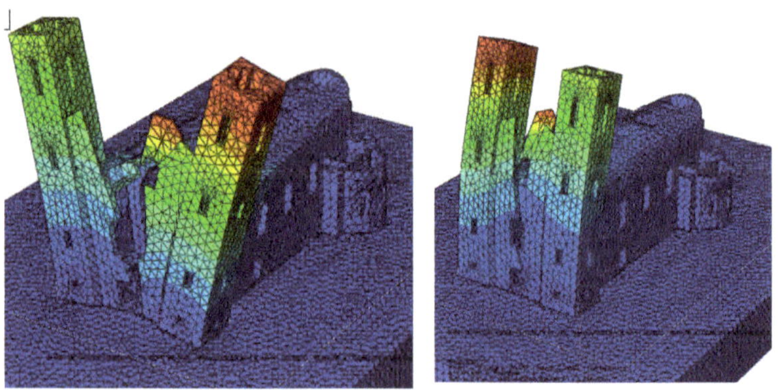

(**a**) First "north-south" mode—0.302 Hz;　　(**b**) First "east-west" mode—0.517 Hz

Figure 17. Modal analysis—first modes in two orthogonal directions.

Calculations have shown (Figures 13 and 14) that the maximum edge pressure is below the southeast and northeast corners of the tower, and is about 816 kPa, which is much greater than the centric stress. The divergence of the towers is about 6.0 mm/m. Although the 3D model is a rough approximation of the actual condition in the structure and soil, the results of static calculation agree well with the results of visual observation of what happens to the object and measurement results, which is a good starting point for determining the methodology of structural rehabilitation.

4. Considered Variants of Foundation Rehabilitation

Based on the analysis of the measurement results and performed calculations, it is concluded that the primary cause of the problem with the Cathedral towers is the unusually high pressure on the layer of compressible clay, which caused large consolidation settlement. Another reason is the eccentric load on the foundation, which caused the towers to tilt. These two quantities are cause-and-effect, because eccentricity causes inclination and inclination increases eccentricity, etc. In addition, the critical height of the centre of gravity makes the structure sensitive and unstable to the effects of relatively small horizontal forces. Starting from the assumption that the cause of the problem is high pressure on the ground and eccentricity of the load, it is logical that the solution is to reduce the weight and pressure on the ground, strengthen the soil and tighten the towers. One "obvious" strategy for tightening the towers is to apply steel ties, anchored in the large hidden anchor blocks that would not disturb or change the appearance of the front facade. However, the lessons from previous, historic experiences with ties within this structure, and the fact that this structure is a high masonry building, not resistant to bending, raises reasonable suspicion that further settlement due to creep would compromise this solution with excessive stretching of ties and horizontal cracks in the tower walls. Additionally, this solution cannot solve the problem of the towers tilting eastwards, away from the nave. Hawing this in mind, it is much more appropriate to tighten the towers with a strong, massive, reinforced concrete (RC) frame that would be hidden beneath the facade, Figure 18.

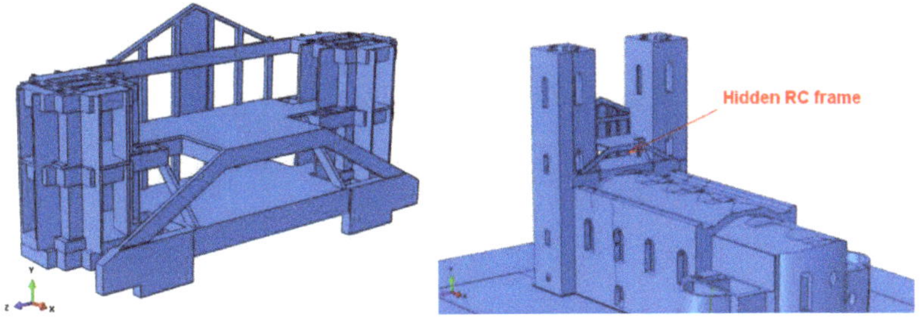

(**a**) Proposed hidden RC frame-detail;   (**b**) Location of the proposed RC frame

Figure 18. Proposed RC frame.

The additional (hidden) RC frame was supposed to be joined with the existing masonry by installing a series of anchors. The size of the RC frame, and the dimensions of interaction surfaces, combined with the anchors, was expected to ensure the joint work of the old and the new structure. With additional structural elements, this RC frame can be joined with the nave, therefore utilizing its mass to counteract the eastward tilting of the towers. On the downside, this strategy raises the issue of increased additional weight and increased pressure on the ground. Hawing in mind all the above, it was concluded that this type of tightening must be combined with partial soil reinforcement, to exclude further soil deformations.

Two possibilities of increasing the bearing capacity of the soil were considered, one of which was "jet-grouting" technology, and the other was the injection of steel piles under the foundation. Both methods belong to invasive construction technologies, since they require temporary and local weakening of foundations and soil through digging, drilling, and grouting. Given the tendency of lime concrete to soften due to fluid drilling, and the significantly higher price, instead of jet-grouting, the preference was given to the installation of jacked-in piles ("mega-piles").

Following the detail analysis of all the above-mentioned aspects of this problem, an additional proposal was formed that would lead to reduced pressure on the ground, while increasing the overall structural strength. This would have to include a partial dismantling of the upper sections of the towers, which would then be reconstructed with all the necessary strengthening elements and reduced weight—see Figure 19. The initial calculations showed that the pressure reduction would be approximatively 15 to 20%. This solution is also supported by the fact that several large structural elements of the front facade were planned for total reconstruction even before the increased settlements raised the question of the structural stability of the towers, such as: the dilapidated roof structure of the tower and the middle section of the facade wall that currently has extremely large cracks and damages. Additionally, new bell carriers, platforms and internal tower stairways need to be built. The major advantage of this strategy is that it allows for the possibility of unhindered construction of the powerful reinforced concrete frame that would connect the towers themselves, and the towers with the nave of the Cathedral. From a structural point of view, this solution is clean, simple, and durable, it does not require soil reinforcement or the installation of steel ties and anchor blocks, and lastly, it is affordable.

Figure 19. Upper sections of the towers and front façade—proposed total reconstructions.

However, since the Cathedral represents a significant cultural heritage monument, this solution was initially put on hold, until other strengthening strategies, however few, are explored.

4.1. Construction of the Jacked-in Piles (Mega-piles) Beneath the Foundation and Its Consequences

The underlaying idea of strengthening the soil with jacked-in piles was based on improving the bearing capacity of the soil beneath the two outer edges of the tower foundations. The calculation showed that in this way, with a relatively small force in the piles, the angular pressure on the ground could be reduced and thus stop the further tilting of the towers. There were four piles planned along the outer edges of the tower foundation, i.e., a total of eight piles under each tower (Figure 20).

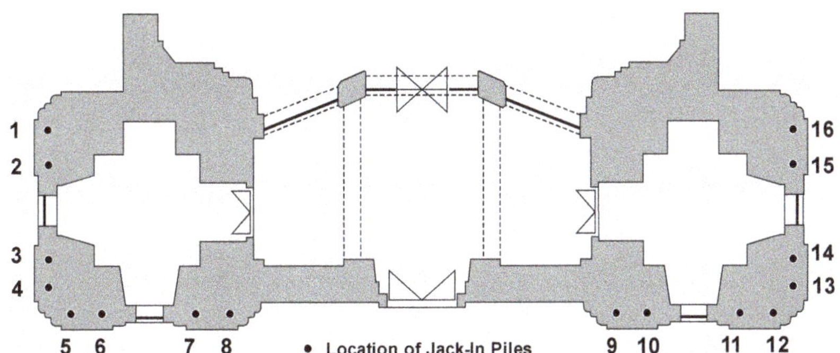

Figure 20. Layout of the piles below the foundations of the towers.

Each pile consists of several steel segments–cylinders with a diameter of 323 mm and a height of 500 mm. In order to install the piles, a narrow trench was excavated next to the foundation. From this trench, the working space ("a gallery") within the foundation was excavated, 1.3 m high, 1.0 m wide and 0.8 m deep. The galleries were excavated in succession—the work on the new pile did not start before the previous one was activated. Relative to the total foundation area of 128 m^3, one such gallery represented less than 1%. Inside the gallery, a steel frame and a hydraulic jack were placed. The hydraulic jack was used to press steel cylinders, one by one, into the ground beneath. The initial steel cylinder of each pile was formed with a conical tip. All segments were welded to each other continuously along the circumference. The piles were pressed all the way down to the layer of sand, where the pressing force was limited to 0.74 MN, so as not to cause a shear fracture in the lime concrete foundation footing above the pile. The resistance diagram of piles being pressed into the ground is shown in Figure 21, and it is very similar to the tip resistance (CPT) in Figure 5.

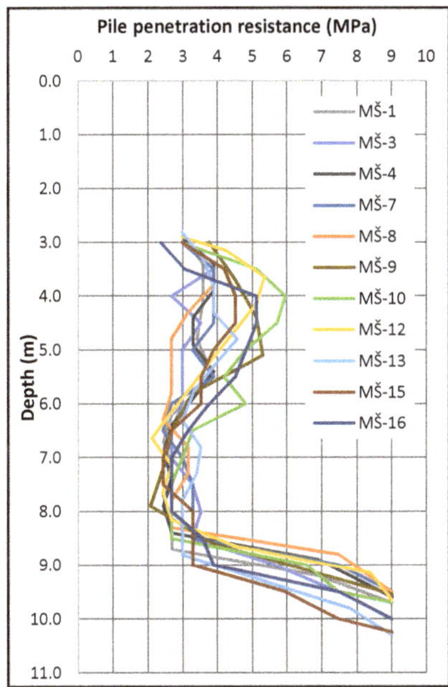

Figure 21. Measured resistances on the hydraulic jacks when pressing piles.

During the works on the pile installation, monitoring of the geodetic markers movement was continuous in order to prevent negative effects that can accompany this foundation strengthening method. Initial, "zero", recording was performed on 6 March 2017 and the first pile was installed on 9 March 2017. After the installation of the last pile segment, the gap between the steel frame and the pile is wedged, the hydraulic jack is removed and the working gallery is concreted. It is of the utmost importance that the wedge is locked adequately, so that the drop in pile force after removing the hydraulic jack is minimized. Due to the geodetic measurements, after installation of a few piles a settlement of 1 to 2 mm order of magnitude was observed. As the work progressed this became increasingly intense with each new pile. The settlement was accompanied by the appearance of new cracks and expansion of the existing ones. Consequently, on 2 April 2017, after the installation of 11 piles, the works were stopped, without piles number 2, 5, 6, 11 and 14 being installed. Geodetic measurements showed that the settlements on both towers reached values between 6 and 8 mm. The inclination of the south tower was increased by 0.4 mm/m and the north by 0.6 mm/m, while the inclination of both towers from the nave were increased by 0.4 mm/m. The total crack width was increased by approximately 4 cm, from 15 to 19 cm. After the cessation of the works, the measurement of the settlement continued on a monthly basis. It was determined that the settlement rate gradually calmed down within the next nine months and, by the end of the year, it fell below geodetic accuracy. In the meantime, possible causes of unsuccessful remediation were analysed, and it was concluded that it was most likely due to a multiple cause, such as sensitivity to weakening of the foundation, vibration of the pneumatic hammer during gallery excavation and slight loosening of the wedges.

4.2. Injection of Polyurethane Resins into the Foundation Soil and Its Consequences

The negative experience with the jacked-in piles ruled out solutions that would include interventions/strengthening on the foundations themselves and reaffirmed the idea of reducing the weight of the towers, as well as the idea of adding massive side apses with pile foundations to stabilize the towers. Figure 1a depicts the present appearance of the front facade, and Figure 22 depicts the remediation proposal with additional massive apses

Figure 22. Remediation proposal with additional massive apses.

However, such solutions were not in accordance with the existing restoration guidelines for cultural monuments protection and therefore were not accepted. At the investors request, an expert team from Budapest proposed, in October 2017, the stabilization of the towers by means of deep grouting of soil with polyurethane expanding resin. During penetration and expansion, the resin exerts pressure on surrounding soil, and creates and penetrates cracks in the form of a branched root. This method was originally developed by the oil industry to increase the yield of wells, while in geotechnics it has been adapted for the purpose of precise remediation of settlement under structures. Based on their experiences on similar objects that were also damaged by excessive settlement, the expert team guaranteed the cessation of the settlement with local uplift of the structure and/or surrounding terrain in the order of magnitude up to 0.5 mm. The injection was performed through long steel pipes with a 12 mm diameter. During the injection procedure, upward movement of the geodetic wall markers is considered to be a proof of successful soil stabilization. The project envisaged the stabilization of the ground beneath the northern and then beneath the southern tower, at 2×92 injection points at a depth of 2, 3, 4, 5, 6, 7, 8, 9 and 10 m, measured from the ground surface. The external injection points were placed in three rows, the first row along the wall, the second and third row at a distance of 1 m and 2 m from the wall, respectively. The internal injection points were placed only in one row along the inner walls—see Figure 23. First, the outer row was injected at a distance of 2 m, then the row at a distance of 1 m, and finally, the rows directly next to the tower wall on the outside and on the inside. The downward direction of injection along the wall is sloping at an 8° angle towards the wall, while the other directions are vertical. The works lasted from February to August 2018. According to an approximate calculation, the volume of injection mass per injection point, after expansion, is about 0.17 m^3, or a total of 15.5 m^3 for each tower. If the average injection radius is about 0.3 m, then in relation to the affected soil volume around the injection pipe, the initial soil porosity of about 0.45 should have been reduced by about 15%.

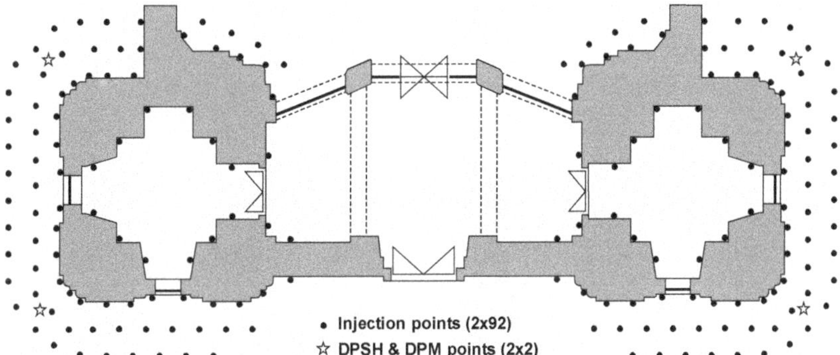

Figure 23. Position of injection points around the foundations of the towers.

To determine the effect of injection, a continuous test of the soil with dynamic penetrometers was performed, at four locations before and after injection, Figure 24.

Based on comparative diagrams, at depths below 3 m, a certain increase in soil resistance is observed beneath both towers as a result of injection. Increasing the penetration resistance means that in a certain region of the soil around the foundation, the strength and stiffness of the soil are increased, which should have favourable consequences on the foundation bearing capacity. However, contrary to expectations, very soon after the beginning of the soil injection, next to north tower, an increase in settlement and the appearance of new cracks in the walls and arches of the Cathedral nave was observed. These effects were significantly less favourable than the ones observed during jacked-in piling. Based on the assumption that the settlement was the result of a large increase in pore pressure and a decrease in effective stresses in the clay, the contractor reduced the injection speed, which prolonged the works. Despite this, the settlement continued at a lower intensity. In order to gain an insight into settlement dynamics, geodetic measurements were performed during and after the injection for 12 h straight, with final measurement conducted 24 h after the injection. The geodetic markers were measured on the south tower, above the injection point next to the wall. Settlement from 10 July 2018, at 08:42, until the next day, 11 July 2018, at 07:49 are shown in Figure 25.

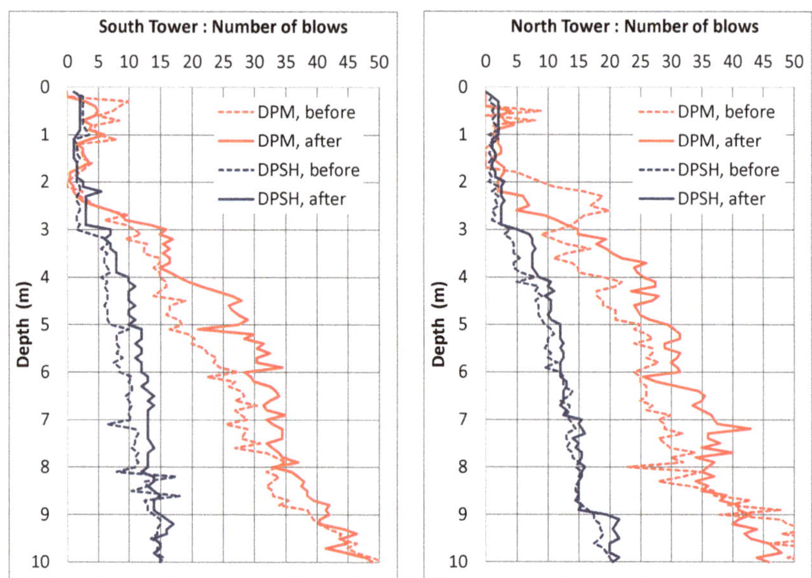

Figure 24. Results of penetration tests before and after soil injection.

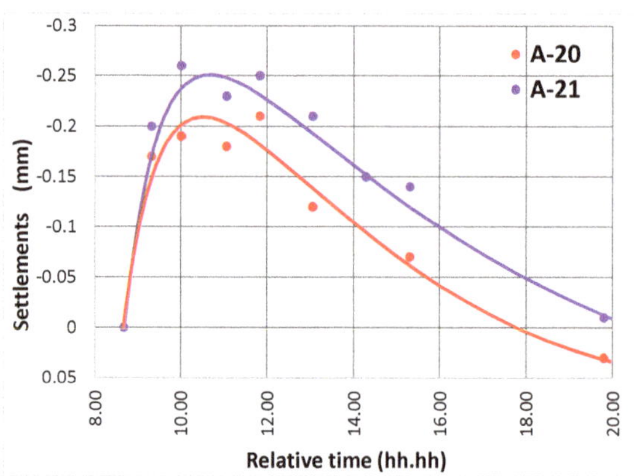

Figure 25. Movements of geodetic markers A-20 and A-21 during and after injection.

The geodetic markers were measured eight times during working hours, and the ninth measurement was on the next day, shortly before the next injection session continued. According to the diagram, the markers showed upward movement that had a peak during the injection of 0.21 to 0.25 mm, but after injection ceased, the markers began to move downward. After 24 h, out of 0.25 mm rise, only 0.01 mm remained on the marker A-21. The marker A-20 showed a downward movement of 0.03 mm, after reaching 0.21 mm of upward movement. Based on the shape of the displacement curve, it can be concluded that, despite the initial rise during injection, both markers eventually have some settlement. Although the individual settlements are very small, when superimposed for each injection point (2 × 92 points), settlements of 8 to 10 mm were observed. In other words, during and after polyurethane resins injection, soil with reduced permeability, instead of increasing the soil volume around the injection zone, showed a permanent reduction [15].

The results of geodetic marker settlement measurements, during and after the completion of rehabilitation works, are shown in Figure 18. Geodetic measurements covered a period of about 2.5 years, after the completion of the rehabilitation works, and as it can be seen, the total settlement of the north tower geodetic markers is between 5 and 27 mm, and the south between 13 and 22 mm. Uneven settlement indicates the inclination of the towers, the direction of which is indicated by the arrows in Figure 26.

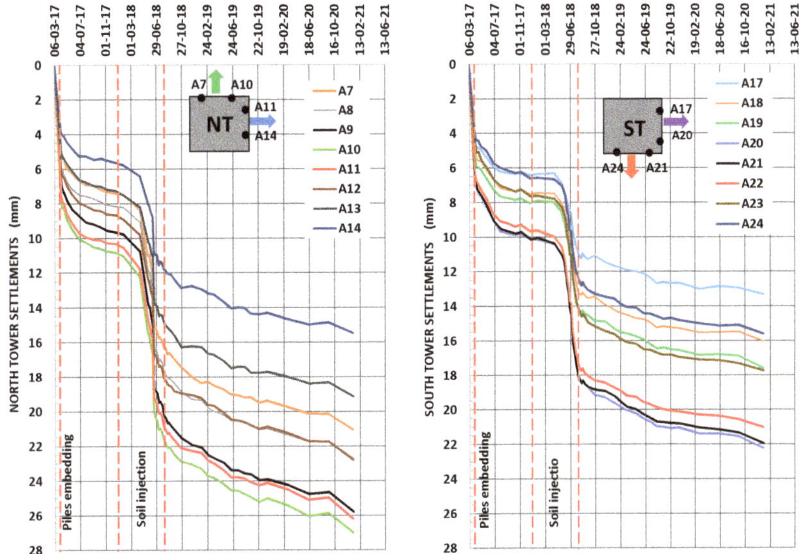

Figure 26. Diagrams of geodetic markers settlement during and after remediation works.

During the installation of the jacked-in piles, the geodetic markers on both towers showed approximately the same level of settlement, 6 to 11 mm. After the soil injection with polyurethane expanding resin was completed, these settlements increased to 11 to 22 mm for the northern, and to 11 to 19 mm for the southern tower. Approximatively 2.5 years after injection, settlements increased by an average of an additional 4 mm for the north tower and 3 mm for the south tower, meaning that neither piles nor grouting stopped the settlements.

Figure 27 shows the average settlement and the average settlement increment of the geodetic markers. It is evident that, as with the settlements (Figure 26), the average settlements of the northern tower, which was founded at a depth of 2.5 m, are about 4 mm larger than the ones of the southern tower, which is founded at a depth of 3.0 m from the surface around the Cathedral.

After the completion of the injection of piles, which was followed by very intensive settlement, the average settlement rate slowed down and practically stabilized at about 0.07 mm/month in both towers. However, when the injection of soil under the north tower began, the settlement of the north tower accelerated, while on the south tower the rate of settlement remained the same. It was only when the injections under the south tower began that its settlement also accelerated. The injection process was accompanied by intense settlement, which began to slow down only after completion of the works. After about a year, the settlement of both towers has stabilized, with the northern tower at about 0.08 mm/month and the southern tower at about 0.05 mm/month. While after the installation of the piles, the settlement rate of both towers remained the same, after the injection, the rate of the north tower was about 50% higher than the settlement rate of the south tower. The last measurement, which was performed in January 2021, showed an increase in the settlement rate, which is more clearly seen in Figure 19, where the settlement increments are shown in two consecutive measurements. The reason for that is the earthquake in Croatia in December 2020, which was also felt in Subotica, in the period between two geodetic measurements. The diagram showing the increments of the average settlement clearly shows that there was a tendency to decrease, which was disturbed by the seismic actions.

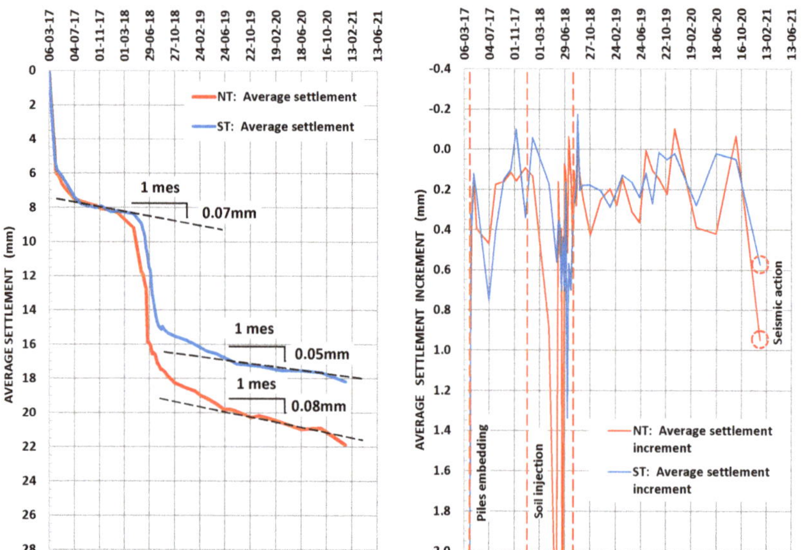

Figure 27. Diagrams of average settlement and average increment of settlement.

In order to explain the measured monthly settlement of towers of 0.05 to 0.08 mm/month (Figure 27), it should be noted that the settlement on clay consists of three components: instantaneous settlement, settlement due to primary consolidation and settlement due to secondary compression (creep). Having in mind the construction time, the first two components were completed, so only the settlement due to creep remained. Based on the time required to complete settlement due to primary consolidation of about t_{100} = 6.0 years and the secondary compression index C_α which can be estimated by compressibility index Cc = 0.18, from the oedometric test (Mesri and Godlewski, 1977), the settlement due to creep from 1912 to 2017, or for about 105 years, is:

$$S_{sc} = \int_0^H C_\alpha \log \frac{t}{t_{100}} dz = \int_0^{13} \frac{0.04 C_c}{1+e} \log \frac{t}{t_{100}} dz = \int_0^{13} \frac{0.04 \cdot 0.18}{1+0.8} \log \frac{105}{6} dz = 0.064 \quad (6)$$

If the settlement due to creep for t = 105 years is about 64 mm, then the average monthly settlement due to creep is about:

$$S_{sc} = \frac{0.064 \cdot 1000}{105 \cdot 12} = 0.051 mm/mes \quad (7)$$

The calculated size of 0.051 mm/month approximately corresponds to the order of magnitude of the measured monthly settlement of the towers, which is shown in Figure 19

4.3. Discussion and Possible Future Research Directions

For the reconstruction and strengthening of the Cathedral structure to be successful, it was extremely important to determine, beyond any doubt, the governing cause of its current damages. By unifying all the findings from the available historic data, and reconstruction efforts since 2015, it was concluded that the damages were caused by extremely high contact stresses in the soil which still cause steady and relatively uniform settlements. This fact directed the remediation proposals towards very invasive measures, but ones that are deemed unavoidable—partial dismantling of the towers followed by their reconstruction. The reconstruction is to be completed in such a way that the outer appearance remains the same but with the lighter and stronger structure. This type of reconstruction measure would reduce the soil pressures, and consequently future settlements, so that they cause no more damage to the structure.

Prior to the final decision regarding the acceptance of the proposed solution, all other aspects of the reconstruction must be explored. In recent years, there were several structures, similar to this one, that have undergone major repair and strengthening works and that could be used as examples of good practice. One of these is the strengthening of the ancient masonry tower in Torre Orsaia (Italy), [16]. The authors of this reconstruction managed to achieve sufficient strengthening without applying very invasive techniques. A similar example is the restoration of the "Carmine Maggiore" bell tower in Naples (Italy), [17]. Apart from the damages caused by the passage of time, most of the damages on this tower were caused by earthquakes, while no damages were observed or attributed to the foundation structure. Both examples are characterized by minimization of the structural interventions, which should always be the goal when dealing with historic buildings. On the other hand, in both cases, there were no severe damages caused by increased and uneven settlements, so this gave more freedom for the strengthening techniques to be chosen.

Future research and investigations that could help with making a final decision related to the strengthening technique should utilize some of the recent analysis tools that were proposed. Since the Cathedral, in its current state, is very sensitive, even to the smallest uneven settlements, the ground water level can induce some negative effects on the dynamic properties of the towers, and therefore reduce its current capacity to resist seismic forces. Provided that current measurement methods are improved with the addition of accelerometers and other long-term monitoring devices, some simplified, low-cost methods to evaluate the dynamic soil–structure interaction for varying ground levels are available [18] that are also non-destructive and based on neural networks.

The numerical analysis shown in this paper was focused on the general understanding of the mechanisms that lead to the current state of the structure and it utilized nonlinear material models only for the soil, while the masonry was modelled with linear materials. Some recent studies, such as the evaluation of the seismic vulnerability of the Trani Cathedral's bell tower (Italy) [19], showed good results by combining sophisticated numerical modelling and experimentally obtained measurements. A seismic risk study shown for the Norman tower of Craco in Matera (Italy) [20] proposes the simplified and versatile procedure for the seismic risk evaluation of historic buildings.

Even though the excessive damage, observed on the masonry, is caused by the weak foundations and not by the masonry itself, it was found that the quality of the masonry and the mortar is very inconsistent throughout the Cathedral. These material properties, combined with construction imperfections, make the estimation of the load-bearing capacity for these masonry elements very difficult. The application of artificial neural networks can be applied for this task as well [21]. Most of the parameters for this approach to be applied can be found in previous analyses and material tests. This approach would require additional tests for masonry stiffness and tensile strength of the masonry, but these tests would be generally beneficial, so they should be conducted.

5. Conclusions

Considering the historic data related to the Cathedral, registered damages, results of geodetic measurements and experiences gained from rehabilitation works, it can be concluded that this is a very complex problem of interaction between a very heavy object on one side and a poorly load-bearing base on the other. From the professional point of view, it is safe to say that, in this case, an inadequate foundation system was applied, and that instead of shallow foundation, it was obviously necessary to apply deep foundation on piles. It is not clear why the builders of that time opted for shallow foundations of relatively heavy and high Cathedral towers, which caused very high contact stresses on the soft water-saturated clay base, when adequate knowledge, experience and means for installing piles existed at that time.

The paper shows that in terms of calculated settlements and low soil compressibility, the centre of gravity of the towers is close to the critical height, making them very sensitive to relatively small side effects. The result of such relations, in the long run, causes uneven settlements and progressive tilting of the towers. This is clearly shown by the continuously increasing width of cracks on the front facade wall, which, in the present form, appeared after the restoration of the facade in 1912, and continued to grow until the rehabilitation work attempts during 2017 and 2018, and to this day.

The results of geodetic measurements and geotechnical calculations also showed that the foundations of the towers are exposed to the process of very small but constant settlements due to soil creep. Such settlements, over a long period of time and eccentric loading, cause progressive tilting of the towers and the propagation of cracks. This process, due to its nature, cannot be stopped by itself, but will only accelerate and worsen the general condition of the building over time.

The inclination of the towers, weak construction of the foundation footings, very high contact stresses in the soil, and a relatively low soil permeability indisputably contributed to the rehabilitation failures, which themselves were also significantly invasive in character. Intensive, continuous measurements during one of the PU resin soil injection sessions showed that the towers initially moved upwards, but this was followed by a slower downward movement, therefore rendering this approach not only ineffective but also damaging.

All of this leads to the conclusion that the solution for rehabilitation should not be sought in the soil, but in reducing the weight of the towers, i.e., reducing the pressure on the ground. Of course, due to the nature of creep, reducing the weight of an object cannot completely stop settlement, but it can significantly slow it down.

Partial dismantling of the towers, their interconnection and connection to the nave of the Cathedral by means of a powerful reinforced concrete frame, followed by their rebuilding with lighter materials, preserving the current appearance, can solve the problem of cracks in the long run with a radical extension of the service life.

Author Contributions: Conceptualization, P.S., D.K., and S.G.; writing–original draft preparation, P.S.; writing–review and editing D.K. and S.G.; software, M.S.; validation, P.S., D.K., S.G., and N.Đ.; data curation, N.Đ.; visualization, M.S.; formal analysis, P.S. and N.Đ.; project administration, P.S., D.K., S.G.; funding acquisition, P.S., D.K., and S.G. All authors have read and agreed to the published version of the manuscript.

Funding: This research received no external funding.

Institutional Review Board Statement: Not applicable.

Informed Consent Statement: Not applicable.

Acknowledgments: The realization of the research presented in this paper, most prominently, experimental research, was supported by companies "Geoexpert Ltd.", "Construction Center Ltd." and "Invest Pro Centar Ltd." from Subotica.

Conflicts of Interest: The authors declare no conflict of interest.

References

1. Csak, B. *A Study on the Assessment of the Condition of the Cathedral of St. Theresa of Avila in Subotica (Tanulmány a szabadkai Avilai szent Teréz székesegyház állapotának értékeléséről)*; Department of Mechanics, Materials and Structures, Budapest University of Technology and Economics: Budapest, Hungary, 2009. (In Hungarian)
2. *Preliminary Repair Design of the Roman Catholic Church of St. Theresa of Avila in Subotica*; Šidprojekt JSC: Šid, Serbia, 2015. (In Serbian)
3. *Technical Report on the Testing of the Mortar and Bricks from the Outer Walls of the Church of the St. Theresa of Avila in Subotica*; Institute IMS, JSC: Belgrade, Serbia, 2015. (In Serbian)
4. *Technical Report, No. 02/15-20, on the Monitoring of the Horizontal and Vertical Displacements*; Geopanonija Ltd.: Novi Sad, Serbia, 2015. (In Serbian)
5. *Technical Report, No. EG-041/15, on the Geomechanical Conditions for the Construction*; Geoexpert Ltd.: Subotica, Serbia, 2015. (In Serbian)
6. *Structural Design: Strengthening of the Foundation and the Foundation Soil, Stiffening the Towers and the Frontage Wall Cathedral St. Theresa of Avila in Subotica*; Geoexpert Ltd.: Subotica, Serbia, 2016. (In Serbian)
7. *Technical Report on the Testing of the Soil Permeability before and after the Soil Injection (Talajtömörség vizsgálata Injektálás előtt es injektálás után)*; Csillagtér Építőipari Ltd.: Budapest, Hungary, 2018. (In Hungarian)
8. *Technical Report on the Stabilization of the Soil beneath the Towers of the Cathedral of St. Theresa of Avila in Subotica (Jelentés a talaj stabilizálódásáról a szabadkai Avilai Szent Terézia székesegyház tornyai alatt)*; ABA Inovator Ltd.: Subotica, Serbia, 2019. (In Hungarian)
9. *Technical Report on the Measurement of the Verticality of the Towers of the Cathedral of St. Theresa of Avila in Subotica*; Geosoft Ltd.: Belgrade, Serbia, 2017. (In Serbian)
10. Monthly Technical Reports on the Settlements of the Geodetic Markers on the Cathedral of St. Theresa of Avila in Subotica; Subotica, Serbia. From 2017 to 2021. Available online: http://www.szt.bme.hu/18-tartoszerkezet-tervezk/a-magyar-statikus-tarsadalom-nagyjai?start=8 (accessed on 27 May 2021). (In Serbian).
11. Mayne, P.W. In-Situ Test Calibrations for Evaluating Soil Parameters. Characterization and Engineering Properties of Natural Soils. In Proceedings of the Second International Workshop on Characterization and Engineering Properties of Natural Soils, Singapore, 29 November–1 December 2006; Volume 3, pp. 1–56.
12. Timoshenko, S.P.; Goodier, J.N. *Theory of Elasticity*, 3rd ed.; McGraw-Hill: New York, NY, USA, 1982; pp. 398–409.
13. Taylor, D.W. *Fundamentals of Soil Mechanics*; John Wiley: New York, NY, USA, 1948.
14. Nonveiller, E. *Soil Mechanics and Foundation Structures*; Školska Knjiga: Zagreb, Croatia, 1980. (In Croatian)
15. Segars, S. Uretek Deep Injection Method—Full Scale Test. Master's Thesis, Faculty of Civil Engineering and Geosciences, TU Delft,, Delft, The Netherlands, 2006.
16. Ferraioli, M.; Lavino, A.A. Seismic Assessment, Repair and Strengthening of a Medieval Masonry Tower in Southern Italy. *Int. J. Civ. Eng.* **2020**, *18*, 967–994. [CrossRef]
17. Nuzzo, M.; Faella, G. The carmine maggiore bell tower: An inclusive and sustainable restoration experience. *Sustainability* **2021**, *1445*, 1–30.
18. Ivorra, S.; Brotóns, V.; Foti, D.; Diaferio, M. A preliminary approach of dynamic identification of slender buildings by neuronal networks. *Int. J. Non-Linear Mech.* **2016**, *80*, 183–189. [CrossRef]
19. Diaferio, M.; Foti, D. Seismic risk assessment of Trani's Cathedral bell tower in Apulia, Italy. *Int. J. Adv. Struct. Eng.* **2017**, *9*, 259–267. [CrossRef]
20. Diaferio, M.; Foti, D.; Sabbà, M.F.; Lerna, M. A procedure for the seismic risk assessment of the cultural heritage. *Bull. Earthq. Eng.* **2021**, *19*, 1027–1050. [CrossRef]
21. Garzón-Roca, J.; Adam, J.M.; Sandoval, C.; Roca, P. Estimation of the axial behaviour of masonry walls based on Artificial Neural Networks. *Comput. Struct.* **2013**, *125*, 145–152. [CrossRef]

applied
sciences

MDPI

Article

Investigation of the Time Dependence of Wind-Induced Aeroelastic Response on a Scale Model of a High-Rise Building

Fabio Rizzo 

Deapartmet of Engineering and Geology, Gabriele D'Annunzio University, viale Pindaro 42, 65013 Pescara, Italy; fabio.rizzo@unich.it

Abstract: Experimental wind tunnel test results are affected by acquisition times because extreme pressure peak statistics depend on the length of acquisition records. This is also true for dynamic tests on aeroelastic models where the structural response of the scale model is affected by aerodynamic damping and by random vortex shedding. This paper investigates the acquisition time dependence of linear transformation through singular value decomposition (SVD) and its correlation with floor accelerometric signals acquired during wind tunnel aeroelastic testing of a scale model high-rise building. Particular attention was given to the variability of eigenvectors, singular values and the correlation coefficient for two wind angles and thirteen different wind velocities. The cumulative distribution function of empirical magnitudes was fitted with numerical cumulative density function (CDF). Kolmogorov–Smirnov test results are also discussed.

Keywords: aeroelastic experiments; experimental uncertainty; singular value decomposition; correlation field

Citation: Rizzo, F. Investigation of the Time Dependence of Wind-Induced Aeroelastic Response on a Scale Model of a High-Rise Building. *Appl. Sci.* **2021**, *11*, 3315. https://doi.org/10.3390/app11083315

Academic Editors: Chiara Bedon, Flavio Stochino and Daniel Honfi

Received: 17 March 2021
Accepted: 29 March 2021
Published: 7 April 2021

Publisher's Note: MDPI stays neutral with regard to jurisdictional claims in published maps and institutional affiliations.

1. Introduction

Uncertainty related to experiments with scale models is an issue which is currently being discussed by the international scientific community. The reason for this is that codes of practice do not give experimental protocols [1,2] and that scientific literature focuses on specific cases. Error propagation due to the small scale of experimental models is known to affect results at the prototype scale [3–8].

The most common experimental errors can be grouped in three families: instrumental error, inaccuracy error and random repetitiveness error. Instrumental error is due to the limits of the instrument and is checked before the experiment is set up [3]. This is an awareness error and is commonly taken into account. Inaccuracy errors can be due to several aspects such as the inaccuracy of the model, inaccuracy when performing experiments and others. The small scale of the models [9] used for wind tunnel tests generally causes the most common inaccuracy errors. Small scales may cause loss of important details or undesired asymmetries or scaling inaccuracy due to the Reynolds number effects. However, the modern technology in the field of the 3D printer and control systems reduces the inaccuracy due to the model geometry. The situation for the Reynolds number effects is different because this is a long-standing problem that should be investigated by repeating tests many times.

The last family of experimental errors, random repetitiveness errors, is the hardest to check or to plan before experiments because it depends primarily on the amount of time and economic resources available for experiments. It is well known that the number of repetitions of experiments affects results at the prototype scale [3–5]. This is particularly relevant in the case of experiments that cannot be totally automated. Rizzo and Caracoglia (2018) [4] showed that the number of repetitions affects flutter critical velocity estimation during wind tunnel aeroelastic tests. Rizzo et al. (2020) [3] showed that the number of experiment repetitions affects dynamic identification of an aeroelastic model and consequently affects the prediction of magnitudes at the prototype scale.

Acquisition time is another important aspect to take into account during dynamic experiments. Acquisition time in the wind tunnel is a relevant issue for both pressure and force tests and for aeroelastic tests. This dependence is discussed in literature with particular relevance to peak statistics. The maxima and minima of a random process depend on the recorded time history length. Most analytical models used to predict peak factors are affected by this aspect [10–20]. In the case of dynamic experiments, where structural response is affected by nonlinearity and vorticity, model vibration is related to acquisition time length even if recording processes in the wind tunnel are mostly stationary.

This paper discusses the acquisition time dependence of wind-induced floor acceleration on a scale model high-rise building. Individual values, eigenvectors and floor acceleration correlations between different levels and wind angles are assumed as significant magnitudes in order to have a measure of variability due to acquisition time length.

The complete recorded processes of floor accelerations at different levels of a high-rise building scale model were subdivided into random sub-processes. The variability of the singular values, eigenvectors and correlation coefficients is discussed.

Sections 2 and 3 discuss the structural setup of the prototype, the experimental model scaling and construction procedure and, finally, the wind tunnel aeroelastic tests. Section 4 discusses the acquisition time dependence of the individual values and of the wind induced floor acceleration correlation coefficient between both different levels of the building and different wind directions.

2. Structural and the Experimental Setup

The building being investigated is a 60-floor high-rise building with a cruciform plan inscribable in a 138.1 × 138.1 m square. Total height is 300 m, and the main structure is made of steel columns and reinforced concrete walls. Floors are made of truss steel structures, and slabs change their dimensions alternately from one floor to another. Structural elements were designed according to [1,2] through a finite element method (FEM) model illustrated in Figure 1a.

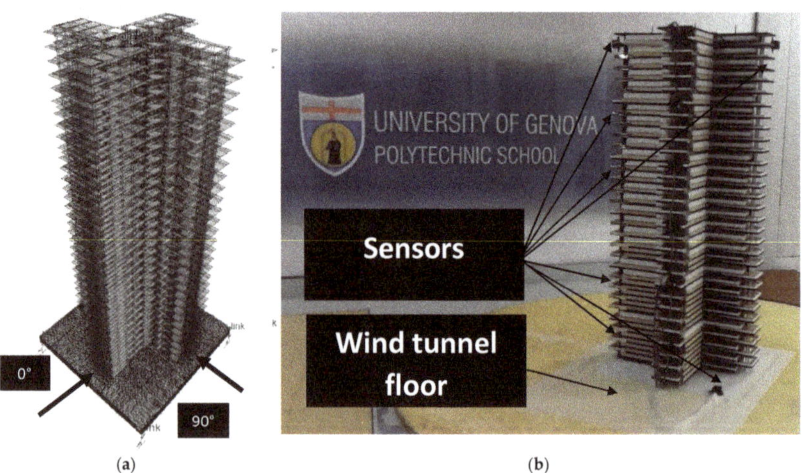

Figure 1. Structural (**a**) and experimental (**b**) setup.

The first two natural frequencies, estimated through modal analyses, are horizontal, along X and Y, and equal to 0.12 Hz (Y) and 0.13 Hz (X). The structural damping ratio was assumed equal to 1.3%. Estimated mean wind velocity at the top of the building, v_m was 48.5 m/s for $T_{R0} = 50$ years [1].

A scaled model to carry out aeroelastic wind tunnel experiments was designed and constructed according to the aeroelastic scaling procedure given by [9]. Velocity scale was calibrated to obtain a value of 5.9 m/s (a prototype scale 48.5 m/s considering the blockage correction) at the top of the model. The first natural frequency and mass were 5.86 Hz and 7.95 kg. Scaling values are: $\lambda_L = 2.50 \times 10^{-3}$ (a geometric scale of 1:400), $\lambda_V = 0.12$ (velocity scale), $\lambda_a = 5.93$ (acceleration scale), $\lambda_\eta = 48.71$ (frequency scale), $\lambda_m = 1.56 \cdot 10^{-8}$ (mass scale), $\lambda_{I,m} = 9.76 \times 10^{-4}$ (inertia scale) and, finally, $\lambda_t = 0.02$ (time scaling). The structural damping ratio was 1.3% for both prototype and model scale [10]. The scale model was made of steel and wood (Figure 1b). The floor slabs were made of 4 mm thick poplar wood. Four $2 \times 30 \times 750$ mm steel plates were used to simulate walls and together with four 4 mm circular steel bars, located in the center of the wings were used to reproduce the total vertical load-bearing structure. The connection between slabs and walls is obtained through steel screws. Steel plates and bars are welded. The model mass is concentrated on six different points.

The first natural frequency and structural damping of the aeroelastic model were identified by experiments on the joint DIST-UNINA and ICD-CNR shaking table at the University of Napoli Federico II, Italy, laboratory. The first modal frequency, structural damping ratio and first modal shape were assessed [3,21,22].

3. Wind Tunnel Experimental Results

Aeroelastic testing was carried out at the DICCA laboratory at the University of Genoa, Italy. The DICCA wind tunnel is a closed-loop subsonic circuit tunnel with a cross-section of 1.70 (width) $\times$ 1.35 (height) m. A total of three different wind angles were investigated, 0° (along wind), 45° and 90° (across wind), according to a suburban Terrain I velocity profile. Correction factors were used to take blockage effects into account. The factors ranged between 1.028 (for 0° and 90°) and 1.04 (for 45°), [23] and were applied to the velocity and acceleration scale factors. Nine accelerometers monitored acceleration signals. Six accelerometers were located on six floors at z/H = 0.17, 0.33, 0.5, 0.67, 0.83; one accelerometer was linked to the wind tunnel basement; and two accelerometers were located on the top floor of the model. A total of thirteen different velocities were tested, ranging from 5.4 to 112 m/s at the building top at prototype scale.

Accelerations were measured at a sampling frequency of 1000 Hz for a time length of 180 s. At the prototype scale, this corresponds to a time step of 0.05 s and a time length of 9000 s.

The entire process, to investigate the variability of magnitudes such as singular values or eigenvectors, was subdivided into fifteen 10 min long recording sessions [24–29]. Figure 2 shows two examples of processes with $U = 49.0$ m/s, along and across wind, subdivided in sub-processes. The along versus across wind acceleration point cloud graph shown in Figure 3 shows that the difference between sub-processes (in the 10 min records) is relevant because it changes significantly as for example with T [0; 9000] (Figure 3c), T [2400; 3000] (Figure 3d), T [7200; 7800] (Figure 3e) and T [840; 9000] (Figure 3f) [24–29].

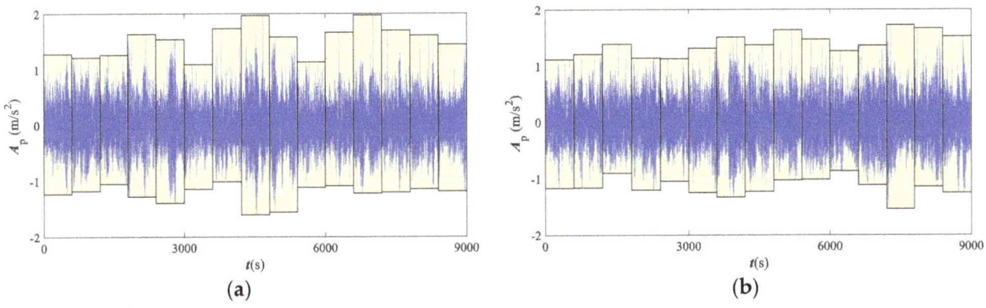

Figure 2. Ten minute accelerometric signal records at the building top with $U = 49.0$ m/s, along wind (0°) (**a**) and across wind (90°) (**b**).

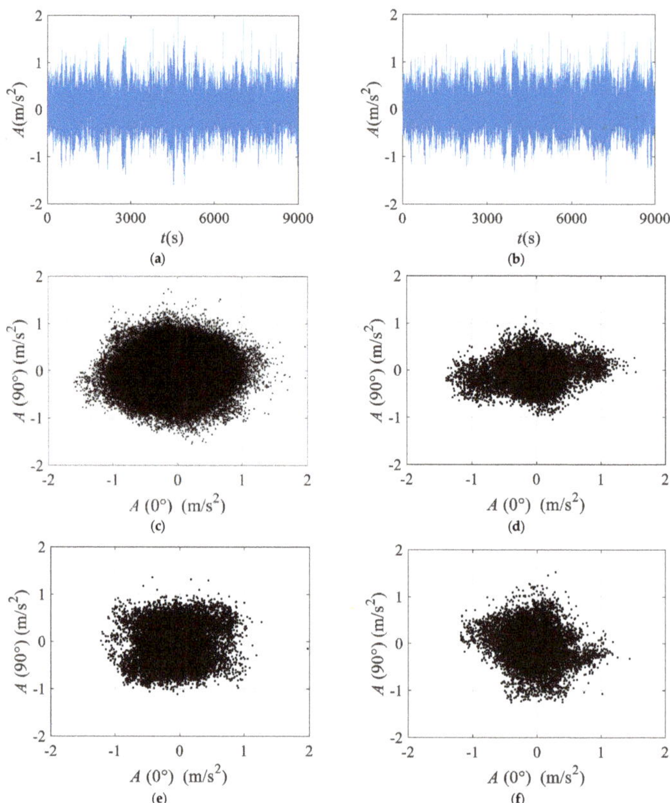

Figure 3. Building top acceleration signal with $U = 49.0$ m/s, along wind (**a**), across wind (**b**), along wind versus across wind with T [0; 9000] (**c**), T [2400; 3000] (**d**), T [7200; 7800] (**e**), T [840; 9000] (**f**).

4. Time Dependence of Wind Induced Floor Acceleration

4.1. Singular Value Decomposition

In linear algebra, the singular value decomposition (SVD) of a matrix is a factorization of that matrix into three matrices. This value has some interesting algebraic properties and conveys important geometrical and theoretical insights about linear transformations. It also has some important applications in data science [30]. Singular value decomposition takes a rectangular matrix of gene expression data (defined as A, where A is a $n \times p$ matrix) in which the n rows represent the genes, and the p columns represent the experimental conditions. The SVD theorem states:

$$A_{n \times p} = U_{n \times n} \times S_{n \times p} \times V_{p \times p}^{T} \tag{1}$$

where the $U_{n \times n}$ columns are the left singular vectors (gene coefficient vectors); $S_{n \times p}$ (the same dimensions as A) has singular values and is diagonal (mode amplitudes); and $V_{p \times p}^{T}$ has rows that are the right singular vectors (expression level vectors). The SVD represents an expansion of the original data in a coordinate system where the covariance matrix is diagonal. Calculating the SVD consists in finding the eigenvalues and eigenvectors of $A_{n \times p} \times A_{n \times p}^{T}$ and $A_{n \times p}^{T} \times A_{n \times p}$. The eigenvectors of $A_{n \times p}^{T} \times A_{n \times p}$ make up the columns of $V_{p \times p}$; the eigenvectors of $A_{n \times p} \times A_{n \times p}^{T}$ make up the columns of $U_{n \times n}$. The singular values in $S_{n \times p}$ are also square roots of eigenvalues from $A_{n \times p} \times A_{n \times p}^{T}$ or $A_{n \times p}^{T} \times A_{n \times p}$. The singular values are the diagonal entries of the $S_{n \times p}$ matrix and are arranged in descending order. The singular values are always real numbers. If matrix $A_{n \times p}$ is a real matrix, then $U_{n \times n}$ and $V_{p \times p}$ are also real.

The singular value decomposition of the $A_{180000 \times 6}$ acceleration matrix was estimated for two different wind angles, $0°$ and $90°$, respectively along and across wind, and thirteen different wind velocities, in order to examine the variability of $V_{6 \times 6}$ and the diagonal of $S_{180000 \times 6}$ computed for different 10 min records.

4.1.1. Singular Value Variability

The diagonal of $S_{180000 \times 6}$ was estimated for each fifteen 10 min record of the wind induced floor accelerations. Figure 4 shows the diagonal of $S_{180000 \times 6}$ variability along wind and across wind for three significate velocities at the building top: a small velocity, $U = 5.4$ m/s; the estimated by CNR DT 207, 2018 velocity, $U = 49.0$ m/s; and a very large velocity, $U = 111.8$ m/s.

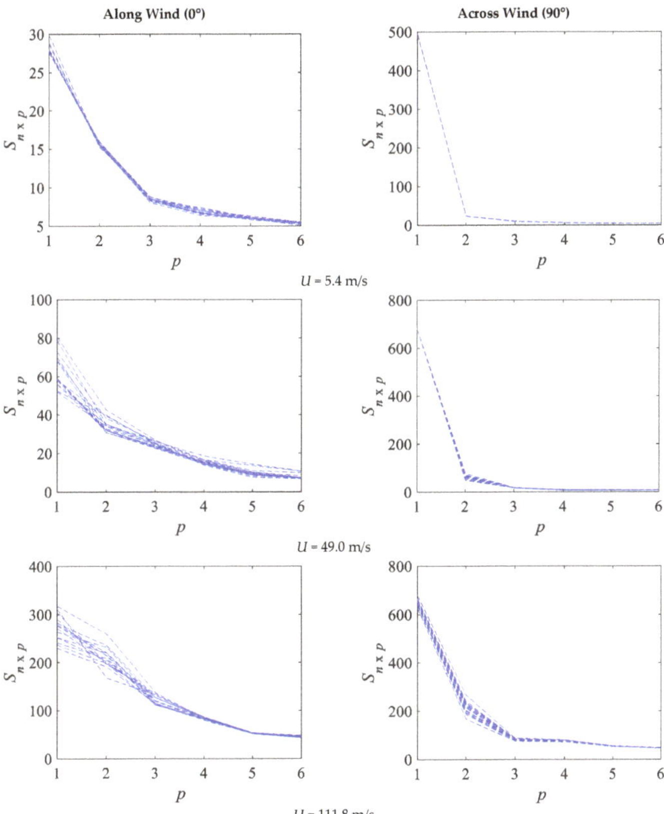

Figure 4. Variability of the singular values $S_{n \times p}$ with different velocities.

Figure 4 shows that the variability is significant for the first three modes (i.e., $p = 1$, 2 and 3) and along wind. Variability increases when velocities increase and is much smaller across wind. This suggests that the across wind singular values are not affected by the random nature of the wind flow. Along wind, by contrast, the singular values significantly depend on the recorded records. They vary up to 40% for the highest velocity. It is very important to specify that the estimated modes through the Property Orthogonal Decomposition (POD) methodology and their discussion only refer to the experimental model. The experimental model was calibrated only on the first mode of the prototype, and for this reason, there is no reliable correspondence for modes higher than the first between the test model and the prototype structure. However, the paper goal is to investigate and to discuss the variability of the test model modes as a function of acquisition time length, and for this reason, the discussion on the higher modes is being reported.

4.1.2. Eigenvector Variability

Eigenvectors were estimated along and across wind with different velocities and for each 10-min record in order to examine the variability at each level (levels from 1 to 6, from base to top). Figure 5 shows the first three eigenvectors along wind (Figure 5a) and across wind (Figure 5b) variability. Different curves overlapping in Figure 5 represent the eigenvector shape for each subinterval. It was observed that the first mode is very similar for both oscillation directions, whereas modes 2 and 3 are quite different between along wind and across wind. The eigenvector variability increases for the first modes for both along wind and across wind, and it increases with increasing velocities as is illustrated in Figure 6 for both oscillation directions. In addition, eigenvectors have a trend which is quite different with increasing velocities. This result seems significant because it means that eigenvector trends are affected by the randomness of the wind flow, and consequently, their variability should be taken into account.

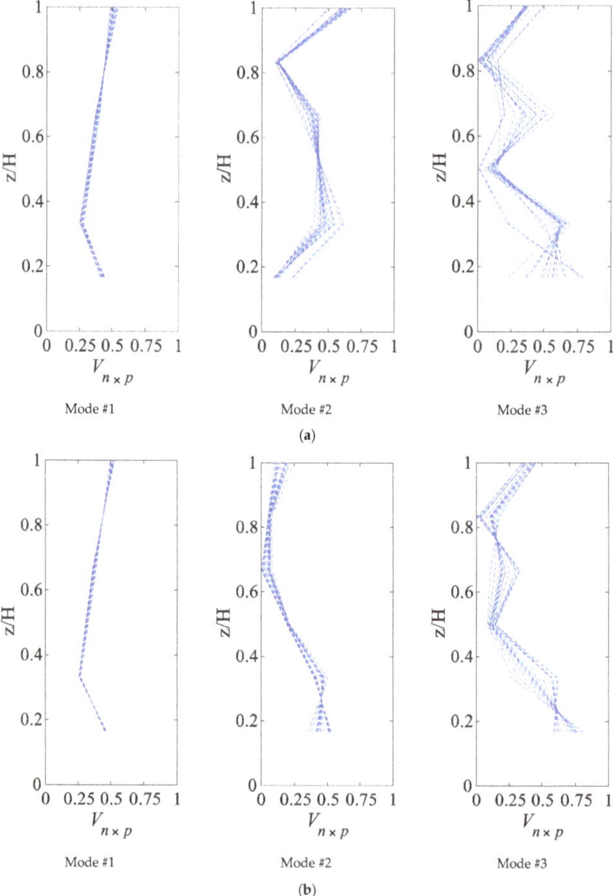

Figure 5. Eigenvectors along wind (**a**) and across wind (**b**) variability with $U = 49.0$ m/s.

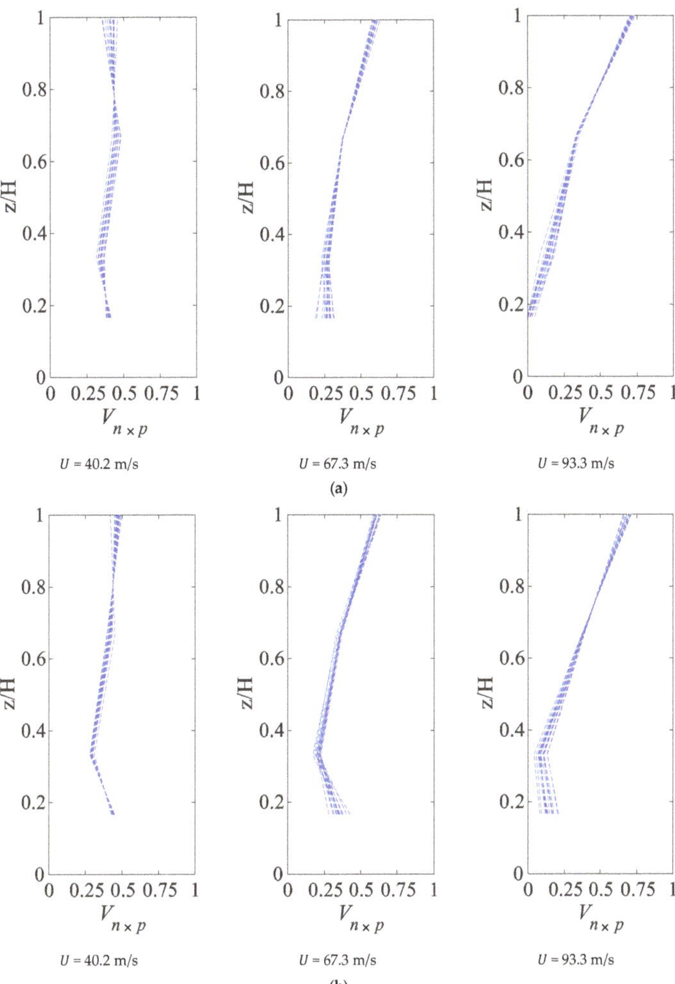

Figure 6. Eigenvector #1 along wind (**a**) and across wind (**b**) variability as a function of the wind velocity.

The eigenvectors were calculated for each subinterval (i.e., fifteen), for each wind velocity (i.e., thirteen), and for each investigated building level (i.e., six). The standard deviation of eigenvector #1 for different subintervals has been plotted as a function of the wind velocity and building floor in Figure 7. Figure 7 shows the three-dimensional variability of the standard deviation of the eigenvectors for Mode #1 along (Figure 7a) and across (Figure 7b) wind. The standard deviation is much larger along wind than across wind for all wind velocities tested, confirming the variability shown in Figures 5 and 6.

The empirical cumulative density function (CDF) of the fifteen different values was estimated for each building level wind velocity and wind angle to give a measure of the probabilistic trends of the eigenvectors. The empirical CDF and numerical CDF are not the same distribution. The Kolmogorov-Smirnov (K–S) test was repeated by varying significance levels from 5% to 50%, and similar results were obtained. The best fitting, in any case, is given by the Generalized extreme value distribution (GEV) distribution even though it is not satisfactory.

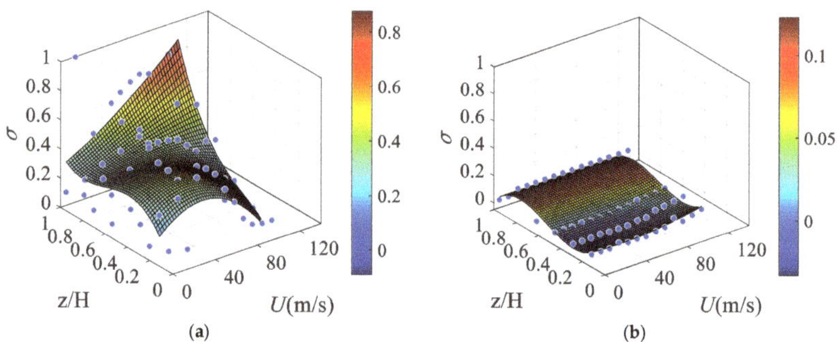

Figure 7. Standard deviation variation for Mode #1 along wind (**a**) and across wind (**b**).

Figure 8 shows some examples. The empirical CDF was fitted through numerical CDF as the Normal, Gamma, Weibull, GEV and Gumbel distribution CDF. The two-sample Kolmogorov–Smirnov (K-S) [31] test was carried out to look for the best fitting. The KS test did not reject (i.e., failed to reject) the null hypothesis of inequality at the 5% significance level.

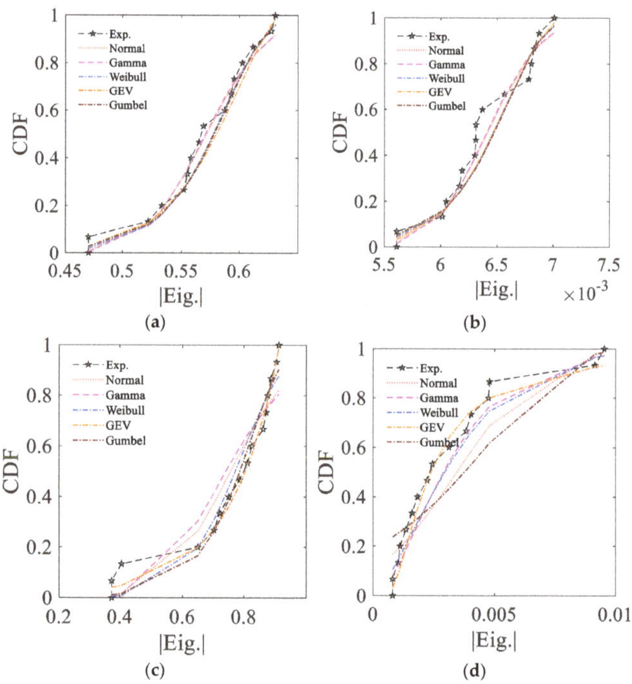

Figure 8. Cumulative density function (CDF) of Mode #1 and z/H = 1 of the eigenvector with U = 49.0 along wind (**a**) and across wind (**b**) and 111.8 m/s along wind (**c**) and across wind (**d**).

4.2. Time Dependence of the Floor Acceleration Correlation Coefficient

The space correlation of two random processes is usually taken as a measure of how much one variable is related to another variable. In the case of a high-rise building, the correlation of floor acceleration at different building levels is a measure of the effects of aerodynamics, such as vortex shedding, and aeroelastics, such as nonlinear structural movements.

A further attempt was made to calculate a measure of response variability due to different acquisition records, examining the correlation coefficients $\rho_{(6,1\ldots5)}$ between the signal on the top, level #6 and on the other five levels, #1, #2, #3, #4 and #5, corresponding to $z/H = 1/6, 2/6, 3/6, 4/6$ and $5/6$, for each 10 min recording session.

Figure 9 shows the variability of the correlation coefficient $\rho_{(6,1\ldots5)}$ estimated for fifteen 10 min recording sessions with two significant velocities ($U = 49.0$ m/s and 111.8 m/s) and two wind angles (0° and 90°). The variability of ρ ranges from 5% to 75% and consequently is non-negligible. Its variability is larger along wind (Figure 9a) than across wind (Figure 9b) and increases with increases in velocity.

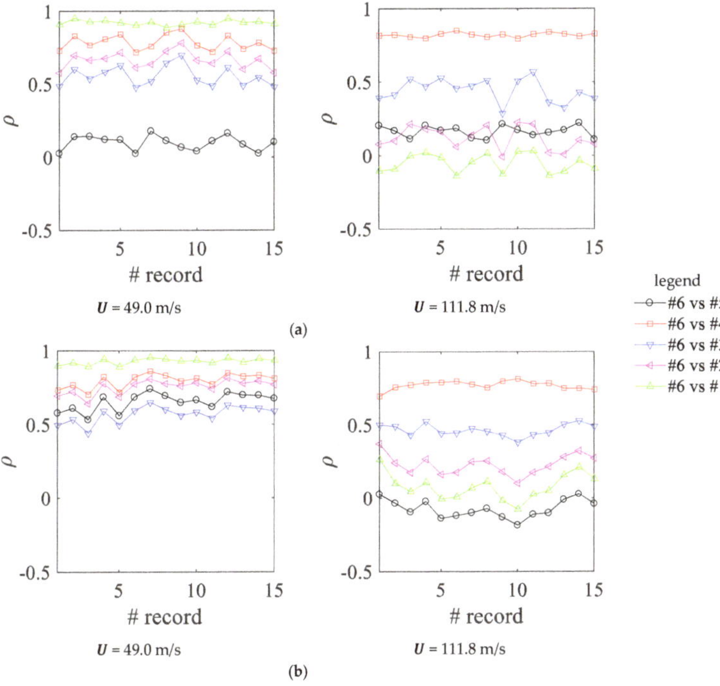

Figure 9. Variability of the correlation coefficient from the building top (i.e., level #6) to levels #1, #2, #3, #4 and #5 along wind (**a**) and across wind (**b**).

It can be seen that the best along wind correlation for all velocities was calculated between levels #6 and #1, while the best across wind correlation was between levels #6 and #4. The correlation between two adjoining intermediate floors is quite low for all velocities and wind directions. This seems to suggest that along wind aerodynamics, in the range between $z/H = 0.33$ and $z/H = 0.83$, gives significant vorticity that induces a structural vibration different from that at the top. This effect is probably due to the not slender shape of the building, to the slabs and to the free vibration associated with the first structural modal shape which, in this case, is not a pendulum motion like that of a simple harmonic oscillator.

The standard deviation of the ρ coefficient estimated for all fifteen along and across wind short (10 min) records ranges between 0.01 and 0.1 as shown in Figure 10. Figure 10 shows that the standard deviation, and consequently ρ variability, depends on flow velocity and on floor position. Values for the correlation between levels #6 and #5 along wind seem anomalous because the correlation is very small. It is reasonable to think that acquisitions were affected by noise during tests. The difference between small and large velocities is relevant but not the same for all levels. The three-dimensional trend shows a valley around z/H = 0.5 for all velocities and two peaks at the top and close to the base for the highest velocities. This irregular trend of the standard deviation as a function of wind velocity and height above ground confirms that structural vibration is very different between the top and the other levels and also that the correlation coefficient is affected by the randomness of the wind flow distribution, consequently varying significantly from one time interval to another.

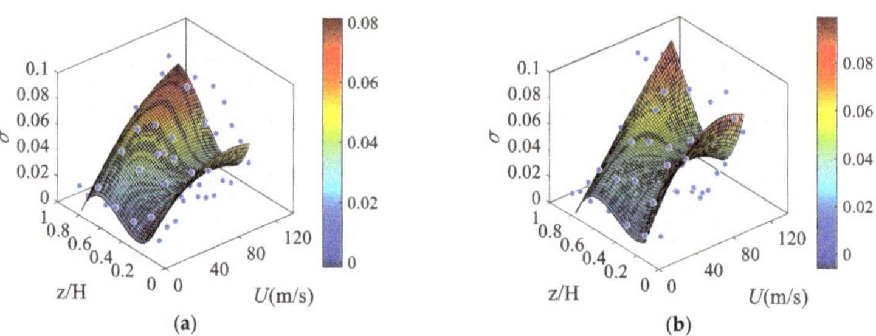

Figure 10. Standard deviation variation of $\rho_{(6,1...5)}$, along wind (**a**) and across wind (**b**).

The correlation coefficient between along and across wind floor acceleration was estimated, and its variability shown in Figure 11a for two significant velocities, $U = 49.0\,\text{m/s}$ and 111.8 m/s. It can be seen that the variability is slightly larger for $U = 49$ m/s than $U = 111.8$ m/s. This effect is evidenced in Figure 11b that shows the three-dimensional trend of the standard deviation of the correlation coefficient for the fifteen records as a function of wind velocity and building levels. It was observed in the range from 40 to 80 m/s. It is reasonable to think that in the range from 40 m/s to 80 m/s, a resonance between oscillation along wind and across wind occurs, although this aspect should be carefully investigated through additional tests.

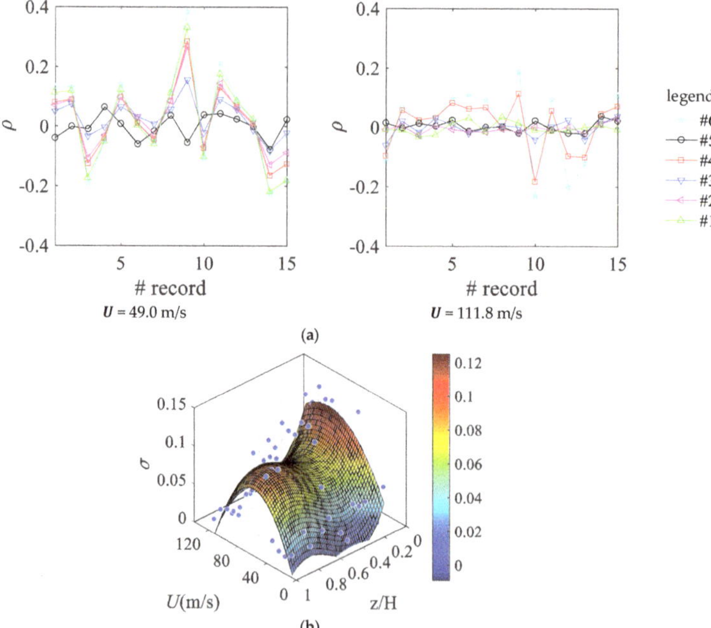

Figure 11. Variability of the correlation coefficient at different levels between along and across wind (**a**) and the standard deviation as a function of wind velocity and building levels (**b**).

The standard deviation ranges from 0.01 to 0.15, taking its maximum value at low wind velocity levels, in the range from 40 m/s to 80 m/s. Figure 12 shows the CDF of the correlation coefficient estimated between level #6 and levels from #1 to #5. Figure 12 shows that, for $U = 49.0$ m/s, the GEV distribution gives the best fitting of the empirical CDF even if, as with the eigenvectors, the KS tests accepted the null hypothesis of inequality for all numerical CDFs.

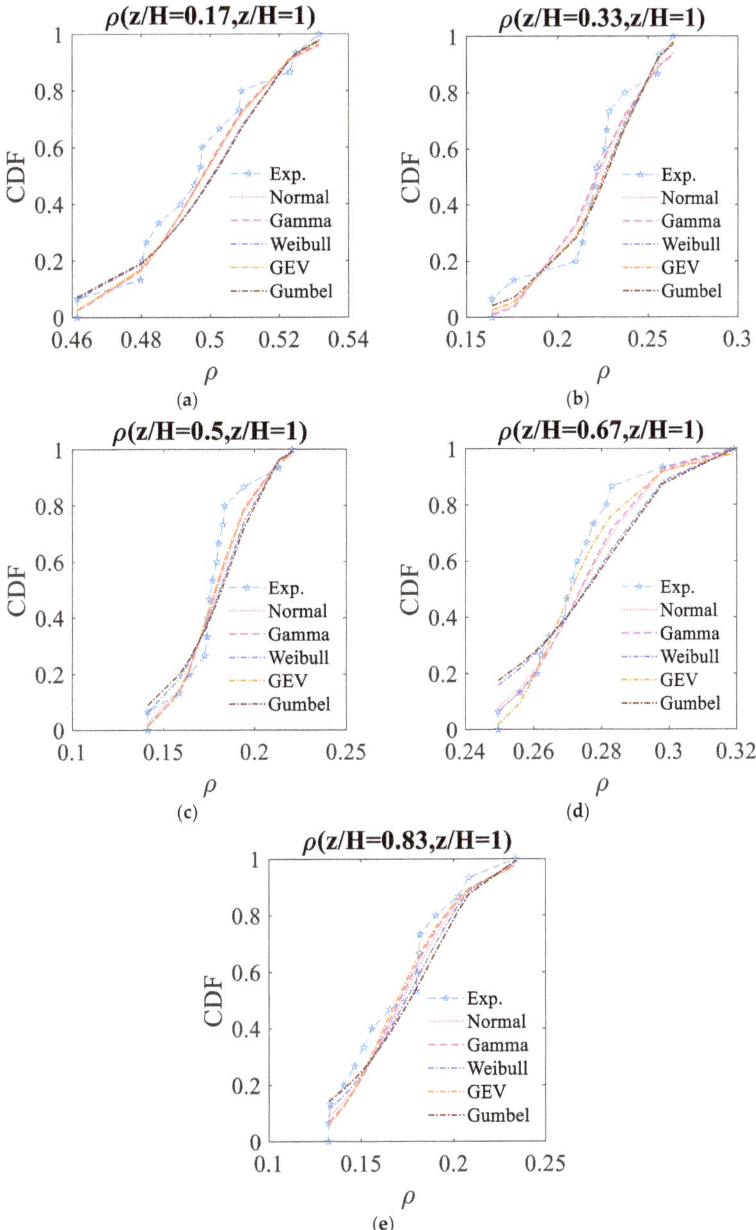

Figure 12. CDF of the correlation coefficient r between level #6 and (**a**) level #1 (**b**) level #2 (**c**) level #3; (**d**) level #4; and (**e**) level #5; along wind with $U = 49$ m/s.

5. Conclusions

Acquisition time dependence of significant magnitudes, such as singular values eigenvectors and correlation coefficients of wind-induced floor accelerometric signals acquired on a scale model of a high-rise building, is discussed by examining their variability from one time interval to another.

The entire wind tunnel data acquisition process was subdivided into fifteen 10 min recording sessions.

The SVD was calculated for two wind angles, along and across wind, and for thirteen wind velocities for all recording sessions. It was found that variability of the singular value increases when the wind velocity increases and that there is a significant difference between along and across wind values. Across wind variability is much smaller than along wind. The same trend was noted for the eigenvectors. The probability density function (CDF) of the eigenvectors for different velocities and at different levels was fitted through numerical CDF. It was found that the GEV distribution gave the best fitting even if it was not satisfactory according to the Kolmogorov–Smirnov test. The three-dimensional trend of standard deviation as a function of building levels and wind velocities shows that the variability of the eigenvectors assumes its maximum along wind value at the building top and for the highest velocities.

Likewise, the correlation coefficient between the top and the other levels was calculated, and its variability for the fifteen 10 min recording sessions was investigated. It was found that these values are closely affected by the randomness of the flow because it varies significantly from one time interval to another. The correlation coefficient at the building top between along and across wind measurements was calculated and was found to vary slightly more for velocities in the range from 40 m/s to 80 m/s and for low levels.

The results discussed here show that uncertainty due to the randomness of wind tunnel wind flow affects structural response during aeroelastic tests. The dependence of results from the acquisition time length was confirmed and nonnegligible variability was observed. In particular, it was observed that the variability of the eigenvectors under wind increases with the wind velocity and distant from the ground along wind, whereas this trend flattens across wind. This uncertainty should be taken into account during experiments and output processing, examining the cumulative probability trends and assuming a reliable level of confidence for the estimated crucial magnitudes, such as the modal shapes, frequencies and damping ratio.

Funding: This research was supported by the Italian Ministry of Education.

Data Availability Statement: The data used to support this research are available from the corresponding author upon request.

Acknowledgments: The Author thanks Andrea Freda and Giuseppe Piccardo for their significant support during the experimentation sessions in the DICCA Wind Tunnel of the University of Genoa and Giuseppe Maddaloni, University of Sannio (Benevento), for his support during shaking table experiments. He is also grateful to Davide Perazzetti for his support during the model construction and to the company H-Sign srl for its support of this research.

Conflicts of Interest: The author declares no conflict of interest.

References

1. CNR (National Research Council of Italy). *Guide for the Assessment of Wind Actions and Effects on Structures*; CNR-DT 207/2008. CNR: Rome, Italy, 2008.
2. CEN (Comité Européen de Normalization). *Eurocode 1: Actions on Structures—Part 1–4: General Actions—Wind Actions*; EN-1991-1-4; CEN: Brussels, Belgium, 2005.
3. Rizzo, F.; Ricciardelli, F.; Maddaloni, G.; Bonati, A.; Occhiuzzi, A. Experimental error analysis of dynamic properties for a reduced-scale high-rise building model and implications on full-scale behaviour. *J. Build. Eng.* **2020**, *28*, 101067. [CrossRef]
4. Rizzo, F.; Caracoglia, L. Examining wind tunnel errors in Scanlan derivatives and flutter speed of a closed-box. *J. Wind Struct.* **2018**, *26*, 231–251.

5. Rizzo, F.; Caracoglia, L.; Montelpare, S. Predicting the flutter speed of a pedestrian suspension bridge through examination of laboratory experimental errors. *Eng. Struct.* **2018**, *172*, 589–613. [CrossRef]

6. Desceliers, C.; Soize, C.; Ghanem, R. Identification of chaos representations of elastic properties of random media using experimental vibration tests. *Comput. Mech.* **2007**, *36*, 831–838. [CrossRef]

7. Schoefs, F.; Yáñez-Godoy, H.; Lanata, F. Polynomial chaos representation for identification of mechanical characteristics of instrumented structures. *Comput. Aided Civ. Infrastruct.* **2011**, *26*, 173–189. [CrossRef]

8. Seo, D.W.; Caracoglia, L. Derivation of equivalent gust effect factors for wind loading on low-rise buildings through Database-Assisted-Design approach. *Eng. Struct.* **2010**, *32*, 328–336. [CrossRef]

9. Isyumov, N. The Aeroelastic Modelling of Tall Buildings. In *Proceedings of the International Workshop on Wind Tunnel Modelling Criteria and Technique in Civil Engineering Applications, Gaithersburg, MD, USA, April 1982*; Timothy, A.R., Ed.; Cambridge University Press: Cambridge, UK, 1982.

10. Huang, M.; Huang, C.M.; Chan, C.M.; ChanWen-juan, L.; Chung-Siu, K. Statistical extremes and peak factors in wind-induced vibration of tall buildings. *J. Zhejiang Univ. Sci. A Appl. Phys. Eng.* **2012**, *13*, 18–32. [CrossRef]

11. Huang, M.F.; Lou, W.; Chan, C.M.; Lin, N.; Pan, X. Peak distributions and peak factors of wind-induced pressure processes on tall buildings. *J. Eng. Mech.* **2013**, *139*, 1744–1756. [CrossRef]

12. Huang, M.F.; Huang, S.; Feng, H.; Lou, W.J. Non-Gaussian time-dependent statistics of wind pressure processes on a roof structure. *Wind Struct.* **2016**, *23*, 275–300. [CrossRef]

13. Huang, G.; Luo, Y.; Gurley, K.R.; Ding Ying Luo, J. Revisiting moment-based characterization for wind pressures. *J. Wind Eng. Ind. Aerodyn.* **2016**, *151*, 158–168. [CrossRef]

14. Huang, G.; Luo, Y.; Gurley, K.R.; Ding Ying Luo, J. Response to Revisiting moment-based characterization for wind pressures. *J. Wind Eng. Ind. Aerodyn.* **2016**, *158*, 162–163.

15. Kareem, A.; Tognarelli, M.A.; Gurley, K.R. Modeling and analysis of quadratic term in the wind effects on structures. *J. Wind Eng. Ind. Aerodyn.* **1998**, *74–76*, 1101–1110. [CrossRef]

16. Kareem, A.; Zhao, J. Analysis of non-Gaussian surge response of tension leg platforms under wind loads. *J. Offshore Mech. Arct. Eng.* **1994**, *116*, 137–144. [CrossRef]

17. Gurley, K.R.; Tognarelli, M.A.; Kareem, A. Analysis and simulation tools for wind engineering. *Probabilistic Eng. Mech.* **1997**, *12*, 9–31. [CrossRef]

18. Gurley, K.; Kareem, A. Analysis, interpretation, modeling and simulation of unsteady wind and pressure data. *J. Wind Eng. Ind. Aerodyn.* **1997**, *69–71*, 657–669. [CrossRef]

19. Kwon, D.; Kareem, A. Peak Factors for Non-Gaussian Load Effects Revisited. *J. Struct. Eng.* **2001**, *137*, 1611–1619. [CrossRef]

20. Sadek, F.; Simiu, E. Peak non-Gaussian wind effects for database-assisted low-rise building design. *J. Eng. Mech.* **2002**, *128*, 530–539. [CrossRef]

21. Rizzo, F.; Maddaloni, G.; Occhiuzzi, A.; Prota, A.; Pagliaroli, A. Earthquake induced floor accelerations on a High-rise building: In Scale Tests on Shaking Table. In Proceedings of the Resilient Technologies for Sustainable Infrastructures (IABSE 2020), Christchurch, New Zeeland, 2–4 September 2020.

22. Rizzo, F.; Maddaloni, G.; Occhiuzzi, A.; Prota, A. High-rise building dynamics identification through shaking table measurements on scale model for multi-hazard experiments. In Proceedings of the ANIDIS 2019, Ascoli Piceno, Italy, 15–19 September 2019.

23. Ross, I.; Altman, A. Wind tunnel blockage corrections: Review and application to Savonius vertical-axis wind turbines. *J. Wind Eng. Ind. Aerodyn.* **2011**, *99*, 523–538. [CrossRef]

24. Cook, N.J.; Mayne, J.R. A novel working approach to the assessment of wind loads for equivalent static design. *J. Wind Eng. Ind. Aerodyn.* **1979**, *4*, 149–164. [CrossRef]

25. Cook, N.J.; Mayne, J.R. A refined working approach to the assessment of wind loads for equivalent static design. *J. Wind Eng. Ind. Aerodyn.* **1980**, *6*, 125–137. [CrossRef]

26. Cook, N.J. Calibration of the quasi-static and peak-factor approaches to the assessment of wind loads against the method of Cook and Mayne. *J. Wind Eng. Ind. Aerodyn.* **1982**, *10*, 315–341. [CrossRef]

27. Cook, N.J. *The Designer's Guide to Wind Loading of Building Structures*; Part 1: Background, Damage Survey, Wind Data and Structural Classification; Building Research Establishment: Butterworths, UK, 1985.

28. Cook, N.J. *The Designer's Guide to Wind Loading of Building Structures*; Part 2: Static Structures; Building Research Establishment: Butterworths, UK, 1990.

29. Harris, R.I. An improved method for the prediction of extreme values of wind effects on simple buildings and structures. *J. Wind Eng. Ind. Aerodyn.* **1982**, *9*, 343–379. [CrossRef]

30. Tamura, Y.; Yoshida, A.; Zhang, L.; Ito, T.; Nakata, S. Examples of modal identification of structures in Japan by FDD and MRD techniques. In *Proceedings of the EACWE4—The Fourth European & African Conference on Wind Engineering*; Naprstek, J., Fischer, C., Eds.; ITAM AS CR: Prague, Czech Republic, 2005.

31. Massey, F.J. The Kolmogorov-Smirnov Test for Goodness of Fit. *J. Am. Stat. Assoc.* **1951**, *46*, 68–78. [CrossRef]

applied
sciences

Article

Short-Term Analysis of Adhesive Types and Bonding Mistakes on Bonded-in-Rod (BiR) Connections for Timber Structures

Jure Barbalić [1], Vlatka Rajčić [1], Chiara Bedon [2,*] and Michal K. Budzik [3]

1   Faculty of Civil Engineering, University of Zagreb, 10000 Zagreb, Croatia; jure.barbalic@unizg.hr (J.B.); vrajcic@grad.hr (V.R.)
2   Department of Engineering and Architecture, University of Trieste, 34127 Trieste, Italy
3   Department of Mechanical and Production Engineering, Aarhus University, 8000 Aarhus, Denmark; mibu@mpe.au.dk
*   Correspondence: chiara.bedon@dia.units.it; Tel.: +39-040-558-3837

Abstract: Bonded-in rods (BiR) represent a structural connection type that is largely used for new timber structures and rehabilitation (repair or reinforcement) of existing structural members. The technology is based on steel / Fiber Reinforced Polymer (FRP) / Glass Fiber Reinforced Polymer (GFRP) rods bonded into predrilled holes in timber elements. The mechanical advantages of BiRs include high local force capacity, improved strength, a relatively high stiffness and the possibility of ductile behaviour. They also offer aesthetic benefits, given that rods are hidden in the cross sections of wooden members. As such, BiR connections are regarded as a solution with great potential, but still uncertain design formulations. Several research projects have dealt with BiRs, but a final definition of their mechanics and a universal design procedure is still missing. This research study explores the typical fracture mechanics modes for BiR connections. A special focus is given to the evaluation of the impact of adhesive bonds under various operational conditions (i.e., moisture content of timber). A total of 84 specimens are tested in pull-out setup, and investigated with the support of digital image correlation (DIC). The reliability of empirical equations and a newly developed analytical model in support of design, based on linear elastic fracture mechanics (LEFM), is also assessed.

Keywords: bonded-in rod (BiR) connections; adhesives; fracture modes; moisture; experiments; linear elastic fracture mechanics (LEFM); analytical model

Citation: Barbalić, J.; Rajčić, V.; Bedon, C.; Budzik, M.K. Short-Term Analysis of Adhesive Types and Bonding Mistakes on Bonded-in-Rod (BiR) Connections for Timber Structures. *Appl. Sci.* **2021**, *11*, 2665. https://doi.org/10.3390/app11062665

Academic Editor: Tore Brinck

Received: 16 February 2021
Accepted: 15 March 2021
Published: 17 March 2021

Publisher's Note: MDPI stays neutral with regard to jurisdictional claims in published maps and institutional affiliations.

1. Introduction

Glued-in rod (GiR) or bonded-in rod (BiR) connections are increasingly used in construction of timber structures [1], and so far several researchers addressed the mechanical performance of specific solutions of technical use in buildings.

Owning to their versatility, BiR connections are used extensively, and thus the need for proper assessment of their mechanical properties and standardized assembling procedures is ever increasing [2,3]. Research efforts have been spent to offer an accurate detail on BiR connections behaviour, but mainly for limited applications that can be hardly generalized. In this framework, most of the literature involving experimental studies have been focused on the axial pull-out strength of a single BiR connection, and its dependency on geometrical and material parameters. Examples can be found in [4–10] for various configurations, with a focus on the experimental assessment of various failure mechanisms [4], test protocols [5] or monotonic and cyclic loading [8]. Often, Finite Element numerical modelling techniques are applied to bonded joints in timber engineering [11–14]. Various experimental studies have been carried out with the additional goal of proposition and validation, as well as assessment of existing methods, of empirical formulations in support of design [15–17], based on curve-fitting of experimental outcomes. In this regard, the current study further explores the mechanical behaviour and properties of BiR connections for timber applications, but with a special focus on the effects due to different adhesive types and their operational condition. It is nowadays well recognized that both the environment conditions and the loading configuration severely affect the adhesive properties and thus the mechanical performance of BiR joints [18–20]. Further, the current investigation aims at finding a link between the proposed fracture mechanics failure modes [21] for BiR connections and the impact of adhesive sensitivity to service conditions.

In this paper, original pull-out experiments are carried out on a total of 84 BiR speci mens, characterized by different adhesive types (epoxy or polyurethane glue), rod-to-grain arrangements (parallel or perpendicular), and average moisture content in timber (9%, 18% or 27% respectively, see also Table 1). The experimental results are discussed, with a focus on the load-bearing mechanism, fracture mechanisms and BiR performance analysis using simple empirical formulations of literature. Later on, a more refined analytical model is presented, and further assessed against the available experimental data.

Table 1. Service classes for timber structures and typical examples.

	Service Class		
	1	**2**	**3**
Climatic condition	20 °C, relative humidity >65% for few weeks/year	20 °C, relative humidity <85% except for few weeks/year	Climatic conditions worse than class 2
Average moisture content in timber	about 12%	always <18%	>18%
Examples	Interiors; warmed and conditioned environments, with limited hygrothermal variations	Covered exteriors; unconditioned environments (shelters, cold roofs, terraces) or humid ones (swimming pools); beam ends on interior walls, well ventilated and drained	Exteriors; bridges, columns, piles; beam ends on exterior walls, also for heated environments

2. Problem Definition

Connections and reinforcements with bonded-in rods have been used in building for several decades. For instance, this solution appeared for the first time in 1980, in French historical monuments [22]. Besides, adhesive bonds for timber applications are notoriously sensitive to several aspects, including:

(1)   wetting ability of the adhesive in relation to the surface;
(2)   bulk properties of adhesive after complete hardening;
(3)   severe environmental conditions.

(Point 1) relates to the substrate (type of surface, its treatment, any kind of ageing and chemical modification, etc.), and to the adhesive in use (viscosity, density, chemical affinity with the substrate, etc.). (Point 2) is particularly relevant when the thickness is high (as usual for structural applications in timber buildings). Finally, environment conditions in 3 can include elevated thermal distortions (due to fires or repeated hygrometric variations) such as wood deformations that induce additional coactive stress at the interface between the adhesive and timber.

From a practical point of view, the applied bond strength design values and the explicit strength modification factors are in most cases not retrievable. This is closely related to the fact that the current lack of worldwide standards or commonly accepted specifications exist for assessing and approving adhesives to be used for BiR applications. It is obvious from a chemistry view point that different adhesive types, which may have rather similar short-term bond strengths, can behave differently under variable climates. The design of BiR joints is implemented in European prestandards and technical documents [23–25], which specify the modification factors for accumulated duration of load in different climates (service classes in Table 1, as described in [26]). The final result takes the form of the well-known k_{mod} factor. On the other hand, this factor is irrespective of the adhesive type.

The current investigation presents experimental tests of BiR connections in different humid climates, in order to explore their actual load-bearing capacity (strength and slip modulus, failure mechanism), as a function of two different adhesive types and rod arrangements.

3. Experimental Investigation

3.1. Test Specimens and Materials

In order to gain a better insight into the behaviour of the joint, an extended series of experimental tests was carried out. In total, 84 specimens were taken into account in the laboratory investigation, with 72 "half-size" and 12 "standard" specimens. Among the half-size specimens, 36 samples were tested parallel and 36 perpendicular to the grains of timber. Furthermore, 12 full-size specimens tested to confirm the observed correlation between half-size and standard specimens (Figure 1).

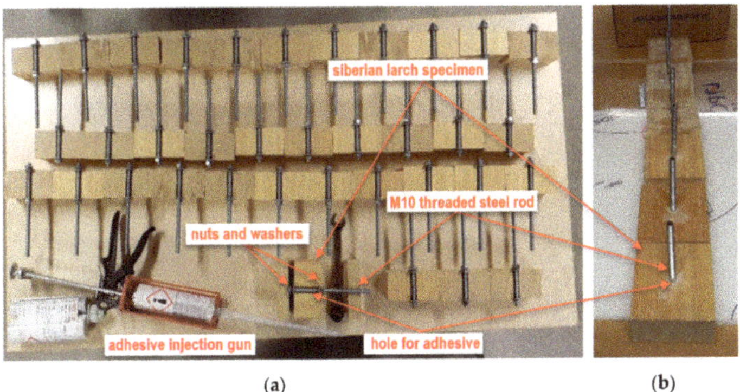

Figure 1. Preparation of bonded-in rod (BiR) connections with M10 steel rod (examples of polyurethane adhesive bonding): (**a**) half-size and (**b**) full-size specimens.

Half-size specimens (with dimensions $B = 120/W = 60/L = 60$ mm) were drilled in their full height L with a concave-notch diameter equal to $d_h = 14$ mm. This hole was placed at the centre of the cross-section of each timber log, in order to introduce both the rod and the adhesive bond. Standard type specimens were characterized by double size (with dimensions $B = 120/W = 120/L = 60$ mm) and prepared with a similar approach. Each timber log was drilled in the full height (60 mm), and a $d_h = 14$ mm hole was drilled at the centre of the wood cross-area.

To get the results for the highest service class, Siberian larch (*Larix sibirica*) wood was used for the timber components [27]. After three weeks in a climate enclosure room with a controlled atmosphere, the moisture content was around 12% (Figure 2). The measured average density was close to 600 kg/m^3, with a standard deviation of 25 kg/m^3.

Figure 2. Preparation of specimens under controlled atmosphere.

One standard metrically threated steel rod with 8.8 strength, nominal diameter $d = 10$ mm and total length of 200 mm was bonded in each wooden specimen ($L = 60$ mm the bonded length). The bonding effect was investigated with two different adhesive types, being represented by a two-component epoxy (KGK EPOCON '88) and a two-component polyurethane (LOCTITE PUREBOND CR 821). Table 2 summarizes the nominal mechanical properties of materials.

Table 2. Nominal mechanical properties for KGK EPOCON '88 (two-component epoxy), LOCTITE PUREBOND CR 821 (two-component polyurethane) and Siberian larch wood (*Larix sibirica*).

		KGK EPOCON '88	LOCTITE PUREBOND CR 821	Wood
Compressive strength	(MPa)	91.5	79.9	61.5
Tensile strength	(MPa)	32.5	27.5	120.5
Bending strength	(MPa)	60.9	-	97.8
Modulus of elasticity	(MPa)	29.5	29.5	8.5

Based on the M10 rod in use and the d_h mm hole, an annular bond-line thickness of 2 mm was created for each sample, and the anchorage length was set equal to $H = 60$ mm. The bonding stage was performed under controlled laboratory conditions (9% humidity in wood and a room temperature of 20 °C, see Figure 1).

3.2. Test Setup and Instruments

After the assembly process, all the specimens were subjected to a controlled atmosphere, so as to achieve different degrees of moisture in wood (at the same temperature of 20 °C). The experiments reported herein comprised three artificial climates being selected as extreme examples of operational conditions for service class 1, 2 and 3 (Table 1). All test series were in fact performed at a constant temperature of 20 °C, with moisture content in wood in the order of 9%, 18% and 27% respectively.

The stiffness and strength characteristics in short-term loading were thus investigated according to EN 15274:2015 recommendations [28]. Further, all tests were carried out based on the EN1382:2016 provisions [29], on a Zwick Roell 50 kN capacity machine, with data recording frequency of 10 Hz. The reference pull-out setup is shown in Figure 3. Each specimen was fixed to the machine with a steel clamping plate. Possible relative displacements of wood logs were restrained by four M8 anchoring bolts. The single rod was hence clamped to the pull machine. To this aim, the nominal cross-head displacement rate was set to ensure a reference value of 0.5–2.0 mm/min (depending on specimen type). The latter was then calibrated (test by test) in order to reproduce a short-term failure mechanism for all the specimens. The axial load F applied to each specimen was recorded and compared with the average relative displacement of the rod with respect to the wood log. In support of these experimental investigations, contactless optical measurements of strain (based on digital image correlation (DIC) techniques), were also implemented to provide full-field strain maps of specimens under load, until failure. The experiments of 24 specimens were further supported by a Canon 700D camera with macro lens and photo recording frequency of 0.5 Hz (Figure 3). All specimens had preprepared surface adapted for recording. The whole postprocessing stage of acquired images was carried out by VIC-2D (Correlated Solutions, University Santa Barbara, Santa Barbara, CA, USA). This approach was used to reveal local aspects of load transfer mechanisms from the rod to the adhesive and wood that could be of importance for further analysis using theoretical or even numerical models.

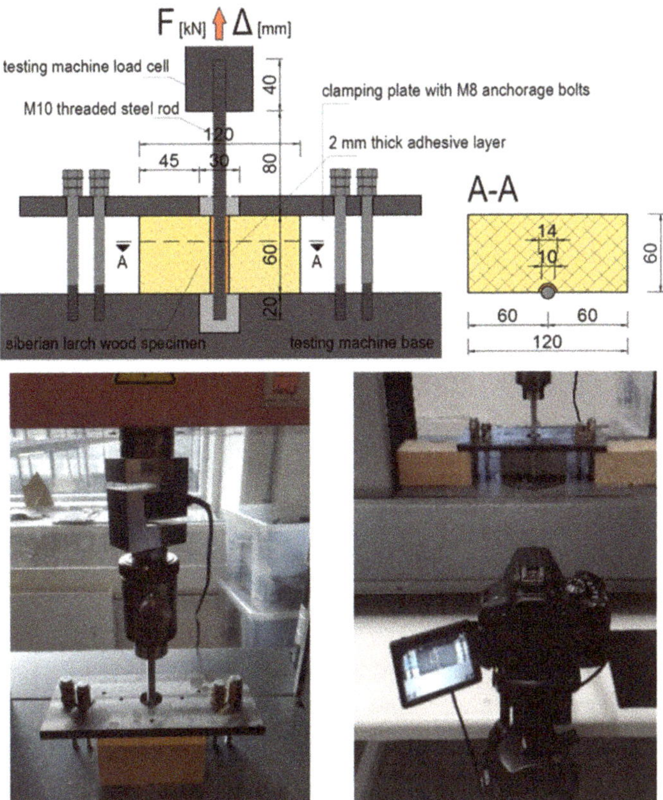

Figure 3. Pull-out test set-up: schematic view and details (nominal dimension in mm).

3.3. Test Results

The analysis of experimental results was carried out at different levels, including (a) qualitative analysis of observed failure mechanisms, (b) measured load-displacement laws, (c) DIC measurement of displacements in the bonded region.

For sake of clarity, the specimens are labelled to detect the type of glue ("E" or "P" for epoxy and polyurethane), the moisture content (9%, 18% or 27%), the rod-to-grain orientation ("0°" or "90°" for parallel or perpendicular arrangement) and the sample number n for each group. The full set of pull-out test results is presented in Appendix A.

Three regimes can be distinguished from the collected load-displacement curves, as shown in the examples of Figures 4 and 5 (specimens under 9% moisture and load parallel or perpendicular to the grain, respectively).

Firstly, a linear elastic stage can be noticed in the load-bearing response of all the specimens, from which the initial stiffness $K_{ser} = F_{ax}/d$ can be estimated from a linear regression procedure.

After the yield point, a progressive decrease of stiffness occurs followed by a sudden failure of the connection. Worth to be noted, in this regard, that the failure path of all tested connections was located in the wood substrate in the vicinity of the wood–adhesive interface, as also emphasized in Figure 6. Thus, the nonlinearity observed in the collected force–displacement responses can be rationally justified in the quasibrittle damage of wood.

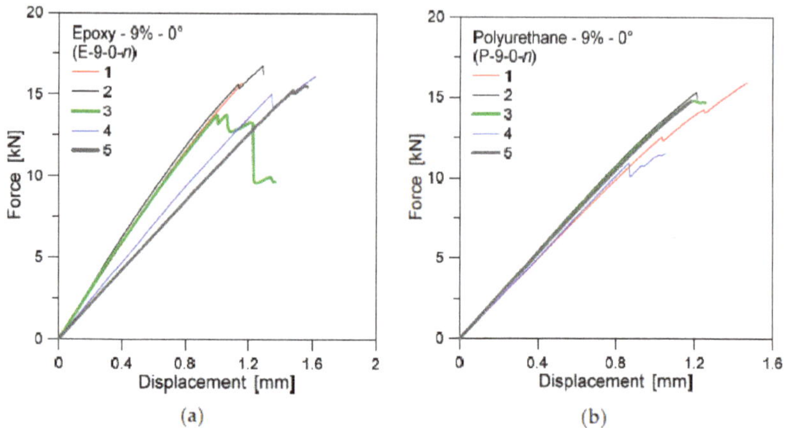

Figure 4. Experimental force–displacement results for specimens with (**a**) epoxy or (**b**) polyurethane glue, under 9% moisture and parallel rod-to-grain arrangement.

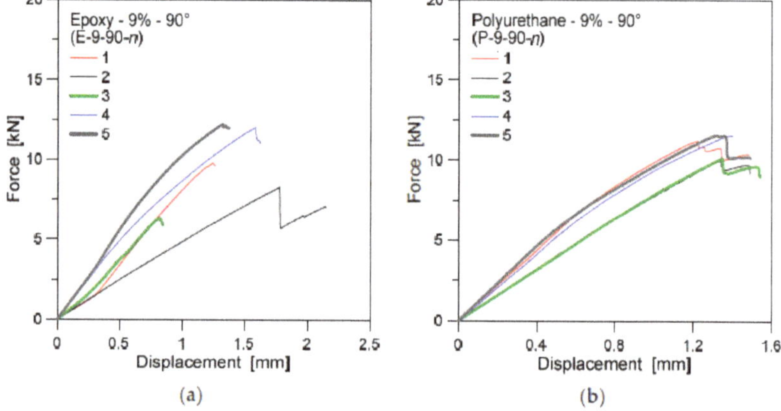

Figure 5. Experimental force–displacement results for specimens with (**a**) epoxy or (**b**) polyurethane glue, under 9% moisture and perpendicular rod-to-grain arrangement.

Finally, the ultimate load measurement for the experimental curves as in Figures 4 and 5 allows estimating the overall shear strength of the examined connections. Following Figure 6 and that the failure mechanism of BiRs is dependent on the mechanical properties of solid wood (strength and stiffness), the typical collapse of BiRs can be assumed as quasi-brittle for general applications.

More in detail, for rods bonded parallel to the grains, wood failure was observed to start in the area around the adhesive matrix, while for rods bonded perpendicularly to the grains, failure typically originated in line between adhesive matrix and wood (Figure 6). The qualitative observations at failure, for a selection of specimens, were further explored by DIC system as in Figure 7.

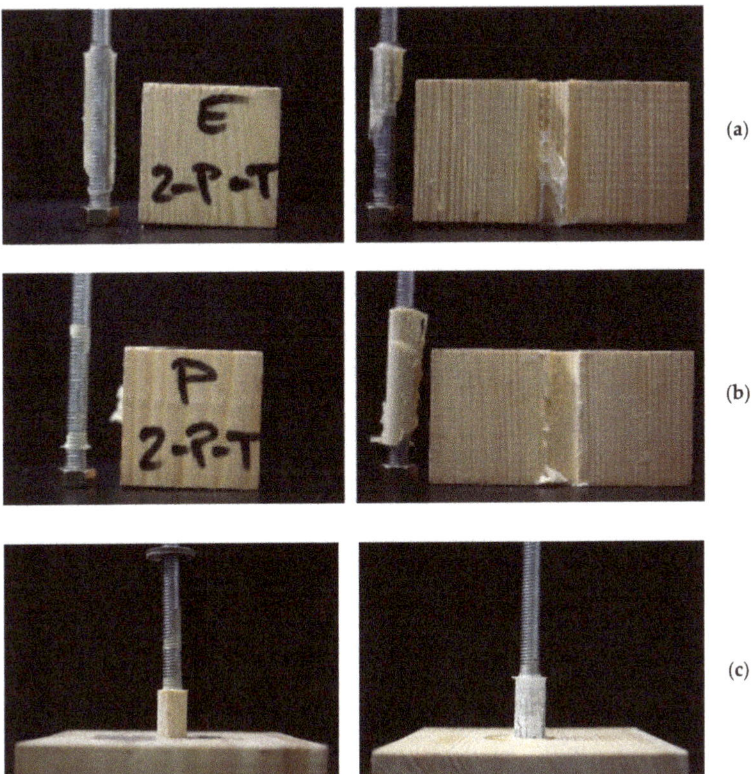

Figure 6. Example of failure configuration for the tested samples: (**a**) epoxy and (**b**) polyurethane bonded rods for half-size specimens (18% moisture, parallel rod-to-grain arrangement) and (**c**) full-size specimens.

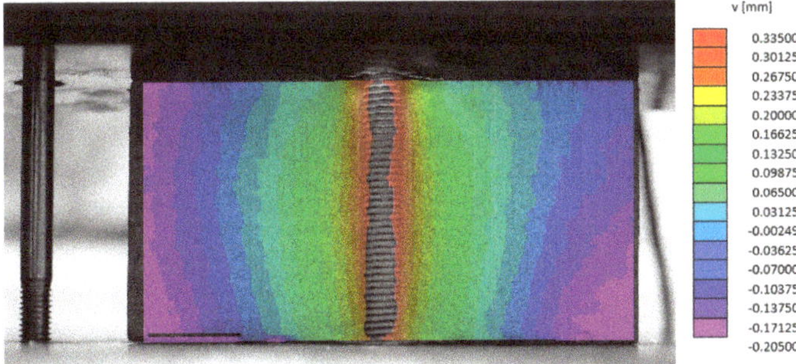

Figure 7. Example of the typical shear strain distribution along the rod, as obtained by digital image correlation (DIC).

Average values of maximum force (F_{max}) and displacement (d_{max}), as well as of slip modulus (shear stiffness K_{ser}) and their corresponding coefficient of variation (CoV.) and standard deviation (St.Dev.) are listed in Tables 3 and 4. Major scatter of grouped predictions is found for the "9%" set in Table 4, which was found characterized by 36% CoV. in terms of slip modulus. On the other side, such an experimental outcome was severely affected by few specimens (like specimen #2 in Figure 5a).

Table 3. Maximum axial force, displacement and slip modulus for specimens with parallel rod-to-grain arrangement (mean experimental values).

		Bond					
		Two-Component Epoxy			Two-Component Polyurethane		
Moisture	Parameter	Avg.	CoV. [%]	St.Dev.	Avg.	CoV. [%]	St.Dev.
	F_{ax} [kN]	15.216	9.1	1.390	15.545	3.3	0.519
9%	d_{max} [mm]	1.222	20.3	0.248	1.362	11.4	0.155
	K_{ser} [N/mm]	14,548.086	10.5	1528.301	12,475.290	9.6	1197.480
	F_{ax} [kN]	12.015	10.8	1.301	11.987	13.1	1.575
18%	d_{max} [mm]	1.233	6.4	0.079	1.227	27.1	0.333
	K_{ser} [N/mm]	6769.256	28.7	1942.825	10,362.893	15.6	1613.003
	F_{ax} [kN]	5.117	11.5	0.591	6.294	20.6	1.293
27%	d_{max} [mm]	0.607	17.1	0.104	0.625	20.9	0.131
	K_{ser} [N/mm]	8341.809	22.1	1843.689	10,728.730	4.1	441.031

Table 4. Maximum axial force, displacement and slip modulus for specimens with perpendicular rod-to-grain arrangement (mean experimental values).

		Bond					
		Two-Component Epoxy			Two-Component Polyurethane		
Moisture	Parameter	Avg.	CoV. [%]	St.Dev.	Avg.	CoV. [%]	St.Dev.
	F_{ax} [kN]	9.722	25.7	2.501	10.908	7.0	0.760
9%	d_{max} [mm]	1.352	26.9	0.364	1.325	4.8	0.064
	K_{ser} [N/mm]	7360.208	36.3	2668.764	9869.502	17.0	1674.434
	F_{ax} [kN]	11.687	3.5	0.408	11.801	6.7	0.792
18%	d_{max} [mm]	1.944	12.9	0.251	1.824	10.6	0.193
	K_{ser} [N/mm]	6943.509	15.9	1105.534	7887.684	3.6	282.079
	F_{ax} [kN]	3.149	25.8	0.814	8.914	6.6	0.587
27%	d_{max} [mm]	0.573	19.4	0.111	1.330	9.3	0.124
	K_{ser} [N/mm]	5675.074	13.9	788.096	7476.200	6.1	456.498

Considering the specimens grouped by adhesive type, from Tables 3 and 4 it is possible to notice a less pronounced sensitivity and scatter of polyurethane bonded rods, compared to the epoxy bonded samples. This effect is even more pronounced for service classes 2 and 3, with higher moisture content.

Mean experimental data can be helpful for the analysis of climate and operational conditions on bonded rods for timber applications. However, an in-depth discussion can be carried out in terms of characteristic mechanical properties that can be obtained from the test observations.

Based on [28], the characteristic axial force at failure for each series of specimens was calculated as:

$$F_{ax,char} = exp\left(\overline{y} - k_s s_y\right) \tag{1}$$

with:

$$\overline{y} = \frac{1}{n} \sum_{i=1}^{n} ln F_{ax,i} \tag{2}$$

$$s_y = \sqrt{\frac{1}{n-1} \sum_{i=1}^{n} \left(ln F_{ax} - \overline{y}\right)^2} \tag{3}$$

where n denotes the number of test repetitions for each series, k_s is a coefficient adopted from [28].

Furthermore, the maximum axial force at failure (both in terms of mean and characteristic values) was correlated with the resisting surface of the bond-line of each specimen, A_{bond}, given that:

$$\sigma_{max} = \frac{F_{ax}}{A_{bond}} = \frac{F_{ax}}{0.5 \, \pi d_h L} \tag{4}$$

where:

$$A_{bond} = 0.5\ \pi d_h L \tag{5}$$

for the half-size specimens.

The estimated results are shown in Figure 8, in terms of stress peak at failure for each set of specimens (under the assumption of uniform stress distribution for the bond-line as a whole). Mean and characteristic values of ultimate stress are grouped by adhesive type and rod arrangement, as a function of the service class/moisture content.

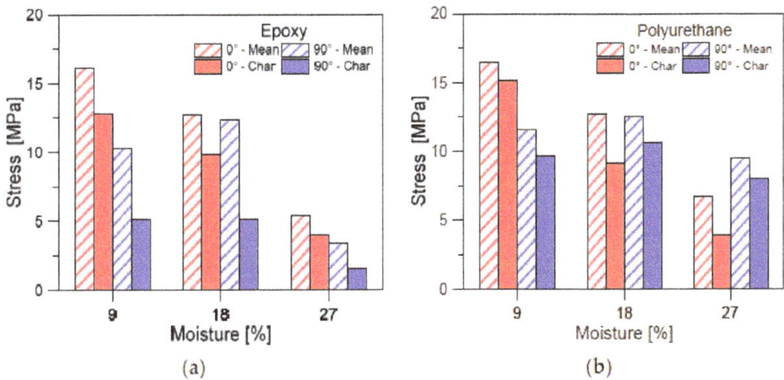

Figure 8. Stress peak at failure, as observed for (**a**) epoxy or (**b**) polyurethane bonded rods in various arrangements. Comparison of mean and characteristic experimental results.

As expected from Tables 3 and 4, a major scatter of mean and characteristic values was observed especially for the epoxy-bonded rods, rather than polyurethane samples. Besides, the global decrease of stress peak can be observed, both in mean and characteristic parameters, as far as the moisture level increases. This is a further confirmation of sensitivity of different adhesive types to operational conditions.

In this regard, it is worth mentioning that the imposed displacement rate (in the range of 0.5–2.0 mm/min), as previously discussed, was adapted test by test. The postprocessing stage of experimental measurements was quantified in average rate values that are summarized in Figure 9, as obtained for each series of specimens. Worth noting are the lower rate values for epoxy or polyurethane specimens under high moisture (27%) and bonding parallel to the grain. This was required by the pronounced viscous response of adhesives used.

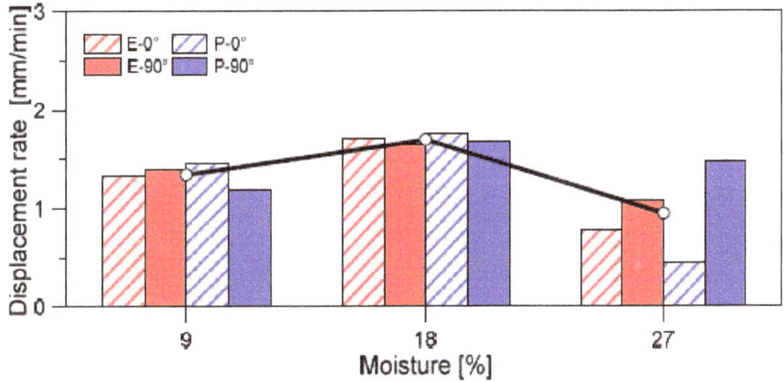

Figure 9. Average experimental displacement rate for the investigated series of specimens.

4. Discussion of Experimental Observations

4.1. Service Class and Adhesive Behaviour

Undoubtedly, the experimental investigations revealed significant differences of the mechanical behaviour of bonded-in rod connections with different adhesives when exposed to wet climate. The test set-up and the support of the DIC system helped in obtaining empirical results of practical use. Rheological behaviour indicates that, in terms of reliability, special attention should be paid to the joints exposed to the extreme climatic conditions. Additional requirements in standards should be included or certification from the adhesive manufacturer should be sought to ensure the safe use of this type of joints.

Tests indicated a significant effect of moisture content on the adhesive stickiness. While only small changes were observed for service class 2, for service class 3 the adhesive stickiness began to recede dramatically. For epoxy specimens, the entire adhesive matrix started to slip smoothly, what is especially characteristic for rods bonded perpendicular to the grains, while on the rods bonded parallel, any pieces of wood grains on the adhesive matrix is not visible. This is indicating that bearing capacity of the joint is defined by shear strength of the interface on exact line between the adhesive matrix and the wood. Polyurethane specimens showed enhanced behaviour, especially for rods bonded perpendicular to the grains. The reason for such behaviour can be justified in higher chemical properties of the new generation of this type of adhesives, which is recommended for use in moist environments. Indicatively, it can be stated that the use of epoxy adhesives is not recommended for service class 3, while polyurethane adhesives can be still used, but with careful consideration for the technical characteristics given by the manufacturer.

4.2. Service Type and Slip Modulus

In Figure 10, the average % variation of slip modulus K_{ser} is shown as a function of the moisture content, the bonding direction to the grain and the type of adhesive.

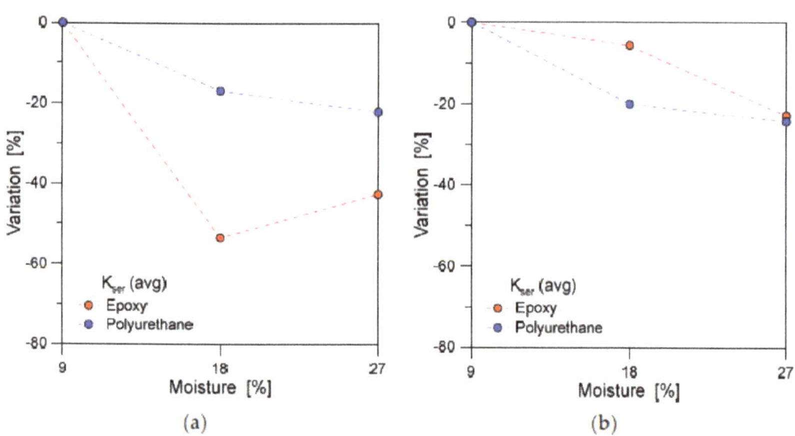

Figure 10. Average experimental variation of slip modulus K_{ser} for specimens with (**a**) parallel or (**b**) perpendicular rod-to-grain arrangement.

The service class, as shown, generally resulted in severe modification of mechanical parameters for the tested specimens, both for epoxy or polyurethane bonded rods, with more pronounced effects for loading parallel to the grain (Figure 10a). In the perpendicular direction, a rather stable variation of average stiffness results can be observed in Figure 10b for both the adhesive types.

4.3. Service Class and Load-Bearing Capacity

The above quantitative comparisons were further supported by the qualitative analysis of experimental outcomes. In general terms, the analysis of BiR performances under different service conditions can be summarized as follows.

For service class 1:

- test results confirmed the assumption of similar load-bearing capacity for both the epoxy and the polyurethane adhesive types;
- specimens with polyurethane showed mild bilinear behaviour and 10% higher capacity then specimens with epoxy, which proved to offer a pure linear behaviour with brittle fracture.

For service class 2:

- specimens showed a $\approx$20% drop in the measured average load capacity;
- the load-bearing behaviour and failure modes were found to closely agree with the experimental observations of specimens in service class 1.

Finally, for service class 3:

- specimens showed large drop of load-bearing capacity. A huge drop was found especially for epoxy bonded specimens, where failure happened on -50% of maximum average force of corresponding specimens in service class 1;
- in any case, the failure modes were still observed in agreement with the previous specimens.

In this regard, the analysis of test results can take advantage of existing empirical formulations that have been proposed for glued-in-rods with parallel or perpendicular rod-to-grain orientation. Among the literature efforts for the analytical analysis and design of BiR connections, the failure load of parallel rod-to-grain arrangement can be for example estimated as [15]:

$$F_{ax,0} = \pi \, L \left(f_v d_{equ} + k(d + e)e \right) \tag{6}$$

The empirical equation has been proposed by Feligioni et al. [15] to predict the pull-out strength at failure (in N), where L is the joint length (mm); f_v is the shear strength of wood (MPa); d is the rod diameter (mm); d_{equ} the smaller between the hole diameter d_h and the rod diameter d multiplied by 1.25 (mm); e is the joint thickness (mm); k is a parameter proposed in 0.086 or 1.213 (based on experimental fitting), for adhesives with brittle or ductile behaviour respectively.

For comparative studies with the current experimental results, the above parameters are set:

$$d_{equ} = min(14; 1.25 \times 10) = 12.5 \text{ mm} \tag{7}$$

$$f_{v,k} = 1.2 \times 10^{-3} d_{eq}^{-0.2} \rho^{1.5} = 10.65 \text{ MPa} \tag{8}$$

with $\rho = 600$ kg/m^3 the average density of wood specimens (±25); $e = 2$ mm, $L = 60$ mm.

From Equation (6), the comparative analysis is carried out towards the average (mean) experimental failure loads earlier discussed, for various configurations of specimens. The findings are summarized in Figure 11, as a function of the analytical vs. experimental failure load, the adhesive type and the moisture/service class. As far as the service class 1 is taken into account, it is possible to see that the analytical to experimental ratio is in the order of the unit. This suggests a certain correlation of literature model with the current experiments. Besides, the analysis of higher moisture content tends to progressively overestimate the analytical failure force for the examined specimens, as shown in Figure 11 for both the adhesive types. Furthermore, the high moisture content reveals also pronounced effects due to the input k coefficients calibrated from [15].

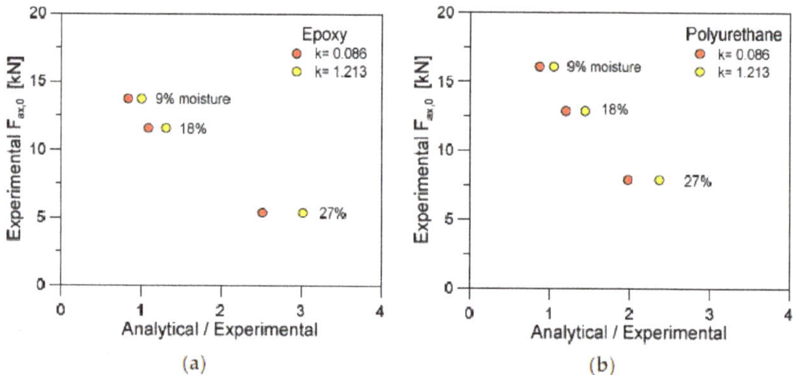

Figure 11. Average experimental failure load versus the ratio of analytical prediction, as obtained for specimens with (**a**) epoxy or (**b**) polyurethane adhesive and parallel rod-to-grain arrangement.

For specimens with perpendicular rod-to-glue arrangement, the analytical model proposed by Yeboah et al. is considered [30]. The model, in particular, assumes that the load-bearing capacity is given by:

$$F_{ax,90,mean} = f_{v,90,mean}\,\pi\,d_h L \tag{9}$$

with the limit applicability condition of $L < 15d_h$.

The empirical model of Equation (9) agrees with the experimental trends earlier discussed in Figure 8. As far as the moisture content increases and the adhesive degradation of mechanical properties increases, Equation (9) itself severely overestimates the expected axial force at failure for the tested joints. In Figure 12, the empirical derivation of material strength is shown for epoxy or polyurethane specimens, as obtained from the mean experimental results.

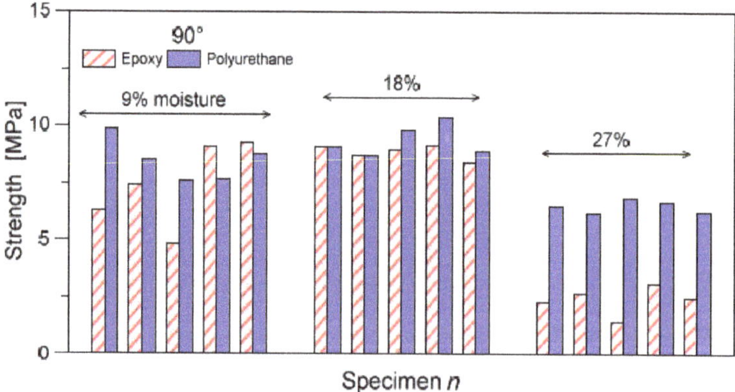

Figure 12. Inverse experimental derivation of $f_{v,90,mean}$ strength, based on Equation (9), for epoxy or polyurethane specimens with perpendicular rod-to-glue arrangement.

5. LEFM-Based Analytical Model

5.1. State-of-Art

Theoretical approaches, based on the stress distribution in the joint, have been used to describe the laws governing the mechanical behaviour of connections by glued-in rods. One of the pioneering works was by Volkersen [31], who developed an elastic analysis of the shear distribution in a single lap joint. However, the substrates were assumed to respond to the load only in tension and the adhesive to respond only in shear. A further development was made by Goland and Reissner [32], who included the influence of a bending moment in the connection in their calculation model. Later, Hart-Smith [33] introduced elastic-plastic stress distribution of anisotropic materials in the analysis. Depending on the ductility of the bond-line, these traditional strength analyses will be more or less accurate.

More recently, the behaviour of glued-in rods was investigated within the framework of fracture mechanics. In this approach, a pre-existing crack in the joint is assumed to lead to a stress singularity so that the traditional maximum stress criterion can be no longer applied. For instance, in accordance with LEFM, Serrano [11] proposed an evaluation model of the load bearing capacity for a single glued-in rod, assuming that failure of the joint could occur when the energy release rate is equal to the fracture energy. In the same year, Gustafsson [17] took into consideration the damage amount preceding the failure of a joint through an approach based on nonlinear fracture mechanics (NLFM), and essentially based on the mode II fracture energy. Thus, many empirical or theoretical design calculations could be found in literature to estimate either the shear strength (especially in studies based on elastic stress analysis) or the fracture energy in mode II. It should be noted that most existing studies were either based on experiments or numerical investigations, but rarely combined both approaches [21].

5.2. Model Definition

In order to check this assumption, the study of the fracture behaviour of BiR connection is proposed within the framework of equivalent LEFM, which is well known to be useful to characterize the quasibrittle failure of load-bearing components [21].

The reference mechanical model is schematized in Figure 13, with evidence of the required geometrical and mechanical parameters in the detailed view.

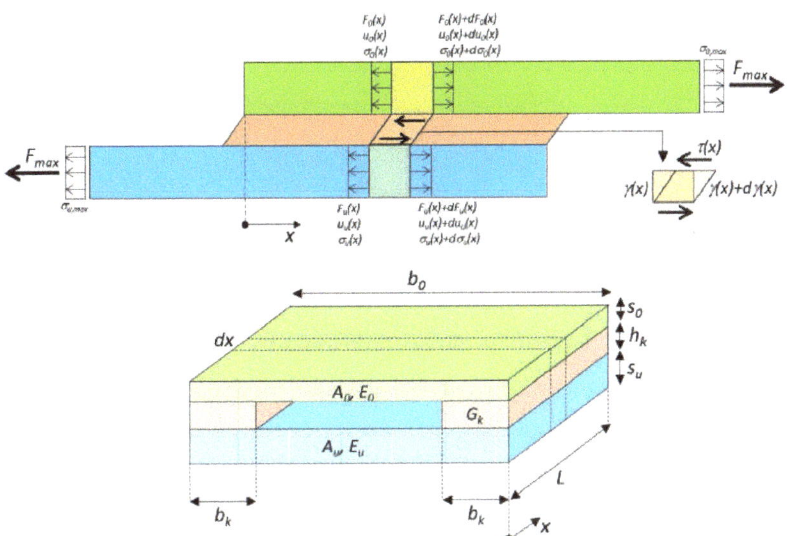

Figure 13. Mechanical model for the analysis of force transmission and deformation behaviour in BiR connections.

Considering that the joint acts as a fibre in the matrix, where the fibre represents a steel glued-in bar and the matrix consists of the wood log as in Figure 13, the following model of shear stresses distribution along the joint is proposed in this study:

$$\tau(x) = \frac{A_s}{b_b}\omega\left(\left(\sigma_{s,max} - q\right)\frac{\cosh(\omega x)}{\sinh(\omega L)} + q\,\frac{\cosh(\omega(L-x))}{\sinh(\omega L)}\right) \tag{10}$$

where A_0 is cross-area of the wooden part; b_k is the mean width of the adhesive layer; $\sigma_{0,max}$ represents the maximum normal stress in the BiR; x is the length coordinate; L denotes the length of the adhesive joint (anchorage length); ω is a correction factor that can be estimated from:

$$\omega^2 = \frac{1}{p} \tag{11}$$

and:

$$p = \left(\frac{E_s A_s E_w A_w}{E_w A_w + E_s A_s} \right) \frac{h_b}{G_b b_b} \quad \left(\text{in mm}^2 \right) \tag{12}$$

$$q = \left(\frac{E_s}{E_w A_w + E_s A_s} \right) F_{ax} \quad (\text{in MPa}) \tag{13}$$

By integrating Equation (10) to get a strain energy release rate, the J-integral method could be implemented directly. Further, the problem analysis and the definition of an accurate behaviour model for BiR connections should be necessarily based on local stress distribution which has been obtained in the framework of DIC system (i.e., Figure 7).

According to the known maximum shear strength of wood, it is possible to predict the bearing capacity of the BiR connection, i.e., maximum pull-out force, as:

$$F_{ax} = \frac{\tau(x) \cdot \frac{b_b}{\omega} \cdot \frac{sinh(\omega \cdot L)}{cosh(\omega \cdot x)}}{1 + \left(\frac{E_s \cdot A_s}{E_w \cdot A_w + E_s \cdot A_s} \right) \cdot \left(\frac{cosh(\omega \cdot (L-x)) - cosh(\omega \cdot x)}{cosh(\omega \cdot x)} \right)} \tag{14}$$

From the developed analytical model, the variation of the shear force, stress and strain in BiR connections can be thus predicted along the bonding length L. Selected examples are proposed in Figure 14.

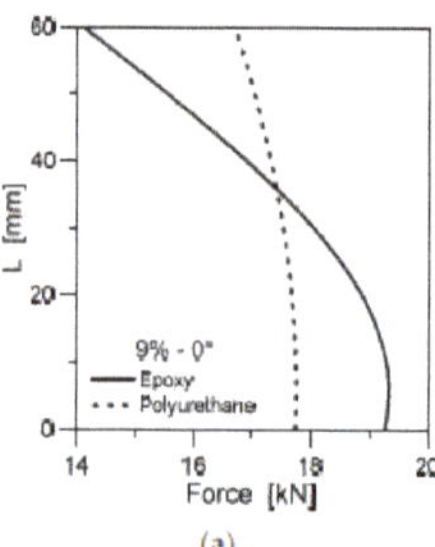
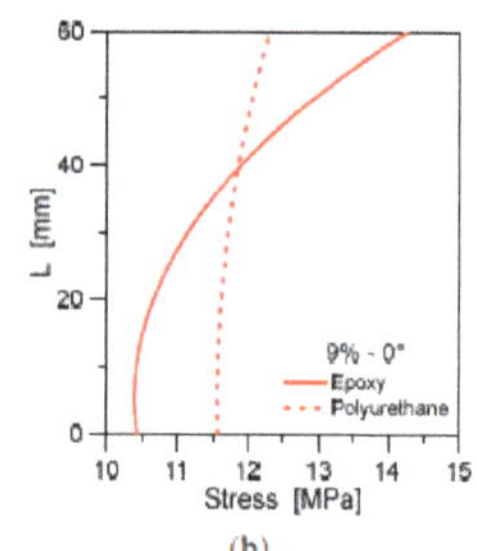
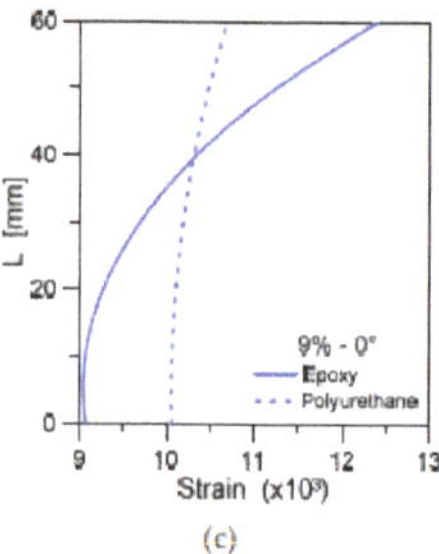

Figure 14. Analytical prediction of shear (**a**) force, (**b**) stress and (**c**) strain in BiR connections with epoxy or polyurethane adhesives (examples for 9% moisture and parallel rod-to-glue arrangement).

5.3. Assessment of Analytical Predictions

The proposed analytical model is further assessed by taking advantage of the available test results and of nominal material properties earlier discussed. Parametric calculations were carried out on the grouped specimens (average estimates) in terms of shear force and stress, by changing the adhesive type, arrangement and environment condition.

In this regard, the analytical model proved to offer reliable estimates for both the parameters of force and stress agreeing with the general trends of Figure 14. Comparative examples are proposed in Figure 15, in terms of force or stress, as obtained for grouped specimens as a function of their service class. It should be noted that the error ratio R is found both to overestimate or underestimate the expected parameter, for all the types of BiR specimens. Most importantly, however, is that the collected R values confirm the rather small scatter for all the analytical predictions, thus confirming the validity and accuracy of the approach.

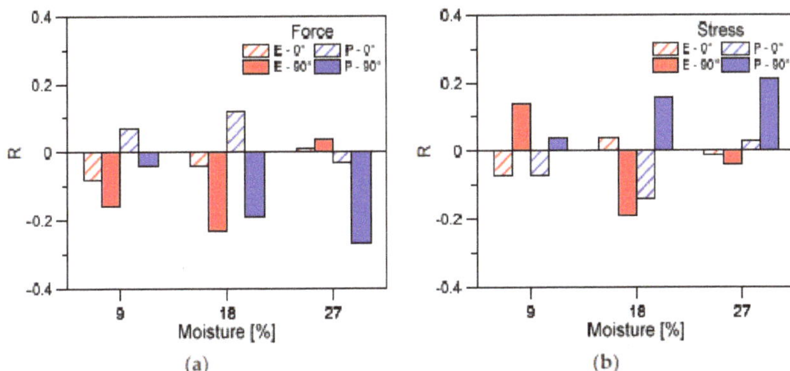

Figure 15. Average experimental failure load versus the ratio of analytical prediction, as obtained for specimens with (**a**) epoxy or (**b**) polyurethane adhesive and parallel rod-to-grain arrangement.

6. Conclusions

In this paper, the mechanical performance of bonded-in rod (BiR) connections for structural timber applications has been explored experimentally and analytically.

The experimental investigations revealed, significant differences in the observed mechanical behaviour of BiR connections, by changing the adhesives type, the bonding arrangement and the wet climate exposure. The pull-out test set-up and the use of a digital image correlation (DIC) system, in particular, helped to obtain results in support of the definition of generalized design tools for this type of connections.

The experimental study, in most of the cases, exhibited a failure mechanism of the connections in the wood, in the vicinity of the wood–adhesive interface. As such, a study of the stress field along this interface was performed with the use of a newly developed linear elastic fracture mechanics (LEFM) formulation. In addition to the shear stress, expected for this kind of connection, the stress field analysis revealed the existence of normal stress (to the interface), which was relevant at the onset of the failure.

The observed rheological behaviour of adhesive types in use further indicates that (in terms of reliability) special attention should be paid to joints exposed to extreme climatic conditions. The current study provided useful information about the short-term behaviour of bonded-in-rods. However, the long-term behavioural analysis of BiR connections requires further investigations, in order to check the mechanical performance of this repair process according to service classes defined in the European timber design codes. Most importantly, additional requirements in standards should be included, or certification from the adhesive manufacturer should be sought, to ensure the safe use of this type of joints in practical applications. In this regard, further research efforts will be dedicated to the in-depth analysis of mechanical parameters and their sensitivity to severe environment conditions, so as to include additional configurations and parameters of technical interest.

Author Contributions: This research paper results from a joint collaboration of the involved authors. More precisely J.B. and M.K.B. carried out the experiments; C.B. and V.R. contributed to the post-processing of results; V.R. supervised all the research activities. All authors have read and agreed to the published version of the manuscript.

Funding: The EU-COST Action CA18120 is gratefully acknowledged for providing financial support to the first and third authors (J.B. visitor at Aarhus University, Denmark and C.B. visitor at University of Zagreb, Croatia), in the framework of CERTBOND Short Term Scientific Missions—STSM grants.

Institutional Review Board Statement: Not applicable.

Informed Consent Statement: Not applicable.

Data Availability Statement: Data supporting this research study will be made available upon request.

Acknowledgments: This publication is based upon work from EU-COST Action CA18120 (CERTBOND—https://certbond.eu/ (accessed on 16 February 2021)), supported by COST (European Cooperation in Science and Technology—https://www.cost.eu/ (accessed on 16 February 2021)). MDPI is also acknowledged for waived APCs (C.B.).

Conflicts of Interest: The authors declare no conflict of interest.

Appendix A

Experimental force–displacement curves for the investigated BiR specimens, grouped by adhesive type, rod-to-glue arrangement and service class.

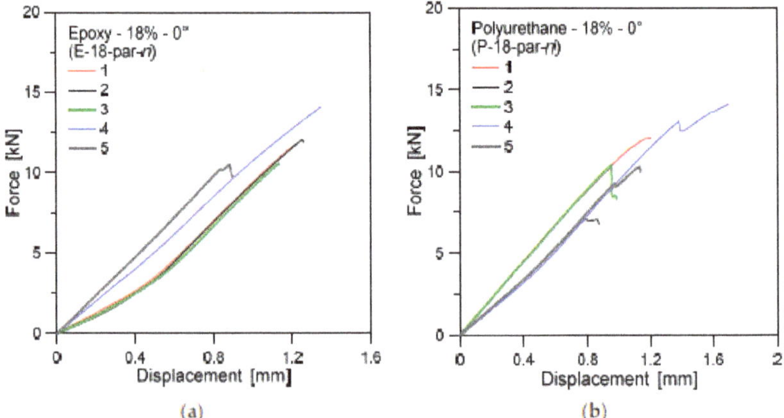

Figure A1. Experimental force–displacement results for specimens with (**a**) epoxy or (**b**) polyurethane glue, under 18% moisture and parallel rod-to-grain arrangement.

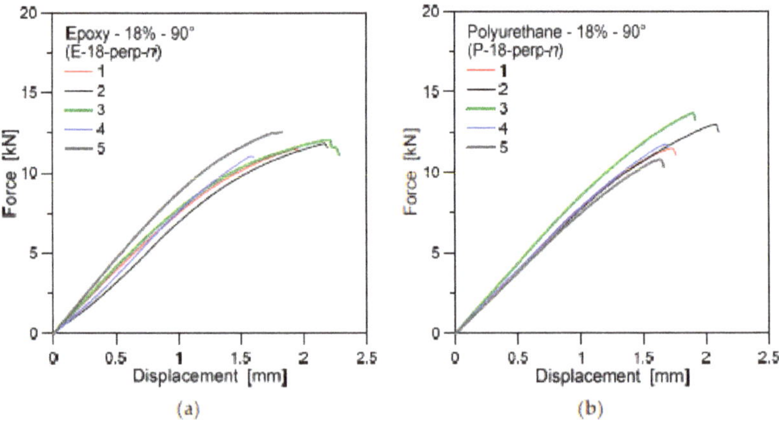

Figure A2. Experimental force–displacement results for specimens with (**a**) epoxy or (**b**) polyurethane glue, under 18% moisture and bonding and perpendicular rod-to-grain arrangement.

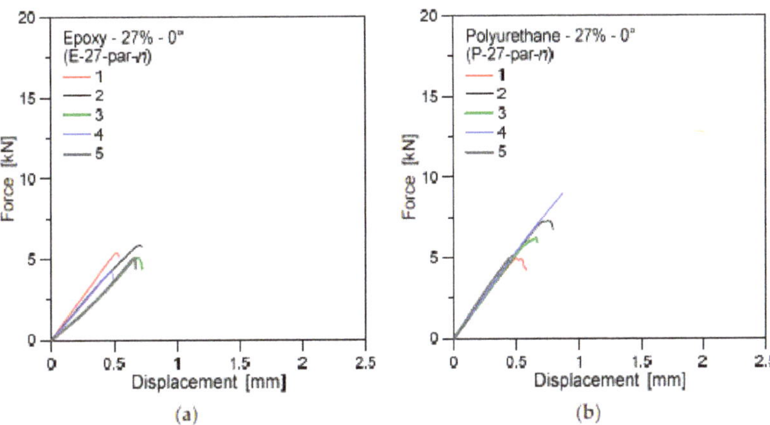

(a) (b)

Figure A3. Experimental force–displacement results for specimens with (**a**) epoxy or (**b**) polyurethane glue, under 27% moisture and bonding and parallel rod-to-grain arrangement.

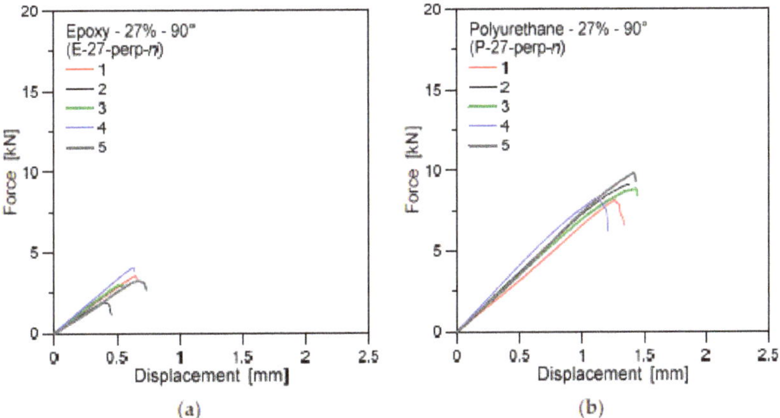

(a) (b)

Figure A4. Experimental force–displacement results for specimens with (**a**) epoxy or (**b**) polyurethane glue, under 27% moisture and bonding and perpendicular rod-to-grain arrangement.

References

1. Tlustochowicz, G.; Serrano, E.; Steiger, R. State-of-the-art review on timber connections with glued-in steel rods. *Mater. Struct.* **2010**, *44*, 997–1020. [CrossRef]
2. Serrano, E.; Gustafsson, P.J. Fracture mechanics in timber engineering—strength analyses of components and joints. *Mater. Struct.* **2007**, *40*, 87–96. [CrossRef]
3. Batchelar, M.; Fragiacomo, M. Timber frame moment joints with Glued-in Steel Rods I: Design. *J. Struct. Eng.* **2012**, *138*, 789–801.
4. Otero Chans, D.; Estévez Cimadevila, J.; Martín-Gutiérrez, E.; Vazquez, J.A. Failure modes in double-sided pull-out test of threaded steel rods glued-in hardwood. In Proceedings of the World Conference of Timber Engineering, Trentino, Italy, 20–24 June 2010.
5. Stepinac, M.; Rajcic, V.; Koscak, J.; Damjanovic, D. Assessment of the pull-out strength of glued-in rods with different test methods. In Proceedings of the World Conference on Timber Engineering, Vienna, Austria, 22–25 August 2016.
6. Steiger, R.; Serrano, E.; Stepinac, M.; Rajčić, V.; O'Neill, C.; McPolin, D.; Widmann, R. Strengthening of timber structures with glued-in rods. *Constr. Build. Mater.* **2015**, *97*, 90–105. [CrossRef]
7. Broughton, J.; Hutchinson, A. Pull-out behaviour of steel rods bonded into timber. *Mater. Struct.* **2001**, *34*, 100–109. [CrossRef]
8. Gattesco, N.; Gubana, A.; Buttazzi, M.; Melotto, M. Experimental investigation on the behavior of glued-in rod joints in timber beams subjected to monotonic and cyclic loading. *Eng. Struct.* **2017**, *147*, 372–384. [CrossRef]

9. Rajcic, V.; Bjelanovic, A.; Rak, M. Comparison of the Pull-out Strength of Steel Bars Glued in GluLam Elements Obtained Experimentally and Numerically. In Proceedings of the International Council for Research and Innovation in Building and Construction "Working Commission W 18—Timber Structures" Meeting Thirty Nine/Goerlacher, Rainer (ur.), Florence, Italy, 28–31 August 2006.

10. Stepinac, M.; Rajcic, V.; Hunger, F.; van de Kuilen, J.W.G. Glued-in rods in beech laminated veneer lumber. *Eur. J. Wood Wood Prod.* **2016**, *74*, 463–466. [CrossRef]

11. Serrano, E. Glued-in rods for timber structures—A 3D model and finite element parameter studies. *Int. J. Adhes. Adhes.* **2000**, *21*, 115–127. [CrossRef]

12. Bedon, C.; Fragiacomo, M. Numerical analysis of timber-to-timber joints and composite beams with inclined self-tapping screws. *Compos. Struct.* **2019**, *207*, 13–28. [CrossRef]

13. Xu, B.; Bouchair, A.; Racher, P. Analytical study and finite element modelling of timber connections with glued-in rods in bending. *Constr. Build. Mater.* **2012**, *34*, 337–345. [CrossRef]

14. Skec, L.; Bjelanovic, A.; Jeleni, G. Glued timber-concrete beams—Analytical and numerical models for assessment of composite action. *Eng. Rev.* **2013**, *33*, 41–49.

15. Feligioni, L.; Lavisci, P.; Duchanois, G.; De Ciechi, M.; Spinelli, P. Influence of glue rheology and joint thickness. *Holz als Roh-und Werkstoff* **2003**, *61*, 281–287.

16. Stepinac, M.; Hunger, F.; Tomasi, R.; Serrano, E.; Rajcic, V.; Van de Kuilen, J.W.G. Comparison of design rules for glued-in rods and design rule proposal for implementation in European standards. In Proceedings of the CIB-W18 Timber Structures, Meeting 46, Vancouver, BC, Canada, 26–29 August 2013.

17. Gustafsson, P.J.; Serrano, E. Predicting the pull-out strength of glued-in rods. In Proceedings of the Sixth World Conference on Timber Engineering, Whistler Resort, BC, Canada, 31 July–3 August 2000; Volume 1.

18. Aicher, S.; Dill-Langer, G. Influence of moisture, temperature and load duration on performance of glued-in rods. In *Stuttgart Rilem Symposium 2001—Joints in Timber Structures*; Springer—International Publisher Science: Berlin, Germany, 2001.

19. Verdet, M.; Coureau, J.-L.; Cointe, A.; Salenikovich, A.; Galimard, P.; Delisée, C.; Toro, W.M. Creep performance of glued-in rod joints in controlled and variable climate conditions. *Int. J. Adhes. Adhes.* **2017**, *75*, 47–56. [CrossRef]

20. Verdet, M.; Salenikovich, A.; Cointe, A.; Coureau, J.-L.; Galimard, P.; Munoz Toro, W.; Blanchet, P.; Delisee, C. Mechanical performance of polyurethane and epoxy adhesives in connections with glued-in rods at elevated temperatures. *BioResources* **2016**, *11*, 8200–8214. [CrossRef]

21. Lartigau, J.; Coureau, J.-L.; Morel, S.; Galimard, P.; Maurin, E. Mixed mode fracture of glued-in rods in timber structures. *Int. J. Fract.* **2015**, *1992*, 71–86. [CrossRef]

22. Taupin, J. Restauration à la résine époxyde de planchers et charpentes au Monastère de la Grande Chartreuse. *Bull. D'informations Tech.* **1980**, *92*, 34–43.

23. *CEN: Timber Structures—Test Methods—Determination of Embedment Strength and Foundation Values for Dowel Type Fasteners (EN 383:2007)*; European Committee for Standardization: Brussels, Belgium, 2007.

24. *CEN: Adhesives, Phenolic and Aminoplastic, for Load-Bearing Timber Structures—Classification and Performance Requirements (EN 301:2017)*; European Committee for Standardization: Brussels, Belgium, 2017.

25. *CEN/TC 250 SC 5: Glued-in Rods in Glued Structural Timber Products—Testing, Requirements and Bond Shear Strength Classification (prEN 17334)*; European Committee for Standardization: Brussels, Belgium, 2018.

26. *EN 1995-1. Eurocode 5: Design of Timber Structures. Part 1-1: General Common Rules and Rules for Buildings*; European Committee for Standardization: Brussels, Belgium, 2005.

27. Bergstedt, A.; Lyck, C. *Larch Wood—A Literature Review*; Forest & Landscape Working Papers no. 23; Forest & Landscape Denmark: Hørsholm, Denmark, 2007.

28. *EN 15274. General Purpose Adhesives for Structural Assembly—Requirements and Test Methods*; European Committee for Standardization: Brussels, Belgium, 2015.

29. *EN1382. Timber Structures—Test Methods—Withdrawal Capacity of Timber Fasteners*; European Committee for Standardization: Brussels, Belgium, 2016.

30. Yeboah, D.; Taylor, S.; McPolin, D.; Gilfillan, J.; Gilbert, S. Behaviour of joints with bonded-in steel bars loaded parallel to the grain of timber elements. *Constr. Build Mate. R* **2011**, *25*, 2312–2317. [CrossRef]

31. Volkersen, O. Die niektraftverteilung in zugbeanspruchten mit konstanten laschenquerschritten. *Luftfahrtforschung* **1938**, *15*, 41–47.

32. Goland, M.; Reissner, E.J. The stresses in cemented joints. *Appl. Mech.* **1944**, *66*, A17–A27. [CrossRef]

33. Hart-Smith, L.J. *Adhesive-Bonded Double-Lap Joints*; Nasa CR-112235: Long Beach, CA, USA, 1973.

applied sciences

MDPI

Article

Experimental Study on the Behavior of Existing Reinforced Concrete Multi-Column Piers under Earthquake Loading

Jung Han Kim [1], Ick-Hyun Kim [2] and Jin Ho Lee [3,*]

[1] Department of Civil Engineering, Pusan National University, Busan 46241, Korea; jhankim@pusan.ac.kr
[2] Department of Civil and Environmental Engineering, University of Ulsan, Ulsan 44610, Korea; ickhyun@ulsan.ac.kr
[3] Department of Ocean Engineering, Pukyong National University, Busan 48513, Korea
*   Correspondence: jholee0218@pknu.ac.kr

Citation: Kim, J.H.; Kim, I.-H.; Lee, J.H. Experimental Study on the Behavior of Existing Reinforced Concrete Multi-Column Piers under Earthquake Loading. *Appl. Sci.* **2021**, *11*, 2652. https://doi.org/10.3390/app11062652

Academic Editor: Chiara Bedon

Received: 8 February 2021
Accepted: 13 March 2021
Published: 16 March 2021

Publisher's Note: MDPI stays neutral with regard to jurisdictional claims in published maps and institutional affiliations.

Abstract: When a seismic force acts on bridges, the pier can be damaged by the horizontal inertia force of the superstructure. To prevent this failure, criteria for seismic reinforcement details have been developed in many design codes. However, in moderate seismicity regions, many existing bridges were constructed without considering seismic detail because the detailed seismic design code was only applied recently. These existing structures should be retrofitted by evaluating their seismic performance. Even if the seismic design criteria are not applied, it cannot be concluded that the structure does not have adequate seismic performance. In particular, the performance of a lap-spliced reinforcement bar at a construction joint applied by past practices cannot be easily evaluated analytically. Therefore, experimental tests on the bridge piers considering a non-seismic detail of existing structures need to be performed to evaluate the seismic performance. For this reason, six small scale specimens according to existing bridge piers were constructed and seismic performances were evaluated experimentally. The three types of reinforcement detail were adjusted, including a lap-splice for construction joints. Quasi-static loading tests were performed for three types of scale model with two-column piers in both the longitudinal and transverse directions. From the test results, the effect on the failure mechanism of the lap-splice and transverse reinforcement ratio were investigated. The difference in failure characteristics according to the loading direction was investigated by the location of plastic hinges. Finally, the seismic capacity related to the displacement ductility factor and the absorbed energy by hysteresis behavior for each test were obtained and discussed.

Keywords: reinforced concrete column; multi-column pier; seismic behavior; lap-splice; transverse reinforcement; plastic hinge; ductility

1. Introduction

Detailed seismic design standards have been introduced recently in countries in low to moderate seismicity regions. In the case of Korea, detailed standards for seismic design were established after recognizing the earthquake damage from the Northridge earthquake in the US and the Kobe earthquake in Japan in the 1990s. Therefore, structures constructed before that time cannot be considered to have been secured against earthquakes. In the case of bridges, the mass is concentrated in the superstructure, therefore when a lateral load such as an earthquake is applied, the bridge should withstand the seismic force through the bending behavior of piers. In the case of a single-column pier, a plastic hinge is generated at the bottom of the pier where the bending moment is at its maximum. This plastic hinge must have sufficient ductile capacity and energy absorption capacity to be safe against earthquake loading. Reinforcement details in the plastic hinge section have a great influence on this flexural behavior. In order to have sufficient ductility capacity, the core concrete must be sufficiently confined by transverse reinforcing bars, and the main reinforcing bar should be connected continuously without a lap-splice. However, when the detailed seismic design criteria are not applied, the longitudinal reinforcing bars at the plastic hinge region were often connected by a lap-splice due to the convenience of construction. Therefore, the behavioral characteristics of such lap-spliced longitudinal bars has emerged as a very important issue in the seismic performance evaluation of the existing bridges.

Most studies about the seismic behavior of bridge piers are about the single-column piers because the number of existing single-column pier bridges is much higher than that of multi-column pier bridges. The multi-column piers may show similar behavior in the longitudinal direction of the superstructure to the single-column piers, but it can be expected that the behavior of the multi-column piers in the transverse direction is different from the single-column piers. Therefore, in this study, one of the aims is to verify the seismic performance of multi-column piers in each direction, as well as the reinforcement details.

Since the superstructure of the bridge has a large mass and is located at the same height, it acts as an inertial force concentrated at the top of the pier when an earthquake load is applied. Therefore, the pier behaves as a deformable body that transmits the inertial force of the superstructure. Most of the piers are reinforced concrete, and how much seismic force these piers can withstand in the horizontal direction is a key consideration in seismic design. If the aspect ratio of the pier, that is, the height compared to the cross-sectional dimension, is low, the shear behavior dominates, and if this ratio is large, the bending behavior occurs [1]. In order to avoid such behavior, reinforcement may be necessary to prevent damage, for example, by seismic isolation, in bridges [2]. The pier can withstand the seismic force with its horizontal strength, but when excessive seismic force is applied, sufficient ductile capacity is required.

In order to increase the ductility of reinforced concrete piers that behave in flexure, it is necessary to prevent compressive failure of the concrete in the cross section by bending. For this purpose, core concrete is confined by a transverse reinforcement bar. The confinement effect at the plastic hinge region of the pier column during earthquakes increases seismic capacity due to ductile behavior [3,4]. In the case of installing sufficient transverse reinforcement, considerable flexural ductility is exhibited while maintaining the load resistance capacity even after the plastic hinge occurs [5,6]. Even if the transverse reinforcement bar ratio is the same, if the spacing is small, the confinement effect increases [7]. Therefore, the horizontal reinforcement ratio is an important variable in seismic detail. This confinement effect can be expressed by transverse reinforcing bars, but can also be implemented through appropriate techniques such as fiber straps [8,9].

The transverse reinforcing bars of reinforced concrete columns contribute to the confinement effect to increase ductility during bending behavior. For this, the transverse reinforcing bar must withstand the tensile force. However, since transverse reinforcing bars are difficult to make into continuous reinforcing bars, they are traditionally composed of columnar hoops and cross ties. In this case, the design details of hoops and cross ties are very important, and an experimental study has been conducted on this [10–12]. In order to improve this, spiral reinforcement detail using a continuous transverse reinforcing bar is applied, so it has very high ductility [13,14].

In addition, the longitudinal reinforcement bar must withstand the tension in the bending section. However, it is difficult to connect the longitudinal reinforcement bars from the foundation to the bottom of the pier continuously during construction; the longitudinal reinforcement bars used to be lap-spliced. Since the bottom of the pier is the part where the plastic hinge occurs, a large tensile force acts on the longitudinal reinforcement bar, and if this part is lap-spliced, bond failure is likely to occur [15–19]. For the occurrence of bonding failure of the lab splice reinforcing bar, the effect of various parameters such as longitudinal bar size and length, amount of transverse reinforcing bars and concrete compressive strength were experimentally studied [20]. In order to prevent the bond failure of the lap-splice, confinement should be increased, and research on improvement methods for this was also conducted [21–23].

Studies on seismic detail in the plastic hinge region have been performed on single-column piers. However, there are many multi-column piers, and their behavior is different from that of single-column piers. Studies on the fracturing of beam–column junctions of multi-column piers with relatively weak cap beams or shear failure of the cap beam [24], a study on the reinforcement method of the cap beam [25], and a study on the design guidelines for multicolumn piers [26] were performed. Since the behavior of multi-column piers is different in both directions, it is necessary to evaluate the behavior characteristics of each direction experimentally. The bidirectional experiment can be performed as a shaking table experiment [27]. However, in order to know the seismic capacity in each direction, an experiment on the horizontal load in each direction is required.

Referring to these existing studies, it is shown that the seismic capacity of the columns with lap-splices is very low and varies according to the details of the transverse reinforcing bar. However, for multi-column piers, seismic performance evaluation is required for these details of the reinforcing bars for each loading direction. Therefore, in this study, the seismic performance of two-column piers is evaluated experimentally.

2. Test Model of Multi-Column Bridge Piers

2.1. Test Specimens

Three sets of test specimens were constructed. They were designed to be one-fourth-scale models of the existing two-column piers on Korean highways. The height of prototype is determined to be 12 m to induce flexural failure for transverse direction loading. The diameter of the column section of scale model is 500 mm and the height is 3000 mm. In the case of the scale model, reinforcement with a diameter of 10 mm was used for the longitudinal reinforcement due to the limitations of size of commercial reinforcement bar. Instead, the reinforcement bar ratio was the same for the full-scale model and the one-fourth-scale model as 1.174%. For the lateral reinforcement, a reinforcement bar with a diameter of 4 mm was fabricated. The concrete strength was 24 MPa, which is the same value applied to the prototype model. Therefore, it is judged that the Young's modulus as determined by the concrete strength was similar between the scale model and prototype. However, the geometrical mix design of the concrete could not be properly scaled because of commercial concrete limitations. The properties of the test specimens are summarized in Table 1, and Figure 1 illustrates the geometry of the test specimen.

Table 1. Properties of the test model.

	Property	Prototype	Model
Material strength (MPa)	Concrete strength (MPa)	24	24
	Reinforcement steel bar strength (MPa)	300	300
Column	Diameter (m)	2.0	0.5
	Height (m)	12.0	3.0
Longitudinal reinforcement	Diameter (mm)	29	10
	Reinforcement ratio (%)	1.174	1.174
Diameter of lateral reinforcement (mm)		16	4
Axial force		$0.052\,\sigma_{ck}A_g$	$0.052\,\sigma_{ck}A_g$

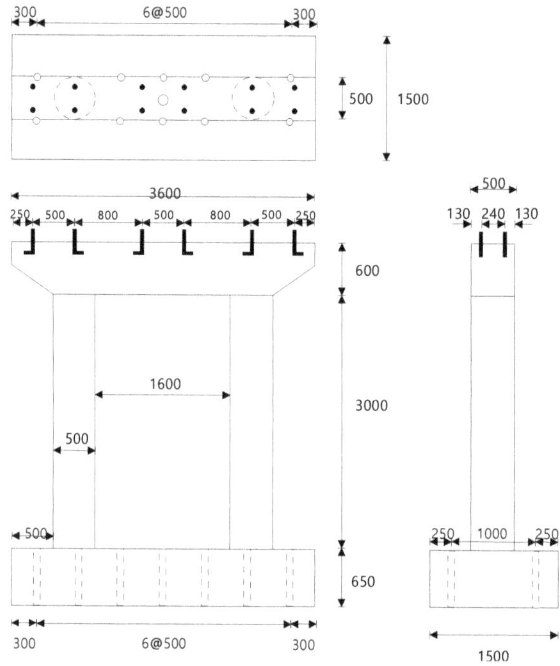

Figure 1. Geometry of the test specimen (units: mm).

The characteristics of 6 test specimens are described in Table 2. The test specimen was composed of two specimens as a set so that the behavior characteristics in the longitudinal direction and the transverse direction to the superstructure could be investigated. Additionally, 3 different reinforcement details were applied for each specimen set. The first specimen (RH-NS) has the same reinforcement details as existing bridge piers. The second specimen (RH-SL) has different reinforcement bar details. For this specimen, the number of lap-spliced longitudinal reinforcement bars in the plastic hinge region was the half of the total number of longitudinal reinforcement bars, and the transverse reinforcing bar was installed with 1/2 of the amount required by the seismic design regulations. The third specimen used continuous longitudinal reinforcing bars without a lap-splice, and a minimal amount of transverse reinforcing bars was installed to prevent local buckling of the main reinforcing bars.

Table 2. Comparison of the characteristics of each specimen.

Specimen	Loading Direction	Lap-Splice of Longitudinal Reinforcement	Lateral Reinforcement Detail	Spacing of Lateral Reinforcement	Volumetric Reinforcement Ratio
RH-NS-L RH-NS-T	Longitudinal Transverse	100% in the lower region	No consideration of seismic performance	37.5 mm	0.268%
RH-SL-L RH-SL-T	Longitudinal Transverse	50% in the lower region	50% of what is required by AASHTO specifications	20 mm	0.503%
RH-SC-L RH-SC-T	Longitudinal Transverse	No lap-splice	Preventing local buckling of the longitudinal bar	30 mm	0.335%

The reinforcement details for each set are shown in Figure 2. Each specimen of the RH-NS set was designed as a prototype without any consideration of seismic performance

For the RH-SL set, the volumetric transverse reinforcement ratio for this set is 50% of what is required by AASHTO (American association of state highway and transportation officials) specifications [28]. The volumetric transverse reinforcement ratio is defined as the ratio of the volume of the transverse reinforcing bar to the volume of the concrete confined by it. In case of the RH-SC set, minimum lateral reinforcement was used to prevent the local buckling of the longitudinal reinforcement.

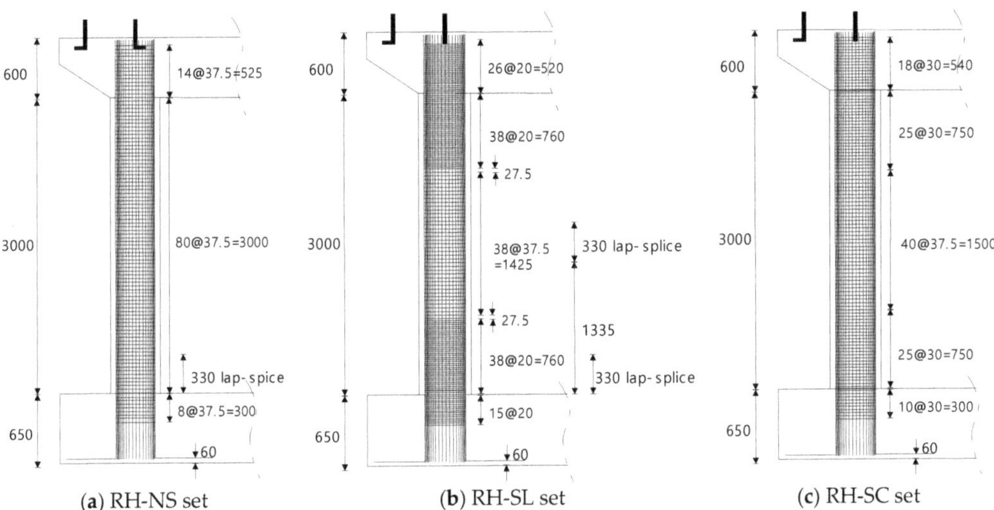

Figure 2. Reinforcement details (unit: mm).

The volumetric transverse reinforcement ratio was 0.268% for the RH-NS set. For the RH-SL set and the RH-SC set, the volumetric transverse reinforcement ratio in the lower and higher regions was 0.503% and 0.335%, respectively. The volumetric transverse reinforcement for these sets was 0.268% except for these regions. For the lateral confinement, two bars of semicircle were used with 135-degree hooks and augmented with two cross ties with a 135-degree hook and a 90-degree hook, as in Figure 3. The cross ties were used in the longitudinal direction and the transverse direction in turns.

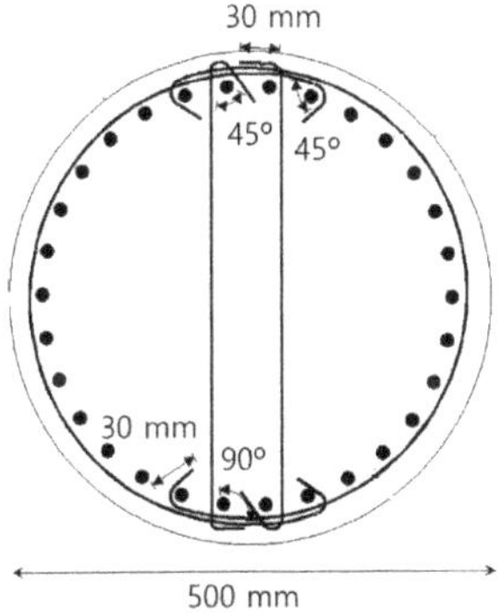

Figure 3. Transverse reinforcement details with hook.

The longitudinal reinforcements were lap-spliced up to 330 mm from the bottom for the RH-NS set. Only 50% of the longitudinal bars were lap-spliced in the lower region and 50% in the middle region for the RH-SL set. In the case of the RH-SC set, the longitudinal reinforcements were extended from the foundation to the top continuously without any lap-splicing. Each set consists of two specimens, one for longitudinal direction loading and the other for transverse direction loading.

2.2. Test Setup

The axial force was applied with a hydraulic pressure jack. Axial force was determined to be 5.2% of the compressive stress of the concrete of the columns, which corresponds with existing bridge structures. The lateral loading was applied with an actuator of 1000 kN (100 tonf) capacity. When longitudinal direction loading was applied, 2 actuators were set and controlled with displacement to prevent torsional behavior. The height at which the horizontal load corresponded to the inertia force of the superstructure was reflected in the existing bridge model. The height from the foundation to the point of loading is 3850 mm, therefore the aspect ratio is 7.7 for longitudinal direction. The aspect ratio for the transverse direction is 3.0 because only column parts are considered to be deformed by rigid joints at both the top and bottom ends. Figure 4 shows the installation of the test specimen for longitudinal loading and transverse loading.

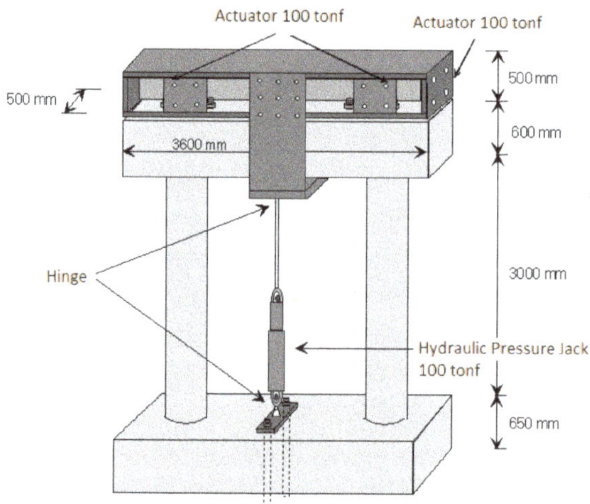

(**a**) Schematic view

(**b**) Longitudinal loading

(**c**) Transverse loading

Figure 4. Test setup for lateral loading and axial loading.

2.3. Loading Protocol

The load patterns are illustrated schematically in Figure 5. The formula proposed in Equation (1) was used to determine the yielding displacement [29].

$$\Delta_y = \frac{\Delta_{+0.75} + \Delta_{-0.75}}{2 \times 0.75} \tag{1}$$

The lateral strength P_i is obtained from analysis at the ultimate compressive strain of 0.003. The displacements $\Delta_{+0.75}$ and $\Delta_{-0.75}$ are displacements when the load $0.75P_i$ is applied in the push and pull directions, respectively. For the comparison of each specimen, the yield displacement was assigned the same value in spite of the differences in reinforcement detail. At the small displacement level of $1.0\Delta_y$~$3.0\Delta_y$, the loading displacement was applied in $0.5\Delta_y$ increments. Two cycles were applied at each loading step. After the $3.0\Delta_y$ displacement level, the load was applied in $1.0\Delta_y$ increments until failure. The displacement was applied at a low speed that was close to the static load. Therefore, during the test, pictures could be taken at each loading step and cracks were checked.

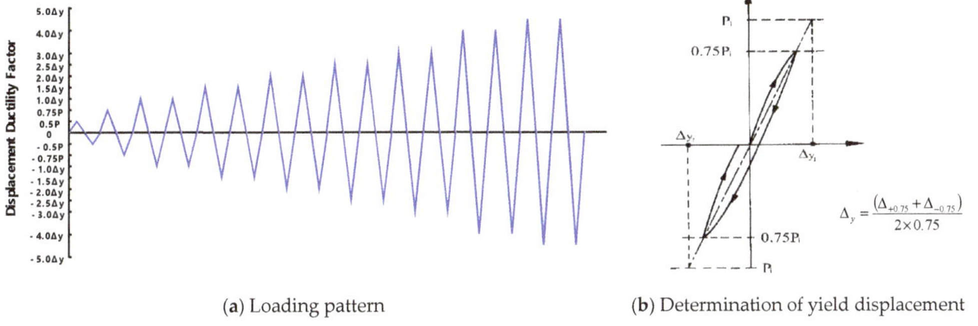

(a) Loading pattern   (b) Determination of yield displacement

Figure 5. Loading patterns of lateral displacement.

Data on load histories were obtained from the load cell in the actuator. The displacement was measured with wire-type LVDT at the cap beam. To measure the curvature of each column, clip gauges were instrumented in pairs on the front and back faces in the loading direction. Since the plastic hinge region is generally up to a height corresponding to the cross-sectional diameter, it was attached to the position corresponding to this height. The heights at which the clip gauges were instrumented were 100 mm, 300 mm, 500 mm, and 700 mm from the foundation or the cap beam. After measuring the strain on both sides of the column using clip gauges, the strain difference was divided by the cross-sectional diameter to convert it into curvature.

3. Seismic Behavior of Specimens

3.1. Longitudinal Loading Test

3.1.1. Failure Mode of Longitudinal Loading Specimen

Figure 6 shows an image of the failure of the longitudinal loading specimens. In the case of the RH-NS-L specimen, concrete cracks hardly occurred up to the height of 330 mm when the loading was applied corresponding to the yield displacement because the ratio of the longitudinal reinforcement was twice that of the other heights by the lap-splicing of the longitudinal reinforcement bars. At $1.5\Delta_y$, the construction joint at the bottom began to crack. At 3.5~$4.0\Delta_y$, the crack in the construction joint was intensified and the concrete adjacent to the foundation was gradually damaged by compression. At $5.0\Delta_y$, the buckling of the reinforcing bar was observed, and spalling occurred. In general, longitudinal reinforcing bars resist tensile forces in flexural members. However, when the surrounding concrete undergoes compression failure, the longitudinal reinforcing bar at the compression section may buckle. In addition, the shear force of the transverse load could cause the longitudinal reinforcing bar itself to bend and contribute to the damage.

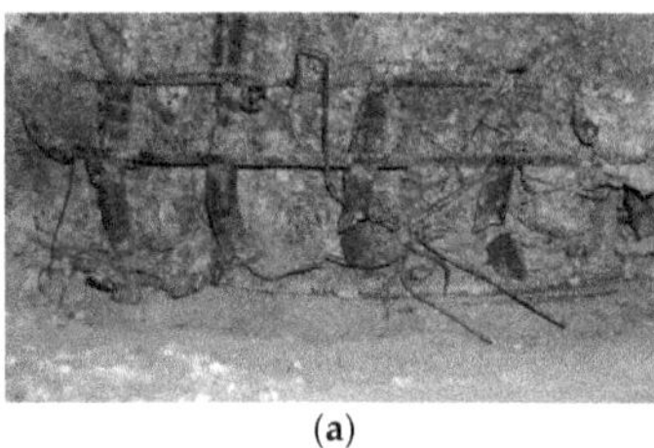  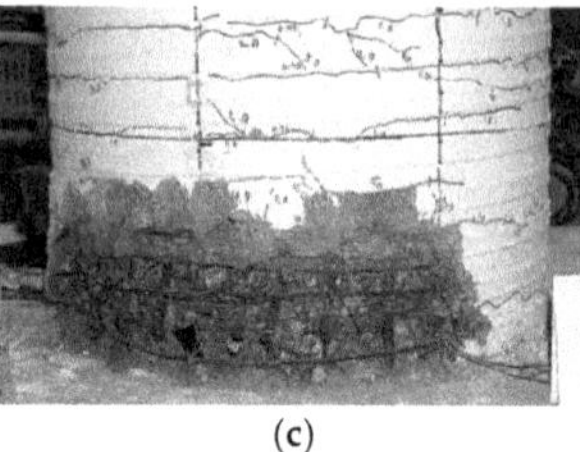

(a) (b) (c)

Figure 6. Failure of the longitudinal loading specimens. (**a**) Buckling and fracture of the longitudinal reinforcement bar of the RH-NS-L specimen ($5\Delta_y$); (**b**) Fracture of the longitudinal reinforcement bar of the RH-SL-L specimen ($7\Delta_y$); (**c**) Buckling and fracture of the longitudinal reinforcement bar of the RH-SC-L specimen ($7\Delta_y$).

In the case of the RH-SL-L specimen, since the longitudinal reinforcement ratio of the lap-splice region was relatively smaller than that of the RH-NS-L specimen, cracks occurred earlier at the bottom. At $6.0\Delta_y$, spalling of the compression part was intensified and the concrete cover began to fall off severely. In the case of the RH-SC-L specimen, several cracks occurred even under 350 mm because the ratio of the longitudinal reinforcement was the same for all sections. At $7.0\Delta_y$, the transverse reinforcement bar at the bottom was deformed due to the buckling of the longitudinal reinforcing bar. However, the gap between the transverse reinforcement bars was determined to prevent buckling of the longitudinal reinforcement bars, so excessive buckling did not occur.

Generally, if the longitudinal bars are lap-spliced in the plastic hinge region, the flexural member is very vulnerable to repeated horizontal loads such as seismic loads. The longitudinal bars do not transmit tensional force and are expected to be damaged by lap-splice bond failure. However, it was observed in this study that the specimens loaded in the longitudinal direction failed through flexural failure at the bottom. The lap-spliced length of longitudinal bars in the scaled specimens was determined in proportion to the diameter of the longitudinal bars, which is larger than the diameter from the scale factor. There was a possibility that the lap-splice was extended to the non-plastic hinge zone, which induced the flexural failure. Another possible reason for flexural failure is the detailing of lateral reinforcements. The minimum volumetric reinforcement ratio for preventing lap-splice bond failure is calculated to be 0.279%. In this study, the volumetric reinforcement ratio of all specimens is almost equal to or larger than this value. The shear friction of confined core concrete by the lateral reinforcements prevented the bond failure of the lap-splice Additionally, the hooks of the lateral reinforcement and cross ties in the specimens can confine the core concrete [30].

Figure 7 shows the distribution of curvature along the height from the bottom of the column for the longitudinal loading test specimen. The lap-splice of longitudinal reinforcements increases the longitudinal reinforcement ratio. Because lap-splices increase the longitudinal reinforcement ratio, the curvature of the corresponding region decreases Therefore, the RH-NS-L specimen has the smallest curvature and the RH-SC-L specimen has the largest curvature. This does not mean that the RH-NS-L specimen does not have small amounts of damage, but it does mean that the damage is concentrated in the lowermost part, where the curvature was not measured.

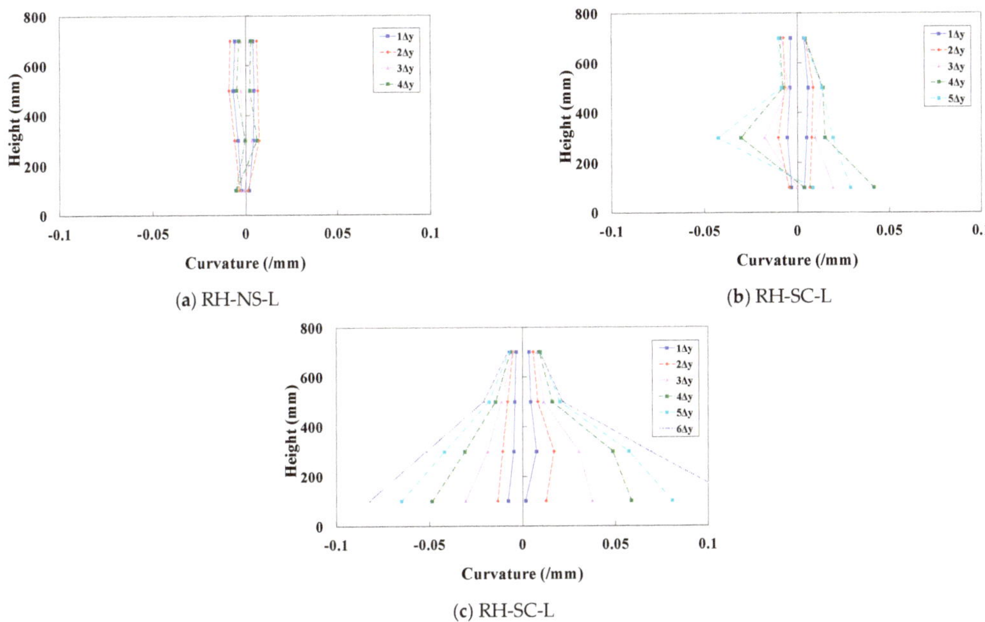

Figure 7. Distribution of curvature of the longitudinal loading test.

3.1.2. Hysteresis Curve of Longitudinal Loading Specimen

Figure 8 shows the load-drift ratio hysteresis curve of the longitudinal loading specimens. For the RH-NS-L specimen, even after the maximum load, it shows a relatively smooth decrease in the hysteresis curve. According to the results of various experiments performed so far, it is expected that the non-ductile behavior will occur due to the bond failure of the reinforcement bar. However, in this test, even if the hysteresis curve goes up to $5.0\Delta_y$, it can be seen that the curve is almost stable. This is because there was no slip between the reinforcement bars raised from the foundation and the column reinforcement bars, and fracturing occurred like with the continuous reinforcement bars. The load-drift ratio curve of RH-SL-L specimen shows a very stable hysteresis up to a ductility of 6.0. It shows more stable behavior than the RH-SL-L specimen because larger a transverse reinforcement ratio affected the confinement of the core concrete. In the load-drift ratio curve for RH-SC-L, displacement ductility at the maximum load capacity was $5.3\Delta_y$ and the fracture ductility was $6.6\Delta_y$. This test specimen shows the most stable hysteresis curve because no lap-splicing of the reinforcing bars was applied.

Figure 9 compares the envelope curve of the hysteresis loop for each longitudinal loading specimen. As can be seen in the figure, there was little difference in the load carrying capacity of each specimen. There was a slight difference in the displacement at failure. This is estimated to be the effect of the curvature according to the lap-splice of the reinforcing bars on the ductility. In general, because lap-slice bond failure occurs, lap-spliced specimens have remarkably low ductility. However, if bond failure does not occur, the lap-spliced section has twice the longitudinal reinforcing bar ratio, resulting in less flexural deformation. Failure will occur at the bottom section of the lap-spliced range. In this section, the longitudinal reinforcing bar ratio is the same without the lap-splice, so the strength of the member does not change significantly.

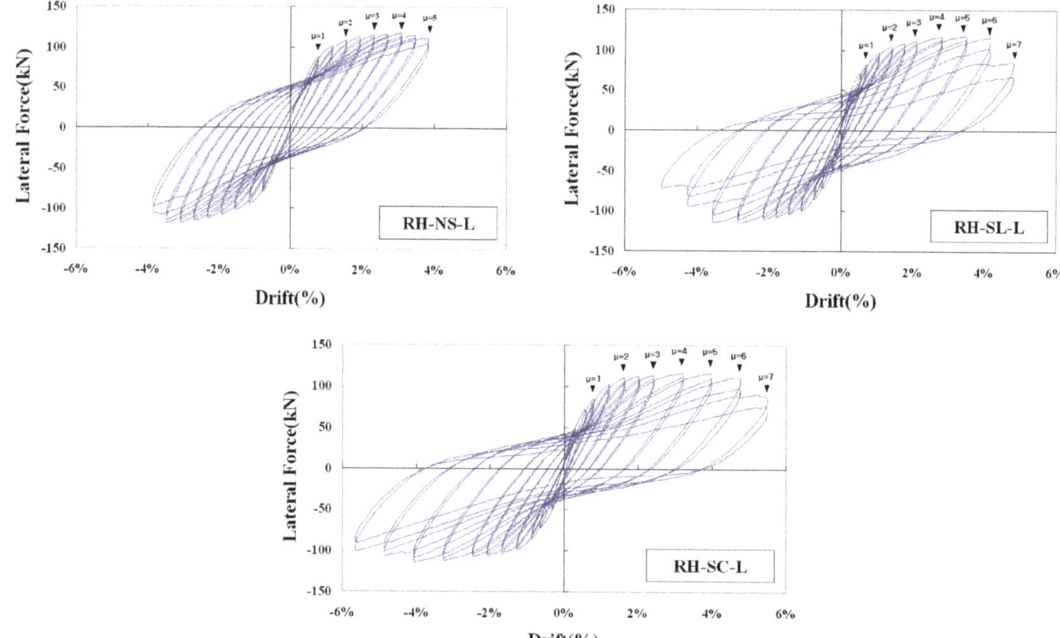

Figure 8. Load-displacement hysteresis curve of the longitudinal loading specimens.

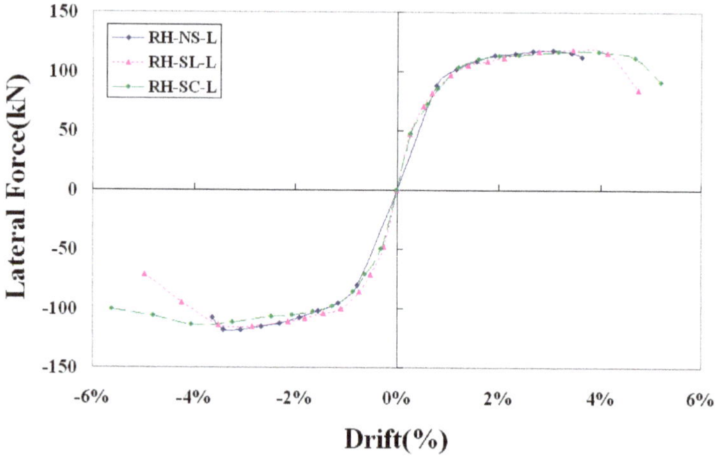

Figure 9. Load-displacement envelop curve of the longitudinal loading specimens.

3.2. Transverse Loading Test

3.2.1. Failure Mode of Transverse Loading Specimen

Since the two-column piers show behavior similar to frames, the behavior of two-column piers in the transverse direction is different from the behavior in the longitudinal direction, as shown in Figure 10. In the longitudinal loading specimens, damage occurs only at the bottom of the column, whereas the transverse loading specimens are damaged at both the top and bottom of the column. The specimens loaded in the transverse direction failed through flexural failure, not through bond-failure of the lap-splices as with the longitudinal loading specimens. Because plastic hinges were formed at both the top and bottom of the column, the ductility is greater than that of the longitudinal loading case. Since the cap beam is not a rigid body, it caused some deformation. Therefore, the moment at the top of the column could be absorbed by the bending deformation of the cap beam, so the failure occurred later than at the bottom, which is fixed to the foundation. The outside of the bottom experienced compression failure and the breakage of the reinforcement bar first, and then failure occurred in the inside of the bottom. Even if the bottom of the column was damaged, the top of the column was not severely damaged. When the resistant moment at bottom was weakened, the resistant moment acted on the top of the column as well, so it could absorb a large displacement. Therefore, in the case of a load in the transverse direction, the displacement ductility could be greater than that in the case of a load in the longitudinal direction.

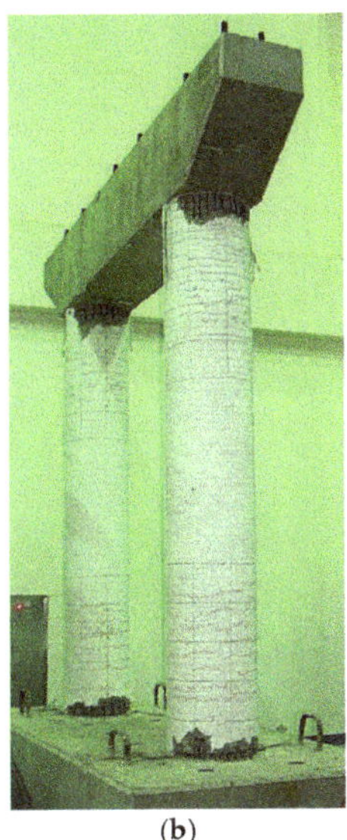

(a)   (b)

Figure 10. Failure of a longitudinal loading specimen and a transverse loading specimen. (**a**) Failure at the bottom of the column of the RH-NS-L specimen; (**b**) Failure at the top and bottom of the column of the RH-NS-T specimen.

Figure 11 shows images of the failure of the RH-NS-T specimen. The cracks were evenly distributed in the upper part of the column from 50 mm to 400 mm when the loading was applied corresponding to the yield displacement. At the bottom of the column, the lap-splice was located up to 330 mm high. In this range, the longitudinal reinforcing bar ratio is doubled, so there is little cracking, and the cracks are evenly distributed in the location from 300 mm to 700 mm at the bottom of the column. At $5.0\Delta_y$, the outside face of the bottom of the column had spalling first by the compressive force, and the concrete cover was damaged and fell off. Buckling of the reinforcing bar was observed at about 5 mm height. At the top of the column, many vertical cracks occurred between about 0 mm and 100 mm in height, and spalling began to appear insignificantly. At $6.0\Delta_y$, the longitudinal reinforcement bar was first broken in a column that was subjected to tensile force during loading. At $8.0\Delta_y$, the longitudinal reinforcement bar was broken at the top of the column at the time of loading. At this time, the longitudinal reinforcement bar at the bottom of the column was almost completely broken where the concrete cover was damaged.

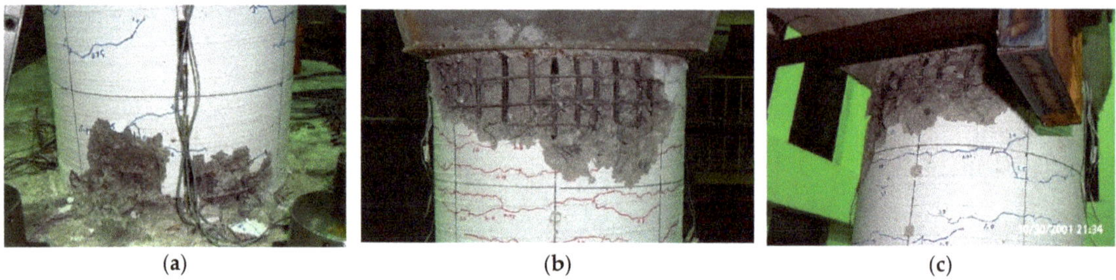

(a)   (b)   (c)

Figure 11. Failure of the transverse loading specimens (RH-NS-T). (**a**) Spalling of the concrete on the outside face of the bottom ($5\Delta_y$); (**b**) Failure of the concrete cover concrete on the outside face of the top ($8\Delta_y$); (**c**) Failure of the concrete cover concrete on the inside face of the top ($8\Delta_y$).

Figure 12 shows images of the failure of the RH-SL-T specimen. The top of the column was evenly cracked in the plastic hinge area, and the bottom of the column had many cracks between 200 mm and 800 mm in height when the loading was applied corresponding to the yield displacement. At the bottom of column, the largest crack occurred at a height of about 350 mm, just above the height of the lap-splice. At $6.07\Delta_y$, the longitudinal reinforcement bar was first broken in a column that was subjected to tensile force during loading, as like the RH-NS-T specimen. At $8.07\Delta_y$, the top of the column was not broken much in spite of the severe fracture of the longitudinal reinforcement bar at the bottom of the column.

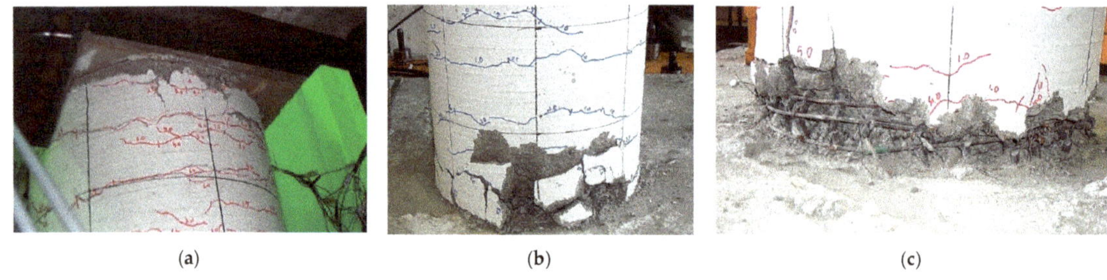

(a)   (b)   (c)

Figure 12. Failure of the transverse loading specimens (RH-SL-T). (**a**) Concrete crack on the inside face of the top ($7\Delta_y$); (**b**) Spalling of the concrete on the inside face of the bottom ($9\Delta_y$); (**c**) Fracture of the longitudinal reinforcement bar on the inside face of the bottom ($9\Delta_y$).

Figure 13 shows images of the failure of the RH-SC-T specimen. Unlike in the other specimens, many cracks occur below the height of 300 mm at the bottom of the column. At $7.07\Delta_y$, the longitudinal reinforcement bar was broken at the outside face of the column subjected to tensile force when the loading was applied. At $9.07\Delta_y$, the longitudinal reinforcement bar at the bottom of the column was severely fractured. The top of the column was not broken much, as like the RH-SL-T specimen.

Figure 14 shows the distribution of curvature along the height for the transverse loading test specimen. Like the longitudinal loading specimen, the RH-SC-T specimen without lap-splicing had the largest curvature at the bottom of the column. On the other hand, at the top of the column, all specimens have relatively the same curvature because there is no lap-splicing. For the transverse loading specimen, double curvature occurred along the column up and down. This means that the aspect ratio is small. Therefore, the height of plastic hinge region that has a large curvature is relatively lower than that of the longitudinal loading specimen.

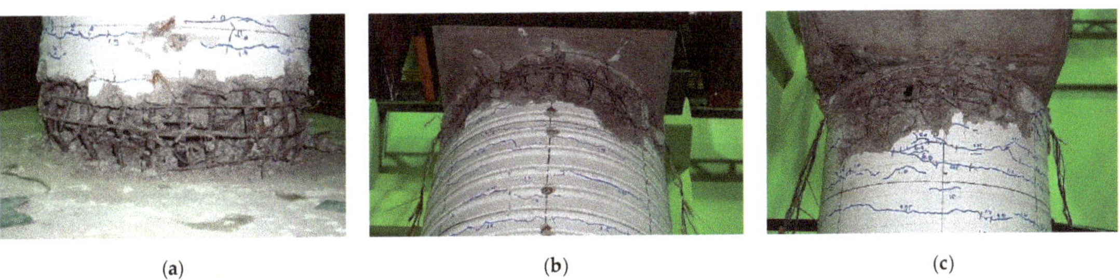

(a)   (b)   (c)

Figure 13. Failure of the transverse loading specimens (RH-SC-T). (**a**) Failure of the concrete and the reinforcement bar on the outside face of the bottom ($9\Delta_y$); (**b**) Failure of the concrete on the inside face of the top ($9\Delta_y$); (**c**) Failure of the concrete on the outside face of the top ($9\Delta_y$).

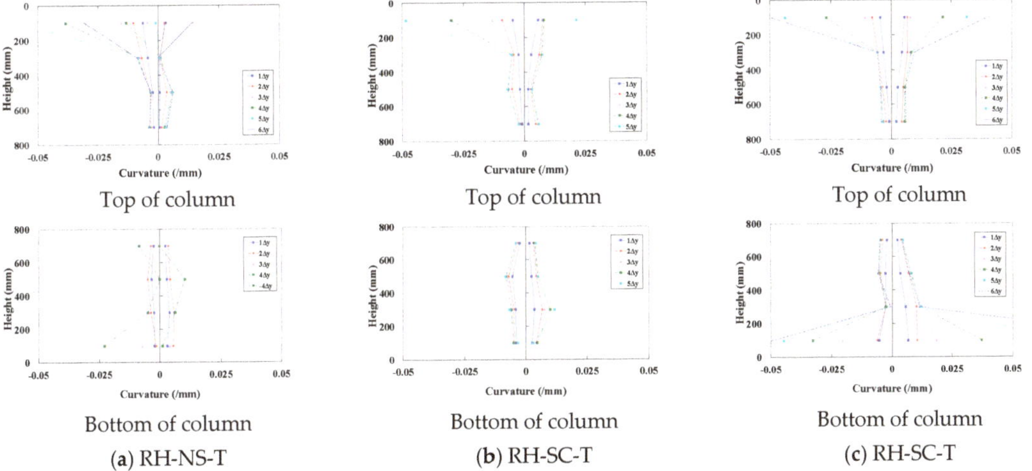

(a) RH-NS-T   (b) RH-SC-T   (c) RH-SC-T

Figure 14. Distribution of curvature of a transverse loading test.

3.2.2. Hysteresis Curve of Transverse Loading Specimen

Figure 15 shows the load-drift ratio hysteresis curve of the transverse loading specimens. For the RH-NS-T specimens, it can be seen that the horizontal load capacity gradually decreases after a stable hysteresis curve up to $5.0\Delta_y$. In contrast to the fact that the resisting force dropped after the breakage of the longitudinal reinforcement bar in the longitudinal loading test, the lateral load was endured more by the moment resistance force, even if the reinforcing bar breaks at the bottom of the column in the transverse loading test. This is because the top of the column can resist the bending moment as the plastic hinge. Thus, it was shown that even if the displacement was applied up to $8.0\Delta_y$, the lateral load carrying capacity was reduced by only about 30%.

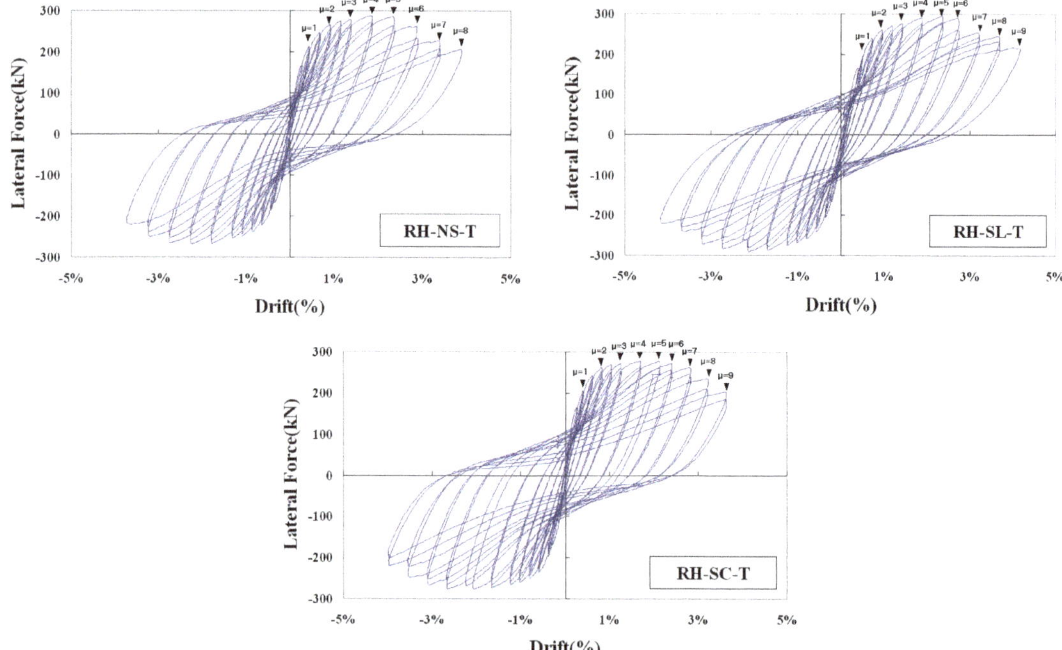

Figure 15. Load-displacement hysteresis curve of transverse loading specimens.

For the RH-SL-T specimens, a stable hysteresis curve was shown up to $6.0\Delta_y$. However, at $7.0\Delta_y$, the resisting force decreased significantly. After $7.0\Delta_y$, the resisting force decreased more slightly. In the case of these test specimens, unlike other transverse loading test specimens, even after $6\Delta_y$, where the fracture progressed to some extent, the lateral load-bearing capacity decreased gradually. This is because the ductility capacity of the top of the column is high, as the transverse reinforcement bar ratio is larger than that of other test specimens.

Figure 16 compares the envelope curve of the hysteresis loops of each transverse loading specimen. As can be seen in the figure, there was little difference in the lateral load carrying capacity of each specimen. This is because bond failure by lap-splice did not occur as in the longitudinal loading test. The reduction of the lateral load-bearing force after the failure of the bottom is determined by the failure of the top of the column. Since there is no lap-splice at the top of the column, the ductility difference due to the transverse reinforcement bar changes the difference in the failure drift ratio.

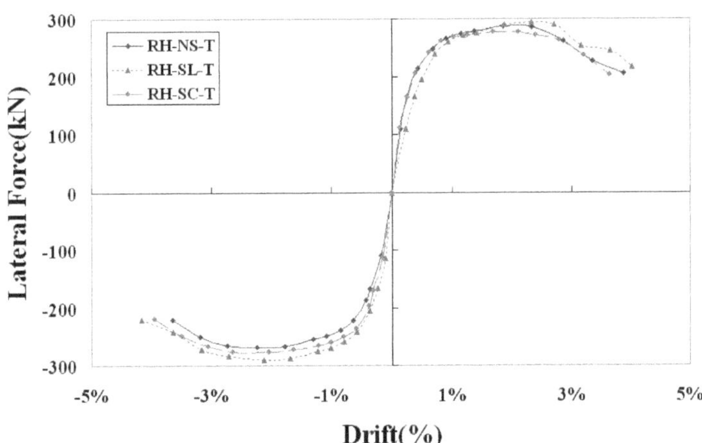

Figure 16. Load-displacement envelope curve of transverse loading specimens.

4. Result of Seismic Capacity

The seismic capacity of a member with nonlinear behavior can be expressed as the energy absorption and ductility factor during hysteretic behavior. The absorbing energy according to the ductility factor was calculated as the area of the hysteresis loop at each loading step. The energy absorption of each specimen is shown in Figure 17. For the case of the longitudinal loading, the energy absorption of the RH-SC-L specimen is higher because the displacement at the failure is slightly larger. For the case of transverse loading, the RH-SL-T specimen is larger than those of any other specimens. It is almost the same until 5.0, before failure occurs at the bottom of the column. After the bottom part of the column fractured, the energy absorption capacity at the top of the column contributed to the seismic performance. It is also confirmed in the failure mode that the top of the column withstands the load in the transverse direction even after damage. The energy absorption capacity at the top of the column is determined by the transverse reinforcing bar ratio at the plastic hinge region. Therefore, even if the specimen RH-SC-T without lap-splicing at the bottom of the column has the largest displacement when the initial damage occurred, the specimen RH-SL-T, which has larger transverse reinforcing bar ratio, has better seismic performance.

The test results are summarized in Table 3. The displacement ductility of the RH-SL-L specimen is smaller than that of the RH-SC-L specimen. There is a difference in ductility between the loading directions. For the longitudinal direction, the pier behaves like a single column. For the transverse direction, the pier behaves as multiple columns which generate multiple plastic hinges. This is consistent with the design value of strength reduction factor defined as three for the single column pier and five for the multi-column pier in the design code. It was expected that there would be a large difference in seismic performance according to the details of the reinforcing bar, but the bond failure in the lap-splice did not occur, so all the specimens were found to have a seismic performance, although there are slight differences. The reason why the lap-splice failure did not occur may be that the plastic hinge region is small because the cross-sectional diameter of the multi-column pier is smaller than single-column pier.

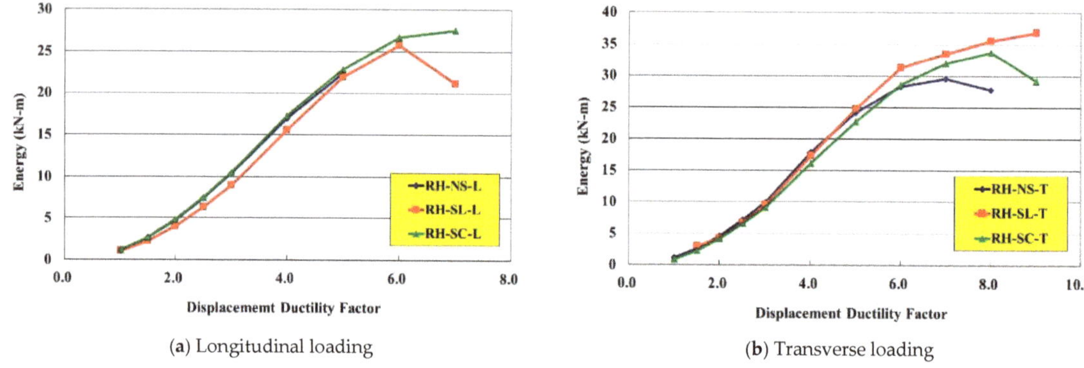

(**a**) Longitudinal loading   (**b**) Transverse loading

Figure 17. Energy absorption with respect to displacement ductility factor.

Table 3. Comparison of failure displacement and ductility.

	RH-NS-L	RH-SL-L	RH-SC-L	RH-NS-T	RH-SL-T	RH-SC-T
Yield displacement Δ_y (mm)		43.67			19.21	
Maximum lateral force, P_m (kN)	117.72	118.13	116.06	287.93	294.06	277.06
Displacement at P_m (mm)	118.34	132.94	153.30	56.25	70.27	63.58
Displacement at failure (mm)		171.47	193.46	93.68	101.69	96.73
Drift ratio at failure		4.45%	5.02%	3.12%	3.39%	3.22%
Displacement ductility at failure		3.92	4.43	4.87	5.29	5.03

5. Conclusions

In this study, six scaled models of two-column piers were constructed and tested to investigate their behavior when seismic load is applied. Test specimens having three different reinforcement details were loaded in the longitudinal direction and the transverse directions.

Initially, bond failure was expected according to lap-splice reinforcement detail. If the longitudinal reinforcements are lap-spliced to enough length and the transverse reinforcement is appropriate, the flexural failure can be expected. Multi-column piers have a smaller cross-sectional size than single-column piers, and the fact that the plastic hinge length is small may be why there was no bond failure at the lap-splice. Accordingly, the effect of the lap-splice resulted in the reduction of the curvature of the corresponding height because the longitudinal reinforcement bar ratio was increased.

The behaviors of two-column piers in the transverse direction is different from the behavior in the longitudinal direction. For the transverse direction, high ductility was produced due to multiple plastic hinges. The ultimate failure occurs due to the bending failure of the top of the column connected to the cap beam, so the ductility is affected by the seismic reinforcement detail of this part.

In this study, lap-splicing did not have a significant effect. However, it should be noted that if there is a lap-splice in the longitudinal reinforcement of an actual pier, bond failure may occur depending on the seismic detail. In particular, since it is a result of a small-scale test, the size effect needs to be additionally considered. Nevertheless, piers with non-seismic details can withstand earthquakes to some extent if adequate details are supported.

Author Contributions: Experiment, J.H.K.; supervision, I.-H.K.; writing—review and editing J.H.L. All authors have read and agreed to the published version of the manuscript.

Funding: This work was supported by the Korea Institute of Energy Technology Evaluation and Planning (KETEP) and the Ministry of Trade, Industry & Energy (MOTIE) of the Republic of Korea (No. 20181510102410) and Basic Science Research Program through the National Research Foundation of Korea (NRF) funded by the Ministry of Education (2019R1I1A3A01058812).

Institutional Review Board Statement: Not applicable.

Informed Consent Statement: Not applicable.

Data Availability Statement: Not applicable.

Conflicts of Interest: The authors declare no conflict of interest.

References

1. Cheok, G.S.; Stone, W.C. Behavior of 1/6-scale model bridge columns subjected to inelastic cyclic loading. *ACI Struct. J.* **1990**, *87*, 630–638.
2. Brito, M.B.; Ishibashi, H.; Akiyama, M. Shaking table tests of a reinforced concrete bridge pier with a low-cost sliding pendulum system. *Earthq. Eng. Struct. Dyn.* **2019**, *48*, 366–386. [CrossRef]
3. Kim, T.; Lee, K.; Yoon, C.; Shin, H.M. Inelastic behavior and ductility capacity of reinforced concrete bridge piers under earthquake. I: Theory and formulation. *ASCE J. Struct. Eng.* **2003**, *129*, 1199–1207. [CrossRef]
4. Mander, J.B.; Priestley, M.J.N.; Park, R. Theoretical stress–strain model for confined concrete. *ASCE J. Struct. Eng.* **1998**, *114*, 1804–1826. [CrossRef]
5. Priestley, M.J.N.; Park, R. Strength and ductility of concrete bridge columns under seismic loading. *ACI Struct. J.* **1987**, *84*, 61–76.
6. Chai, Y.H.; Priestley, M.J.N.; Seible, F. Seismic retrofit of circular bridge columns for enhanced flexural performance. *ACI Struct. J.* **1991**, *88*, 572–584.
7. Hoshikuma, J.; Kawashima, K.; Nagaya, K.; Taylor, A.W. Stress-strain model for confined reinforced concrete in bridge piers. *J. Struct. Eng.* **1997**, *123*, 624–633. [CrossRef]
8. Sheikh, S.A.; Uzumeri, S.M. Strength and ductility of tied concrete columns. *ASCE J. Struct. Div.* **1980**, *106*, 1079–1102. [CrossRef]
9. Saadatmanesh, H.; Ehsani, M.R.; Li, M.W. Strength and ductility of concrete columns externally reinforced with fiber composite straps. *ACI Struct. J.* **1994**, *91*, 434–447.
10. Moehle, J.P.; Cavanagh, T. Confinement effectiveness of crossties in RC. *ASCE J. Struct. Eng.* **1985**, *111*, 2105–2120. [CrossRef]
11. Lukkunaprasit, P.; Sittipunt, C. Ductility enhancement of moderately confined concrete tied columns with hook-clips. *ACI Struct. J.* **2003**, *100*, 422–429.
12. Lee, T.; Chen, C.; Tsou, P.; Lin, K.; Lee, W.; Pan, A.D.E. Experimental evaluation of crossties in large reinforced concrete columns under pure compression. *Adv. Struct. Eng.* **2012**, *15*, 1717–1727. [CrossRef]
13. Sheikh, S.A.; Toklucu, M.T. Reinforced concrete columns confined by circular spirals and hoops. *ACI Struct. J.* **1993**, *90*, 542–553.
14. Tanaka, T.H.; Park, R. Behavior of axially loaded concrete columns confined by elliptical hoops. *ACI Struct. J.* **1999**, *96*, 967–971.
15. Jaradat, O.A.; McLean, D.I.; Marsh, M.L. Performance of existing bridge columns under cyclic loading-Part1: Experimental results and observed behavior. *ACI Struct. J.* **1998**, *95*, 695–704.
16. Ichinose, T. Splitting Bond Failure of Columns under Seismic Action. *ACI Struct. J.* **1995**, *92*, 535–542.
17. Sultana, N.; Ray, T. Implications of bond-slip in inelastic dynamic responses of degrading structural systems. *Int. J. Civ. Eng.* **2020**, *18*, 1039–1052. [CrossRef]
18. Mousa, M.I. Flexural behaviour and ductility of high strength concrete (HSC) beams with tension lap splice. *Alex. Eng. J.* **2015**, *54*, 551–563. [CrossRef]
19. Rakhshanimehr, M.; Reza Esfahani, M.; Reza Kianoush, M.; Ali Mohammadzadeh, B.; Roohollah Mousavi, S. Flexural ductility of reinforced concrete beams with lap-spliced bars. *Can. J. Civ. Eng.* **2014**, *41*, 594–604. [CrossRef]
20. Azizinamini, A.; Pavel, R.; Hatfield, E.; Ghosh, S. Behavior of lap-spliced reinforcing bars embedded in high strength concrete. *ACI Struct. J.* **1999**, *96*, 826–835.
21. Hamad, B.S.; Najjar, S.S. Evaluation of the role of transverse reinforcement in confining tension lap splices in high strength concrete. *Mater. Struct.* **2002**, *35*, 219–228. [CrossRef]
22. Bournas, D.A.; Triantafillou, T.C. Bond strength of lap-spliced bars in concrete confined with composite jackets. *Asce J. Compos. Constr.* **2011**, *15*, 156–167. [CrossRef]
23. Hadhood, A.; Agamy, M.H.; Abdelsalam, M.M.; Mohamed, H.M.; El-Sayed, T.A. Shear strengthening of hybrid externally-bonded mechanically-fastened concrete beams using short CFRP strips: Experiments and theoretical evaluation. *Eng. Struct.* **2019**, *201*, 109795. [CrossRef]
24. Sexsmith, R.; Anderson, D.; English, D. Cyclic behavior of concrete bridge bents. *ACI Struct. J.* **1997**, *94*, 103–113.
25. Mander, J.B.; Kim, J.H.; Ligozio, C.A. *Seismic Performance of a Model Reinforced Concrete Bridge Pier before and after Retrofit*; Technical Report NCEER-96-0009; NCEER: New York, NY, USA, 1996.
26. Wua, R.; Pantelides, C.P. Seismic evaluation of repaired multi-column bridge bent using static and dynamic analysis. *Constr. Build. Mater.* **2019**, *208*, 792–807. [CrossRef]

27. Kar, D.; Roy, R. Seismic behavior of RC bridge piers under bidirectional excitations: Implications of site effects. *J. Earthq. Eng.* **2018**, *22*, 3030–3331. [CrossRef]
28. AASHTO. *Standard Specifications for Highway Bridges*, 17th ed.; American Association of State Highway and Transportation Officials: Washington, DC, USA, 2010.
29. Prestley, M.J.N.; Seible, F.; Calvi, G.M. *Seismic Design and Retrofit of Bridges*; John Wiley & Sons, Inc.: New York, NY, USA, 1996.
30. Lee, J.H.; Seok, S.G.; Yun, S.G.; Kang, S.K. Seismic performance of circular columns considering transverse steel details. In Proceedings of the EESK Conference, Seoul, Korea, 4 March 2000; Volume 4, pp. 259–266.

applied sciences

Article

Application of Component-Based Mechanical Models and Artificial Intelligence to Bolted Beam-to-Column Connections

man Faridmehr [1], Mehdi Nikoo [2], Raffaele Pucinotti [3] and Chiara Bedon [4,*]

[1] Institute of Architecture and Construction, South Ural State University, 454080 Chelyabinsk, Russia; faridmekhri@susu.ru

[2] Young Researchers and Elite Club, Ahvaz Branch, Islamic Azad University, Ahvaz 61349-37333, Iran; m.nikoo@iauahvaz.ac.ir

[3] Department of Mechanics and Materials, University of Reggio Calabria, 89124 Reggio Calabria, Italy; raffaele.pucinotti@unirc.it

[4] Department of Engineering and Architecture, University of Trieste, Via Valerio 6/1, 34127 Trieste, Italy

* Correspondence: chiara.bedon@dia.units.it; Tel.: +39-040-558-3837

Abstract: Top and seat beam-to-column connections are commonly designed to transfer gravitational loads of simply supported steel beams. Nevertheless, the flexural resistance characteristics of these type of connections should be properly taken into account for design, when a reliable analysis of semi-rigid steel structures is desired. In this research paper, different component-based mechanical models from Eurocode 3 (EC3) and a literature proposal (by Kong and Kim, 2017) are considered to evaluate the initial stiffness ($S_{j,ini}$) and ultimate moment capacity (M_n) of top-seat angle connections with double web angles (TSACWs). An optimized artificial neural network (ANN) model based on the artificial bee colony (ABC) algorithm is proposed in this paper to acquire an informational model from the available literature database of experimental test measurements on TSACWs. In order to evaluate the expected effect of each input parameter (such as the thickness of top flange cleat, the bolt size, etc.) on the mechanical performance and overall moment–rotation (M–θ) response of the selected connections, a sensitivity analysis is presented. The collected comparative results prove the potential of the optimized ANN approach for TSACWs, as well as its accuracy and reliability for the prediction of the characteristic (M–θ) features of similar joints. For most of the examined configurations, higher accuracy is found from the ANN estimates, compared to Eurocode 3- or Kong et al.-based formulations.

Keywords: top-seat angle connections (TSACW); component-based models; initial stiffness; ultimate moment capacity; moment-rotation relation; artificial neural network (ANN); sensitivity analysis (SA)

Citation: Faridmehr, I.; Nikoo, M.; Pucinotti, R.; Bedon, C. Application of Component-Based Mechanical Models and Artificial Intelligence to Bolted Beam-to-Column Connections. *Appl. Sci.* **2021**, 11, 2297. https://doi.org/10.3390/app11052297

Academic Editors: César M. A. Vasques and Alkiviadis Paipetis

Received: 1 February 2021
Accepted: 2 March 2021
Published: 5 March 2021

Publisher's Note: MDPI stays neutral with regard to jurisdictional claims in published maps and institutional affiliations.

1. Introduction

Bolted top-seat angle connections without (TSAC) or with web angles (TSACW) have been extensively used in steel and composite structures, because of their relatively high moment capacity and easy construction. These types of connections are mainly designed to resist gravity loads of determinate steel beams. A schematic layout is proposed in Figure 1. Based on the AISC standard [1], a given TSACW is classified as a semi-grid beam-to-column connection, in which the moment–rotation (M–θ) behavior needs to be considered in the overall force distribution and structural analysis. Accordingly, there are no doubts on the necessity to accurately predict the (M–θ) behavior of TSACW joints.

The (M–θ) behavioral trends of steel connections represent essential features of their response, and in particular with regard to the initial stiffness $S_{j,ini}$, maximum moment capacity M_n, and maximum rotation θ_u. Once these features are correctly estimated, a reliable simulation of the actual behavior toward the optimum design of structures would be possible. The behavior of TSACs and TSACWs has been investigated and equivalent equations have been proposed to simulate their mechanical behavior for applications to steel frames [2–6]. In parallel to TSACWs, a similar solution has a key role also for "modular buildings", namely, off-site prefabricated building modules in which the primary structure consists of steel frames, whose construction has become increasingly popular due to the advantages, including reduced resource wastage and improved quality [7–10]. The typical connection takes the form of a bolted connection with a plug-in device or a bolted connection with a rocket-shaped tenon. Being classified as "bolted", these connections have the advantages of reduced site work and demountability.

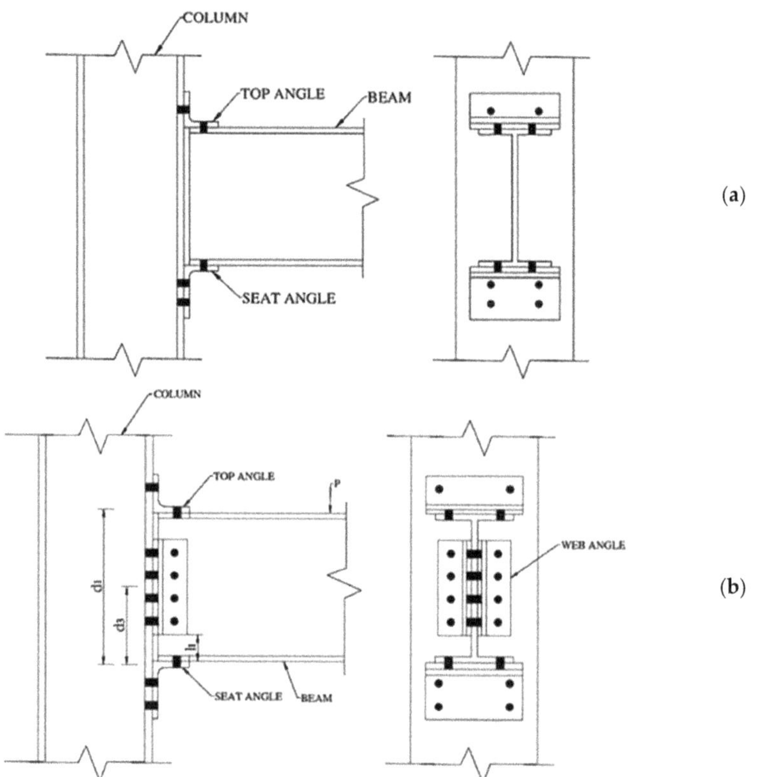

Figure 1. Reference configurations for (**a**) top-seat angle connections without (TSAC) or (**b**) with web angles (TSACW).

In both the cases, the mechanical properties of connections have a great influence on the strength and stability of the structural system to which they belong. An unrealistic design of connections may negatively affect the serviceability of the structure as a whole, due to the occurrence of large deflections. Deformations in connection components as a result of local buckling phenomena is also a common reason for the failure of these connections. The key role of connections for modular buildings has been recognized to affect the overall structural performance of modular systems. Despite this importance, the design of reliable connections is still identified as a major challenge, and prevalent knowledge of their structural performance is still limited [11,12].

In this paper, an original study is presented for TSACWs. An artificial neural network (ANN) model able to capture the underlying mechanism and to accurately estimate both the initial stiffness $S_{j,ini}$ and the ultimate moment capacity M_n for TSACWs is proposed. To this aim, two reference calculation approaches are taken into account from the literature, and investigated in detail, namely, (i) the mechanical component-based model approach proposed by Eurocode 3 ("EC3" [13]) and (ii) the analytical formulation proposed by Kong and Kim in [14] ("KK"). Since the $S_{j,ini}$ and M_n features for a given TSACW are the most influential parameters that affect the overall mechanical characterization of a given bolted beam-to-column connection, these two selected properties are preliminarily calculated based on the selected "EC3" and "KK" formulations, and compared against each other, with the support of experimental data from the literature (77 specimens in total). Successively, a comprehensive ANN approach, combined with a metaheuristic artificial bee colony (ABC) algorithm, is developed in this paper to extract an informational model for TSACWs. A major advantage for the definition of the proposed model is given by the support of experimental data from the literature, and sensitivity analysis to assess the collected comparative results. The potential and limits of each scheme, as well as the original ABC-ANN for TSACW joints, are hence discussed in detail.

2. State of the Art

Different parameters contribute to the (M–θ) behavioral characterization of bolted connections, as shown in Figure 2 [15–20], and it is thus first necessary to use reliable modeling techniques. In the case of modular steel buildings, non-linear link elements are generally incorporated for computational efficiency, where the bending moment, shear, axial force, and corresponding moment–rotation (M–θ) curves are simplified and included as spring elements. Nevertheless, since the knowledge of the structural behavior of modular steel buildings is limited, the simplified models for the force–displacement and moment–rotation curves are not well established. Therefore, the intrinsic uncertainty can lead to excessively conservative design practices, requiring additional investigations on the topic.

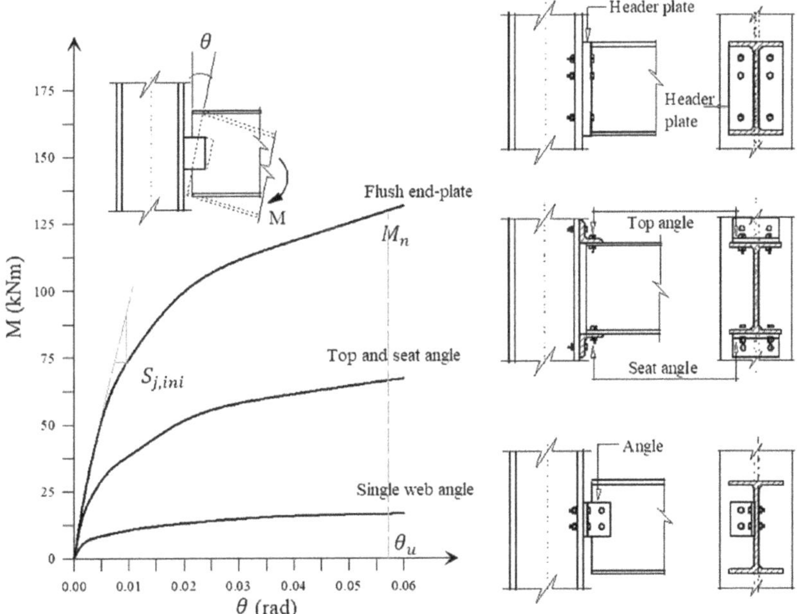

Figure 2. Moment–rotation curve of common types of beam-to-column connections (adapted from [5] under the terms and conditions of a Creative Commons Attribution License (CC BY)).

Extended literature studies are indeed available for TSACs and TSACWs. To simulate the beam-to-column connection behavior, apart from experimental tests, three different modeling options are introduced in the form of analytical or empirical models [21], advanced finite element (FE) models [22–24], and component mechanical models [25,26].

Analytical and empirical models naturally extract simple mathematical expressions, and thus typically facilitate their application in practice, with a reasonable level of accuracy. On the other hand, by applying advanced FE methods, it is possible to simulate more precisely the complex non-linear beam-to-column connection response. Some limitations are unavoidably introduced in terms of high computational cost and sophisticated preparation process (assembly, calibration, etc.). Component-based mechanical models, finally, consider an assembly of rigid elements and equivalent springs relay on the two previous modeling approaches, considering computational complexity and reliability.

Several research studies have documented the $(M-\theta)$ performance of TSACWs, especially through extended experimental investigations. In 1985 and 2000, respectively, Azizinamini and Radziminski [25] and Calado et al. [15] experimentally investigated the $(M-\theta)$ behavior of TSACW specimens under monotonic and cyclic loading. The hysteresis behavior of TSACW specimens was investigated in several cited projects. The presented results indicated that the main sources of plastic deformation are typically located in the top and bottom cleats, which represent the governing components. In addition, the same studies proved a limited influence of the column size on the overall hysteretic performance of the examined connections. A considerable number of literature studies has also been published in support of the prediction of the $(M-\theta)$ behavior, initial stiffness, and ultimate strength of TSACW systems through various mechanical and analytical models. In 1987, Azizinamini et al. [27] proposed a set of relevant equations to predict initial stiffness. Kishi et al. [28,29] presented an analytical model to estimate both the initial stiffness and ultimate moment capacity of TSACWs through cantilever beam theory. In more detail, a three-parameter power model was discussed in [29], in which the elastic stiffness and ultimate moment capacity of connection of TSACWs were estimated by a simple analytical method. Pucinotti [30] proposed a simplified mechanical model along with relevant initial stiffness and ultimate moment equations. A three-linear $(M-\theta)$ constitutive law based on the fiber element formulation was proposed by Shen and Astaneh-Asl [31]. In that study, two possible mechanisms were considered, according to the strength of the angles relative to the bolt. They concluded that plastic hinges were crated at the leg of the angles in thinner top cleats, while the plastic hinge was generally found to appear at the central line of the column bolts, along with plastic deformations of the bolts. A mechanical model for estimating the inelastic cyclic $(M-\theta)$ relationship of TSACWs was proposed by De Stefano et al. [32]. Currently, by the advances in computer technologies, there is a large volume of published studies on $(M-\theta)$ behavior of beam-to-column connections through the FE method. For example, Danesh et al. [33] investigated the effect of shear force on the expected initial stiffness. Pirmoz et al. [34] focused on the effect of web angle on the $(M-\theta)$ behavior. Salem et al. [35] conducted parametric research on the prying action of bolts. Their study indicated that the prying force of top cleat and column flange bolts is found to increase by decreasing the vertical leg of the top cleat. Collectively, all the mentioned studies (and others) outline a set of different methodologies in support of a reliable $(M-\theta)$ modeling and mechanical characterization of beam-to-column connections.

Among the multitude of research studies that have been done on TSACs or TSACWs, traditional experimental or numerical methods are used in most of the cases. The current investigation, in this regard, applies ANN techniques for the analysis of this type of connection. ANNs are largely used for solving complex civil engineering (and other) problems, and their application has increased significantly in the last few years. ANNs can be effectively applied for predictive modeling in different engineering fields, especially in those cases where some prior (experimental or numerical) analyses are already available. The origins of ANNs can be found in the field of biology, where the biological brain consists of billions of highly interconnected neurons, forming a neural network.

A validated informational method is developed in this paper, to provide an alternative tool that could be used to model the complex structural and material behavior in TSACWs that cannot be easily approximated and generalized by conventional approaches. The information about the underlying mechanics for TSACWs is extracted from the observed experimental data of the literature, which are processed in neural networks and subsequently trained with optimization techniques.

3. Data Bank Development

Having recognized that different parameters contribute to the $(M\text{–}\theta)$ behavioral characterization of bolted beam-to-column connections, it is first fundamental to develop a consistent data bank of test results for the examined connection typology. Experimental investigations and their results—when correctly extracted—notoriously allow a more robust and accurate classification of different behavioral features for beam-to-column connections, including $S_{j,ini}$ and M_n, but also hardening, non-linearities, progressive damage and degradation of mechanical parameters, rotation capacity, failure mechanism, and sequence. The same database is also strictly necessary for the development of the ANN model herein proposed.

In this paper, the preliminary verification of the component-based mechanical model for bolted TSACWs, as well as the training data to develop the ANN model, is carried out with the support of experimental data from the literature. Table 1 presents a short summary of the major geometrical and material properties for the examined TSACWs. The full data of TSACW specimens are available in [27,36,37].

Table 1. Geometrical and mechanical characteristics of selected TSACW specimens.

Test	Beam	Column	Size of Bolts (mm)	Top Cleat (mm)	Web Cleat (mm)	Yield Stress of Angle (N/mm^2)
8S1	H210 × 134 × 6.4 ×10.2	H310 × 254 × 9.1 × 16.3	19.1	L152 × 89 × 7.9	L102 ×89 × 6.4	285.4
8S2	H210 × 134 × 6.4 ×10.2	H310 × 254 × 9.1 × 16.3	19.1	L152 × 89 × 9.5	L102 × 89 × 6.4	285.4
8S3	H210 × 134 × 6.4 × 10.2	H310 × 254 × 9.1 × 16.3	19.1	L152 × 89 × 7.9	L102 × 89 × 6.4	285.4
8S4	H210 × 134 × 6.4 × 10.2	H310 × 254 × 9.1 × 16.3	19.1	L152 × 152 × 9.5	L102 × 89 × 6.4	285.4
8S5	H210 × 134 × 6.4 × 10.2	H310 × 254 × 9.1 × 16.3	19.1	L152 × 102 × 9.5	L102 × 89 × 6.4	285.4
8S6	H210 × 134 × 6.4 × 10.2	H310 × 254 × 9.1 × 16.3	19.1	L152 × 102 × 7.9	L102 × 89 × 6.4	285.4
8S7	H210 × 134 × 6.4 × 10.2	H310 × 254 × 9.1 × 16.3	19.1	L152 × 102 × 9.5	L102 × 89 × 6.4	285.4
8S8	H210 × 134 × 6.4 × 10.2	H310 × 254 × 9.1 × 16.3	22.2	L152 × 89 × 7.9	L102 × 89 × 6.4	277
8S9	H210 × 134 × 6.4 × 10.2	H310 × 254 × 9.1 × 16.3	22.2	L152 × 89 × 9.5	L102 × 89 × 6.4	277
8S10	H210 × 134 × 6.4 × 10.2	H310 × 254 × 9.1 × 16.3	22.2	L152 × 89 × 12.7	L102 × 89 × 6.4	277
14S1	H358 × 172 × 7.9 × 13.1	H323 × 310 × 14 × 22.9	19.1	L152 × 102 × 9.5	L102 × 89 × 6.4	285
14S2	H358 × 172 × 7.9 × 13.1	H323 × 310 × 14 × 22.9	19.1	L152 × 102 × 12.7	L102 × 89 × 6.4	365
14S3	H358 × 172 × 7.9 × 13.1	H323 × 310 × 14 × 22.9	19.1	L152 × 102 × 9.5	L102 × 89 × 6.4	285
14S4	H358 × 172 × 7.9 × 13.1	H323 × 310 × 14 × 22.9	19.1	L152 × 102 × 9.5	L102 × 89 × 9.5	285
14S5	H358 × 172 × 7.9 × 13.1	H323 × 310 × 14 × 22.9	19.1	L152 × 102 × 9.5	L102 × 89 × 6.4	277
14S6	H358 × 172 × 7.9 × 13.1	H323 × 310 × 14 × 22.9	19.1	L152 × 102 × 12.7	L102 × 89 × 6.4	277
14S8	H358 × 172 × 7.9 × 13.1	H323 × 310 × 14 × 22.9	19.1	L152 × 102 × 15.9	L102 × 89 × 6.4	277
14S9	H358 × 172 × 7.9 × 13.1	H323 × 310 × 14 × 22.9	19.1	L152 × 102 × 12.7	L102 × 89 × 6.4	277

4. Component-Based Mechanical Models for TSACWs

4.1. Initial Stiffness

The initial stiffness $S_{j,ini}$ of bolted connections is one of the most important parameters for the characterization of their $(M\text{–}\theta)$ response. There is a large volume of published studies investigating $S_{j,ini}$ of bolted beam-to-column connections. The first deep discussion and analysis about $S_{j,ini}$ emerged in the 1980s with the research studies by Azizinamini et al. [27]. The authors proposed a beam model dividing the leg of the top cleat flange into two types of beam segments, flexible and rigid sections. Relevant equations of practical use have been presented to estimate $S_{j,ini}$ for TSACWs. Later, Kishi and Chen [29] presented a new equation that was developed according to the cantilever beam model.

The EC3 standard document [13]—Annex J—implemented a component method to estimate the $S_{j,ini}$ of TSACs. In this method, the connection behavior is simulated by a series of different components, each one represented in the form of an elastic spring with specific stiffness and strength. The overall initial stiffness of the connection is hence notoriously estimated by assembling all the springs in a parallel series configuration, as shown in Figure 3.

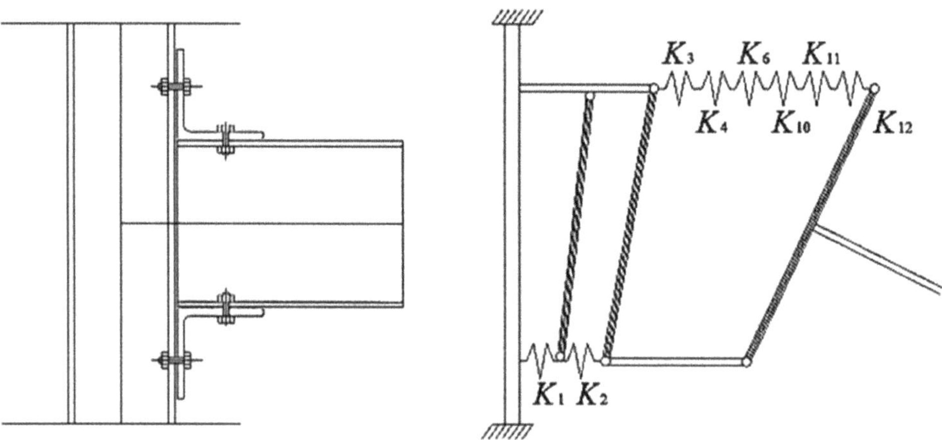

Figure 3. Component-based modeling of TSAC.

In the EC3, the input stiffness coefficients that must be considered are associated with the column shear panel zone (K_1), the column flange in tension (K_3), the column flange in compression (K_2), the flexural stiffness of column flange (K_4), the top cleat flexural stiffness (K_6), the tensile stiffness of bolts (K_{10}), and, for non-preloaded bolts, their shear stiffness (K_{11}) and their bearing stiffness (K_{12}). In conclusion, $S_{j,ini}$ of TSACs is given by:

$$S_{j,\,ini} = \frac{E\,z^2}{\sum_{i=1}^{n} 1/K_i} \tag{1}$$

where E is Young's modulus, z is the lever arm, K_i is the i-th component stiffness coefficient, and n is the number of joint components, as in Figure 3. Finally, z should be taken as the distance from the bolt-row in tension and mid-thickness of the leg of the seat cleat on the compression flange.

While the above formulation is widely used in practice, the EC3 formulation does not include a mechanical model for TSACWs. Accordingly, an extension of EC3 for TSACWs is proposed in this paper. For a bolt-row in a web cleat, the stiffness illustrated in Figure 4 should be considered.

The overall stiffness of basic components illustrated in Figure 4 is represented by a single equivalent stiffness coefficient k_{eq}, that can be calculated as:

$$k_{eq} = \frac{\frac{S_{j,\,ini}}{Ez} + \sum k_{eff,\,r} z_r}{z_{eq}} \tag{2}$$

where $S_{j,ini}$ is again the initial stiffness for TSACs, while z_r represents the distance between the center of the compression cleat and the bolt-row r of the web cleat.

Moreover:

$$k_{eff,\,r} = \frac{1}{\sum_i 1/k_{i,\,r}} \tag{3}$$

where $k_{i,r}$ is the stiffness coefficient representing component i relative to bolt-row r, and:

$$z_{eq} = \frac{\frac{S_{j,\,ini}}{E} + \sum_r k_{eff,\,r} z_r^2}{\frac{S_{j,\,ini}}{Ez} + \sum_r k_{eff,\,r} z_r} \tag{4}$$

In conclusion, the initial stiffness of TSACWs can be calculated as:

$$S_{j,\,ini} = \frac{E z_{eq}}{k_{eq}} \tag{5}$$

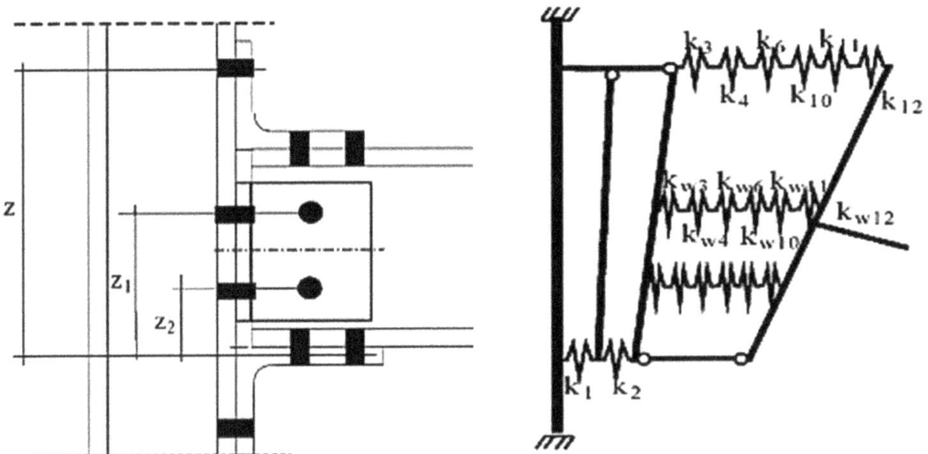

Figure 4. Proposed extension of component-based model for TSACW.

Kong and Kim [14] have used curve fitting software to obtain the effects of top and seat angles on the $S_{j,ini}$ value of TSACWs. Past literature studies also recognized that $S_{j,ini}$ correlates well with correlates the angle thickness, the length of the top angle, the height of the beam, the thickness of the column flange, the fillet size of the top angle, the thickness of the beam web, the gauge distance, and the diameter of bolts. In [14], therefore, a semi-empirical equation has been proposed as:

$$S_{j,\,ini} = \frac{0.49 E l_t t^3_t \left(d + \frac{t_t}{2} + \frac{t_s}{2} + 2k_t\right)}{\left(g_t - t_t - \frac{d_b}{2}\right)^2} \frac{t_{cf}}{t_t} \frac{t_{bw}}{t_t} \left(\frac{d}{t_t}\right)^{0.3} + \frac{0.312 n \alpha E l_p^2 t_a}{(1+v) g_c} \tag{6}$$

In Equation (6), E is the previously defined Young's modulus; l_t is the length of top angle; t_t is the top angle thickness; t_s is the thickness of seat angle; d is the height of beam; k_t is the fillet size of top angle; t_{cf} is the thickness of column flange; t_{bw} is the thickness of beam web; g_t is the distance from top angle heel to the center of bolts; d_b is the diameter of bolts; n is the number of bolts; $\alpha = 1.0$ mm; l_p is the angle length of the web; ta is the angle thickness of web; v is Poisson's ratio; and $g_c = g_1 - t_a$, as shown in Figure 5.

4.2. Ultimate Moment Capacity

The ultimate moment capacity M_n is one of the well-known fundamental parameters that affects the overall response of bolted beam-to-column connections. Several literature studies on TSACW specimens suggest that a rotation of 0.03 rad is sufficient to experience the full plastic moment capacity M_p of the connected beam [38]. The American Institute of Steel Construction [39], in this regard, recommends considering a relatively low rotation during the analysis and design process for this connection typology. Accordingly, the amplitude of 0.03 rad is defined as the maximum rotation capacity of TSACWs.

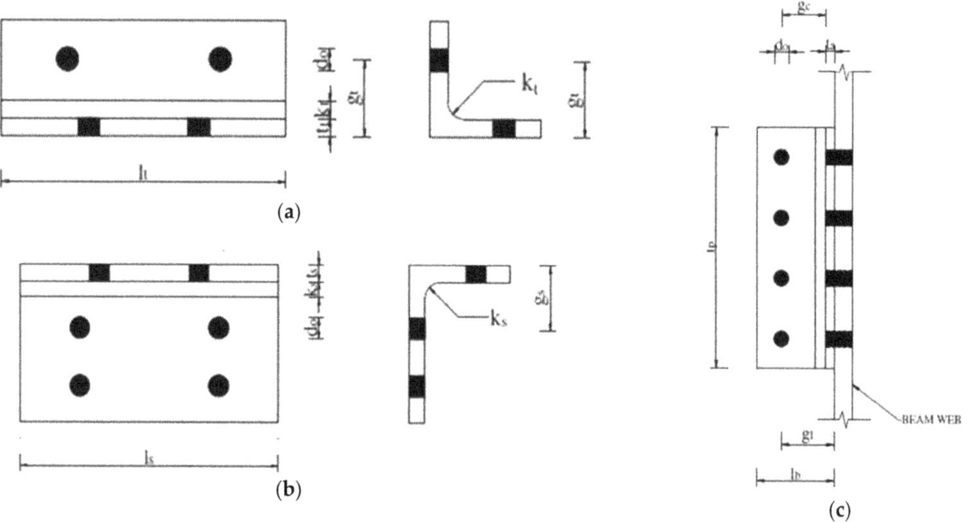

Figure 5. Geometrical parameters of the reference TSACW: (a) top angle, (b) seat angle, (c) web angle. Reproduced from [14] with permission from Elsevier, copyright agreement with license number 5011231057466, February 2021.

The component-based method can be extended to estimate the ultimate moment capacity M_n of TSACWs, as far as the contribution of influencing components is properly considered. These all affect the overall TSACW flexural resistance and are represented by: the column web panel in shear, the column web in compression, the column web in tension, the column flange in bending, the top cleat in bending, the web cleat in bending, the bolts in tension or in shear, the beam web in tension, and the beam flange and web in compression. Past research studies proposed different mechanical models to calculate M_n [30,40,41]. In most of those proposals, the failure of the top, seat, and web cleat's leg under maximum moment was only recognized, while other failure mechanisms (such as the failure of one or more bolts) were fully disregarded. In 2005, the EC3 standard [13] defined the required ultimate moment capacity M_n of TSACs and TSACWs as:

$$M_n = F_{Rd}Z \tag{7}$$

where Z is the lever arm and F_{Rd} is the design resistance of weak joint components. The second term can be governed by one of the following mechanisms: top cleat in bending ($F_{tc,Rd}$), bolts in tension, beam flange in tension and compression, and beam web in tension.

For a TSAC with a single-web cleat, Kishi and Chen [29], in 1990, elaborated a model proposal for the detection of collapse mechanisms. Kong and Kim [14] further extended the method from [29], and presented a formulation for TSACWs:

$$M_n^{top-seat} = M_{os} + M_p + V_{pt}d_2 + V_{pa}d_4 \tag{8}$$

In Equation (8), M_{os} is the plastic moment capacity of the seat angle; M_p is the plastic moment capacity of the top angle; V_{pt} is the ultimate shear force acting on the top angle; d_2 is a parameter related to the depth of the beam and thickness of top and seat cleats; V_{pa} is a parameter that depends on the ultimate shear force at the upper and lower edges of the web cleat; and d_4, finally, is the distance between plastic shear at the lower edge of the web cleat to the center of compression. These parameters are schematized in Figure 6.

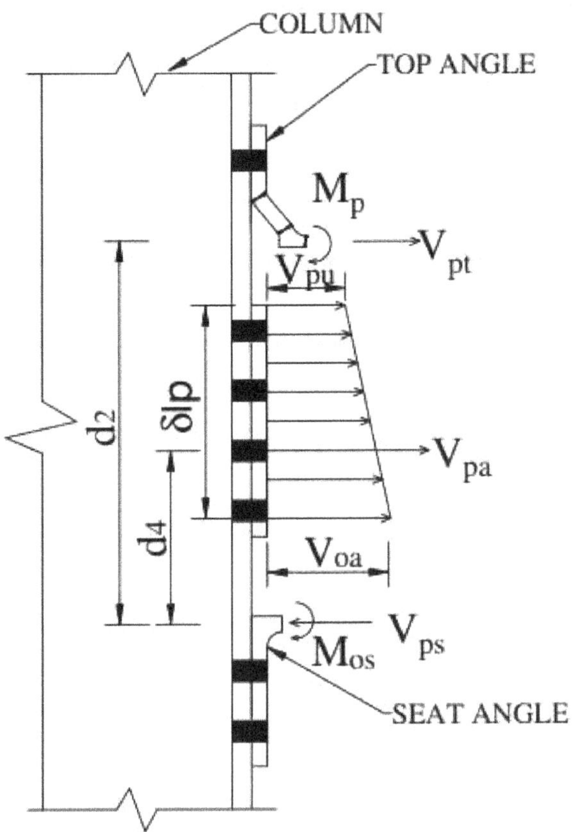

Figure 6. Collapse mechanism for TSACW. Reproduced from [14] with permission from Elsevier, copyright agreement with license number 5011231057466, February 2021.

5. Informational Based Modeling

An informational-based method recognizes, as an alternative method to simulate the complex structural and material behavior, that conventional methods are not simply estimated. This alternative method simulates the behavior by the information provided by selected tested specimens. Accordingly, this is an essential evolution from mathematical equations to preserve data that contain the necessary information on mechanical characteristics. In this approach, the underlying mechanics information is extracted from the experimental test data and processed in the ANN program.

Among numerous structures of ANNs that have been studied, the most widely used one is represented by the multi-layered feed-forward (FF) network, which is the structure implemented in the current investigation. Among others, FF is the oldest and simplest of existing ANNs, representing the most popular class of ANNs for its computational efficiency [42]. The main structure of a neural network is usually made up of three distinct layers, and never cycles. According to the literature, three-layer FF networks are found to be sufficient in civil engineering practices [43]. The input layer is where the data are introduced to the model, the hidden layer is where the data are processed, and the output layer is where the model results are generated. Each layer is made up of nodes called neurons [44]. Apart from the neurons in the input layer, which only receive and transmit incoming signals to other neurons in the hidden layer, each neuron in the other layers consists of three main components, weights, bias, and an activation function, which can be linear or non-linear. After determining the network topology, including the number of layers, the number of neurons, the type of transmission function, and the network learning algorithm, modeling for data can be started [45]. During the analysis of the model using learning algorithms, the number of errors can be reduced by adjusting weights and bias in each neuron. For this purpose, the Levenberg–Marquardt algorithm has been used in modeling the neural network due to its high speed and accuracy. The training algorithm distributes the network error to achieve optimal or minimum error [46,47].

5.1. Artificial Bee Colony (ABC) Algorithm

The artificial bee colony (ABC) algorithm is an optimization algorithm based on the bee population's collective intelligence and intelligent behavior in finding food [44]. In its primary model, the algorithm performs a neighborhood search combined with a random search, and can be used for either combined or functional optimization [48].

In the ABC algorithm, the colony consists of three groups of bees: employed, onlookers and scout bees. The first and the second half of the colony consist of employed artificial bees and onlookers, respectively. There is only one employed bee for every food source. In this algorithm, moreover, a scout is representative of an employed bee of an abandoned food source. Overall, the search process in ABC can be summarized as follows [49,50]:

- A food source is determined by employed bees in their memory within the neighborhood
- The collected information of food sources by employed bees is shared with onlookers within the hive, and subsequently, the optimum food sources will be selected by onlookers.
- A food source will be selected by onlookers themselves within the neighborhood of the food sources.
- An employed bee becomes a scout once the food source has been abandoned and starts to search for a new food source randomly.

5.2. Training the ANN and Methodology

To train the artificial neural network, a total number of 77 specimens from [36] and [37] was considered in this study, in order to determine two outputs of $S_{j,ini}$ and ($M_n/M_{p,beam}$). Up to $\approx 80\%$ of the selected samples (62) was used for training, while 20% (15 specimens) was used for testing the network.

Accordingly, several variables were introduced as input parameters for the model, including:

- the moment inertia ratio of the column to the connected beam (I_{col}/I_b);
- the thickness of the top (th_{tc}) and bottom (th_{bc}) flange cleat;
- the maximum thickness of right or left web cleat ($Max\text{-}th_{wc}$);
- the bolt size (d_b);
- the ratio of column to beam yield strength ($f_{y,c}/f_{y,b}$).

The statistical characteristics of these variables are summarized in Table 2.

Table 2. Characteristics of input and output parameters for artificial neural network (ANN) training.

Statistical Index		Type	Max	Min	Avg.	STD
I_{col}/I_b		Input	20.00	0.29	2.59	3.95
th_{tc}	(mm)	Input	15.90	0.00	8.10	4.79
th_{bc}	(mm)	Input	15.90	0.00	8.70	4.45
$Max\text{-}th_{wc}$	(mm)	Input	15.00	0.00	6.17	4.50
d_b	(mm)	Input	24.00	16.00	19.51	1.69
$f_{y,c}/f_{y,b}$		Input	1.13	0.79	0.99	0.08
$S_{j,ini}$	(kNm/rad)	Output	36,365.00	1633.00	12,021.75	9108.08
$M_n/M_{p,beam}$		Output	0.95	0.13	0.43	0.19

The number of hidden layers and the total number of neurons in the hidden layers in an ANN depend on the nature of the problem [51]. Generally, the trial and error method is used to obtain the ideal architecture that best reflects the characteristics of laboratory data. In this paper, an innovative method for calculating the number of neurons in hidden layers is taken into account, namely:

$$N_H \leq 2N + 1 \tag{9}$$

where N_H represents the number of neurons in the hidden layers and N_I is the number of input variables.

Since the number of influential input variables for the current study is equal to 6, the empirical Equation (9) shows that the number of neurons in hidden layers can be less than 13. Therefore, several networks with different topologies (with a maximum of two hidden layers and a maximum of 13 neurons) were trained and studied in this paper.

The hyperbolic tangent stimulation function and Levenberg–Marquardt training algorithm were used in all networks. The statistical indices used to evaluate the performance of different topologies are the root mean squared error (*RMSE*), the average absolute error (*AAE%*), the model efficiency (*EF*), and the variance account factor (*VAF%*) that are defined as follows [52]:

$$RMSE = \left[\frac{1}{n} \sum_{i=1}^{n} (P_i - O_i)^2 \right]^{\frac{1}{2}} \tag{10}$$

$$AAE = \frac{\left| \sum_{i=1}^{n} \frac{(O_i - P_i)}{O_i} \right|}{n} \times 100 \tag{11}$$

$$EF = 1 - \frac{\sum_{i=1}^{n} (P_i - O_i)^2}{\sum_{i=1}^{n} (\overline{O}_i - O_i)^2} \tag{12}$$

$$VAF = \left[1 - \frac{var(O_i - P_i)}{var(O_i)} \right] \times 100 \tag{13}$$

After examining different topologies of networks, it was found that the network with a 6-7-6-1 topology is characterized by the lowest value of error in *RMSE*, *AAE%*, *EF*, and *VAF%*, and by the highest value of R^2 to estimate the two output parameters $S_{j,ini}$ and $(M_n/M_{p,beam})$, as also shown in Table 3. It is necessary to mention that the error criteria for training and testing the selected data are calculated in the main range of variables and not in the normal range.

In this study, as it is for several structural engineering practice applications, the ABC algorithm is used as a new metaheuristic algorithm to determine the weight optimization of each ANN model. In more detail, the $(M_n/M_{p,beam})$ output has a numerical range of 0–1 while the $S_{j,ini}$ output is characterized by a numerical range of 1600–37,000 (kNm/rad). This means that there is a big difference between the two target outputs. For modulation, two separate ANNs were hence used in this study, each one with one output. Their basic characteristics are summarized in Table 4.

Figure 7 shows the optimal topology of an FF network with two hidden layers, six input variables (neurons), and one output parameter. The ABC has been also used to provide the least prediction error for the trained structure, in order to optimize the weights and biases of the ANN model. The ABC parameters are also presented in Table 5.

Table 3. Statistical indices of ANN with best 6-7-6-1 topology, as combined with artificial bee colony (ABC).

Type	Statistical Index	$S_{j,ini}$ (kNm/rad)	$M_n/M_{p,beam}$
Train	R^2	0.922	0.955
	y = ax + b	y = 0.8107x + 2,037.8	y = 0.9319x + 0.033
	RMSE	3542.657	0.058
	AAE %	0.292	0.108
	EF	0.848	0.911
	VAF %	0.849	0.912
Test	R^2	0.939	0.954
	y = ax + b	y = 0.9406x + 2860.1	y = 0.8749x + 0.0303
	RMSE	3790.584	0.065
	AAE %	0.422	0.109
	EF	0.817	0.892
	VAF %	0.878	0.908
All	R^2	0.918	0.953
	y = ax + b	y = 0.8287x + 2257.6	y = 0.9142x + 0.0348
	RMSE	3592.297	0.059
	AAE %	0.317	0.108
	EF	0.842	0.908
	VAF %	0.843	0.908

Table 4. Structure and topology of the feed-forward (FF) neural network.

No.	Name	Features of Neural Network						
		Number of Input	Number of Output	Neural Network	Hidden Layer	Node	Learning Role	Transfer Function
1	ABC-ANN-s	6	1	FF	2	7-6	Levenberg–Marquardt	tansig
2	ABC-ANN-m	6	1	FF	2	7-6	Levenberg–Marquardt	tansig

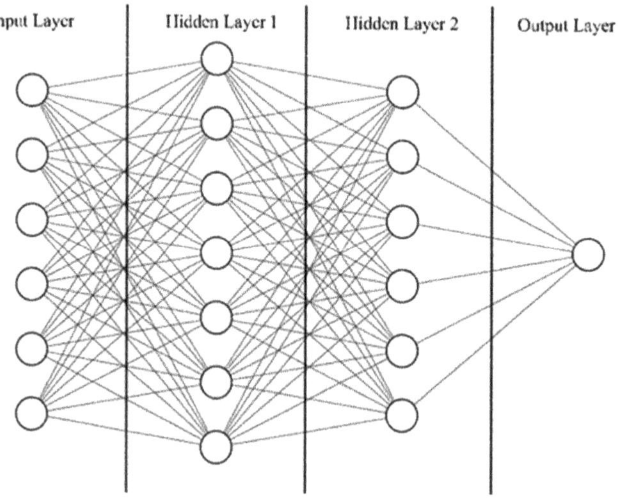

Figure 7. Feed-forward ANN with 6-7-6-1 structure.

Table 5. Features of ABC algorithm in the proposed ANN model.

No.	Name	Features of ABC Algorithm			
		Number of Bees	Source Number	Onlooker Number	Max Number of Cycles
1	ABC-ANN-S	50	25	25	100
2	ABC-ANN-M	30	15	15	150

6. Results and Discussion

6.1. Accuracy of Proposed ABC-ANN Model

Table 6 shows the comparison between different models with experimental data for estimating $S_{j,ini}$ and M_n. Concerning the calculated average ("Avg.") and standard deviation ("STD") values, the results in Table 6 indicate that the ABC-ANN model provides more reliable predictions for both, compared to the EC3 or KK formulations described earlier. Using the existing empirical models, in particular, the $S_{j,ini}$ and M_n predictions for some specimens are either underestimated or overestimated, and this suggests the limitation of mechanical models to capture the underlying mechanism that governs both the parameters. The ABC-ANN predictions, conversely, are characterized by minimum deviation.

Table 6. Comparison of different models with literature test data, as obtained in terms of $S_{j,ini}$ and M_n.

$S_{j,ini}$ (kNm/rad)				M_n (kNm)			
Test	Test/EC3	Test/KK	Test/ABC-ANN-S	Test	Test/EC3	Test/KK	Test/ABC-ANN-M
6000	0.62	0.81	1.32	43.6	1.11	1.22	0.91
13,846	0.44	1.49	0.57	44.9	0.93	0.95	0.92
10,099	0.49	1.03	0.78	54.2	1.11	1.22	0.73
1633	1.32	1.34	1.23	21.7	1.17	1.21	1.9
8089	1.65	1.26	0.98	43.3	1.02	1.09	0.95
4490	1.80	1.13	1.33	33.1	1.25	1.37	1.2
4638	1.17	0.96	1.7	47.4	1.34	1.47	0.87
6060	1.50	1.43	1.32	50.4	1.87	2.07	1.6
10,029	1.61	1.94	0.98	54.6	1.56	1.67	0.97
30,222	2.74	4.09	0.99	74.7	1.35	1.37	0.95
21,623	1.74	0.99	0.88	83.7	1.08	1.11	1.19
26,919	1.05	0.87	0.88	168.8	0.75	1.12	1.01
11,022	0.87	0.51	0.66	80.9	1.30	1.31	1.25
23,852	1.67	1.07	0.8	101.3	1.03	1.06	0.99
22,672	1.78	0.97	0.84	119.9	1.22	1.57	0.81
25,247	0.97	0.76	0.94	127.4	1.00	1.03	0.99
58,679	1.43	1.27	1.14	186.9	1.045	1.07	0.96
24,169	0.93	0.72	0.99	123.8	0.97	1.00	1.02
Avg.	1.32	1.26	1.01		1.17	1.27	1.02
STD	0.56	0.78	0.27		0.25	0.28	0.19

Figures 8 and 9 show the scatter graph that provided the relationship between test results and the proposed ABC-ANN model for estimating the $S_{j,ini}$ and M_n parameters, respectively. In this case, the comparative results also indicate that the ABC-ANN model offers a reliable value for the ratio of experimental to computational predictions (R^2), for both the examined mechanical parameters, and thus confirming further the high potential and accuracy of the proposed model.

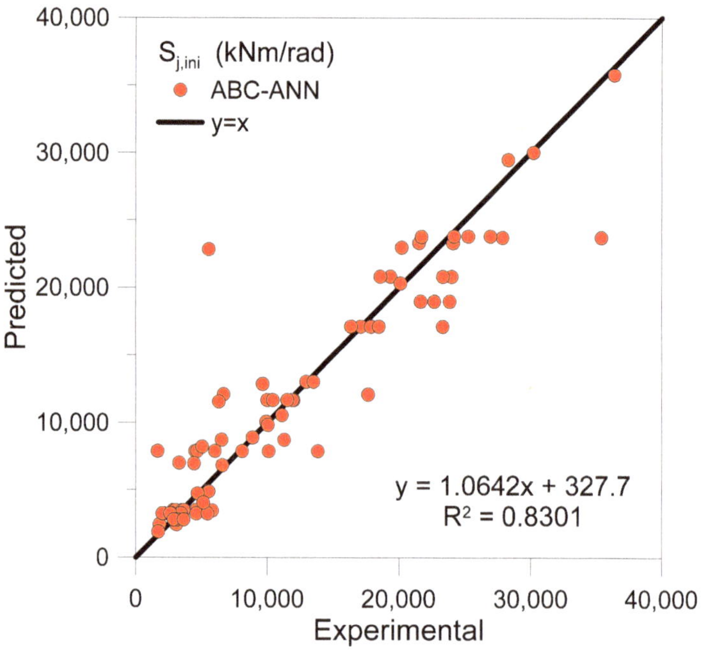

Figure 8. ABC-ANN calculated versus experimentally observed values for the parameter $S_{j,ini}$.

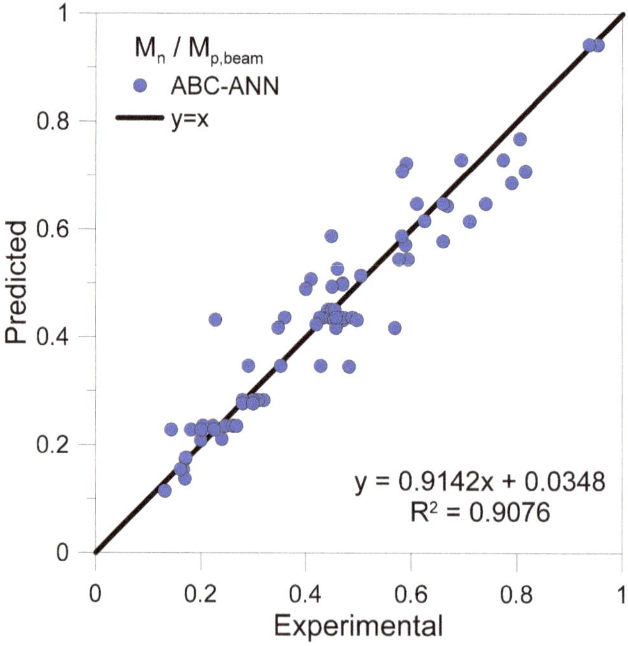

Figure 9. ABC-ANN calculated versus experimentally observed values for the parameter $(M_n/M_{p,beam})$.

In conclusion, Tables 7 and 8 provide the final weights and biases for both hidden layers, as estimated by the ABC-ANN model. Using the values of the weights and biases between the different ANN layers, the two output parameters ($S_{j,ini}$ and $M_n/M_{p,beam}$) can be determined and predicted.

Table 7. Final weights and bias values of the ABC-ANN model for $S_{j,ini}$ (kNm/rad).

IW							b_1
0.4817	0.1907	−0.2197	−0.4365	0.5490	−0.5109		0.9004
−0.3810	0.0292	0.5892	−0.2042	−0.7789	0.7893		0.5127
0.0253	0.0070	−0.7712	0.5666	−0.1401	0.8548		0.3646
0.8274	−0.7504	−0.8258	0.9904	−0.4542	−0.4942		−0.1958
0.5154	−1.0000	0.3092	0.4735	−0.5747	0.0010		0.0809
−0.8149	−0.0035	−0.6034	0.3425	1.0000	−0.0460		−0.4586
−0.4572	1.0000	0.3426	0.9226	−0.1067	−0.9320		1.0000

LW1							b_2
−0.8055	−0.7453	0.8586	−0.3097	0.5595	0.4411	−0.8149	−0.2254
0.9575	−1.0000	−0.7036	0.8996	−0.2134	−0.8109	−0.2879	0.8898
−0.6931	−0.0147	0.1303	−0.3631	0.3113	−0.3478	−0.5636	0.0838
-0.4409	−0.7401	0.4323	−0.9174	0.3017	−0.6847	−1.0000	0.5783
-0.4875	−0.0611	0.3553	−0.8939	1.0000	−0.5234	−0.5076	0.2355
0.5268	−1.0000	−0.7456	−0.1620	0.1855	−0.1735	−0.2715	0.2381

LW2						b_3
−0.8122	−0.2114	0.4500	0.1232	−0.9455	−0.1440	−0.8268

Table 8. Final weights and bias values of the ABC-ANN model for ($M_n/M_{p,beam}$).

IW							b_1
0.5052	0.026	−0.048	−0.457	−0.601	0.880		−0.263
0.0611	−0.002	0.231	1.000	−0.039	−0.041		−0.096
0.8044	0.613	0.636	−0.597	0.068	−0.707		0.830
−0.0309	0.840	1.000	0.972	0.503	0.902		0.328
−0.0211	−0.668	0.193	0.190	0.841	0.214		0.582
−0.8067	−0.706	0.351	−0.533	−0.137	0.048		0.332
0.6398	0.751	0.515	−0.311	0.908	−1.000		−0.874

LW1							b_2
−0.8170	0.552	−0.907	−0.003	0.750	−0.879	−0.423	−0.870
−0.8032	0.822	−1.000	−0.435	0.052	−0.235	0.438	−0.286
−0.7108	−0.702	−0.572	−0.039	0.144	0.154	0.653	0.159
−0.3023	−0.827	0.142	0.368	0.149	0.385	0.467	−0.822
0.6028	0.679	−0.656	−0.584	−0.243	−0.078	0.546	0.176
0.9051	−0.554	−0.576	−0.672	0.981	−0.221	−0.992	0.303

LW2						b_3
−0.4997	−0.239	0.875	−0.930	−0.243	−0.643	−0.930

IW: weight values for input layer; LW1: weight values for first hidden layer; LW2: weight values for second hidden layer; b_1: bias values for first hidden layer; b_2: bias values for second hidden layer; b_3: bias values for output layer.

To formulate ANN results, weights and biases from Tables 7 and 8 should be normalized as:

$$X_n = \frac{2 \times (X - X_{min})}{X_{max} - X_{min}} - 1 \tag{14}$$

By substituting the normalized values of Table 3 for each one of the six input parameters, they are represented by a 6×1 vector that is herein labeled as $a^{(1)}$. Then, by using the following equations, the values of $S_{j,ini}$ and $(M_n/M_{p,beam})$ can be calculated from:

$$a^{(2)} = tanh\left(IW \times a^{(1)} + b_1 \right) \tag{15}$$

$$a^{(3)} = tanh\left(LW1 \times a^{(2)} + b_2 \right) \tag{16}$$

$$Y_{S \ or \ M}^{Predic(Normalize)} = tanh\left(LW2 \times a^{(3)} + b_3 \right) \tag{17}$$

$$Y_{S \ or \ M}^{Predict(Actual)} = \frac{Y_{S \ or \ M}^{Predict(Normalize)} + 1}{2} \times (Y_{max} - Y_{min}) + Y_{min} \tag{18}$$

The parameters *IW*, *LW1*, *LW2*, b_1, b_2, and b_3 are shown as vector matrices in Tables 7 and 8. Furthermore, "*S*" and "*M*" in Equations (15)–(18) stand for $S_{j,ini}$ and $(M_n/M_{p,beam})$, respectively.

Another visual measure that can be taken into account for comparing the performance of the ABC-ANN model against the component-based mechanical models (EC3 or KK) is the Taylor diagram, see Figures 10 and 11. This diagram depicts a graphical illustration of the adequacy of each investigated model, based on the root mean square-centered difference, the correlation coefficient, and the standard deviation.

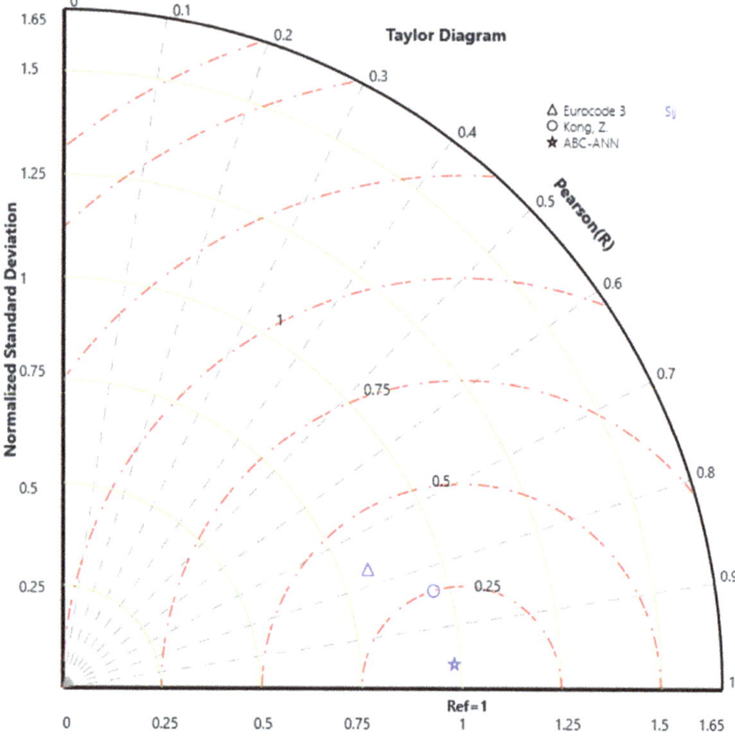

Figure 10. Taylor diagram visualization of model performance, in terms of $S_{j,ini}$ predictions.

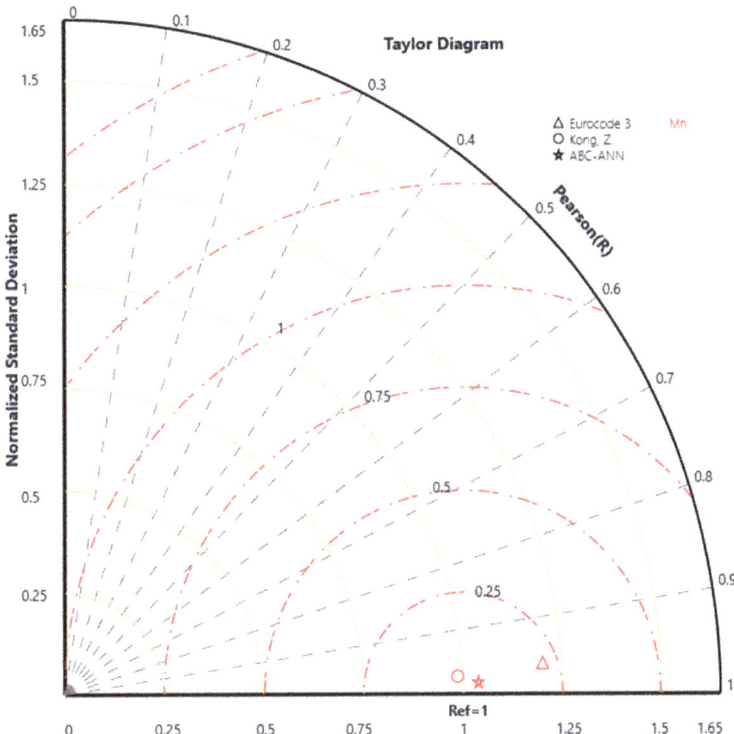

Figure 11. Taylor diagram visualization of model performance, in terms of M_n predictions.

The results shown in Figures 10 and 11 clearly indicate that the closest prediction for both the $S_{j,ini}$ and M_n input parameters, to the point representing the experimental data of the literature, are provided by the herein developed ABC-ANN model. The EC3 component-based model, as shown, also results in high values of root mean square-centered difference and standard deviation, thus further suggesting a good accuracy of the formulation over the selected experimental data. The same comparative parameters, conversely, are rather low regarding the application of the KK model to the selected experimental specimens.

6.2. Sensitivity Analysis

As previously discussed, the accuracy and potential of the proposed ABC-ANN model for the estimation of the $S_{j,ini}$ and M_n parameters in TSACWs was acknowledged by comparison with the EC3 and KK component-based formulations. In order to investigate in more detail the effect of all the required input parameters, a sensitivity analysis (SA) was carried out for the selected TSACW specimens. The SA reveals how significantly the model output can be affected by changes within input variables. There are two main types of SA, known as "global" and "local" sensitivity analyses, where the local sensitivity analysis (LSA) concentrates on the local impact of individual input parameters on the overall performance. The global sensitivity analysis (GSA), on the other hand, evaluates the influence of individual input parameters over their entire spatial range and measures the uncertainty of the overall performance (output) caused by input uncertainty, over interaction with other parameters, or also taken individually. Therefore, considering the nature of the complex non-linear behavior and variation of $S_{j,ini}$ and M_n parameters in the current study, the GSA was selected as much more rational for investigating the impact of input parameters on the overall performance.

Amongst GSA methods, a variance-based approach has been primarily considered in past literature for SA. The method provides a specific methodology for defining total and first-order sensitivity indices for each input parameter of the ANN model. Assuming a model in the form $Y = f(X_1, X_2, \ldots, X_k)$, where Y is scalar, the variance-based technique takes a variance ratio to evaluate the impact of individual parameters using variance decomposition as:

$$V = \sum_{i=1}^{k} V_i + \sum_{i=1}^{k}\sum_{j>i}^{k} V_{ij} + \ldots + V_{1,2,\ldots,k} \tag{19}$$

where V is the variance of the ANN model output; V_i is the first-order variance for the input X_i; V_{ij} to $V_{1,2,\ldots,k}$ correspond to the variance of the interaction of the k parameters

V_i and V_{ij}, which denote the significance of the individual input to the variance of the output, are a function of the conditional anticipation variance:

$$V_i = V_{x_i}[E_{x_{\sim i}}(YX_i)] \tag{20}$$

$$V_{ij} = V_{x_i x_j}\left[E_{x_{\sim ij}}(YX_i, X_j)\right] - V_i - V_j \tag{21}$$

where the suffix $x_{\sim i}$ designates the set of all input variables apart from X_i.

The first-order sensitivity index (S_i) represents the first-order impact of an input X_i on the overall output provided by:

$$S_i = \frac{V_i}{V(Y)} \tag{22}$$

The abovementioned methodology for calculating the first-order sensitivity index was used in this research study. Major results of the SA are presented in Figure 12.

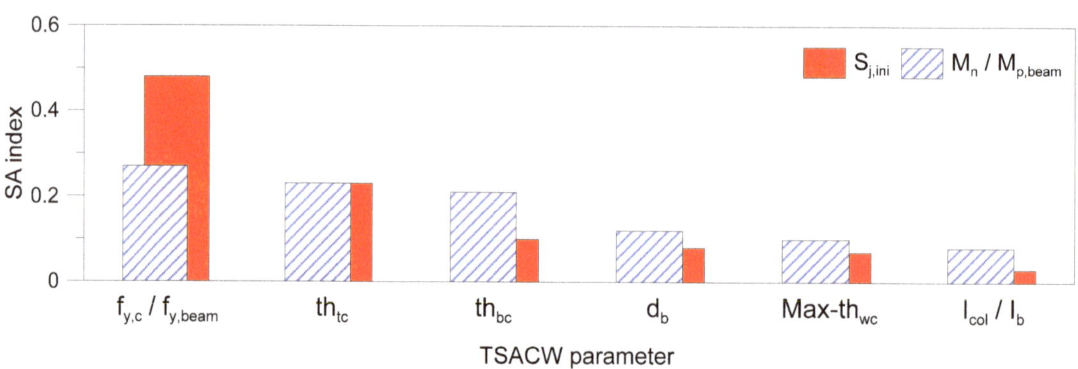

Figure 12. Sensitivity indices of variables $S_{j,ini}$ and ($M_n/M_{p,beam}$).

Apart from the yield strength (f_y) that depends on material properties, the comparative results indicate that the thickness of top flange (th_{tc}) has the most influence, while the moment inertia ratio of column to beam (I_{col}/I_b) has the least effect on both the output parameters, $S_{j,ini}$ and M_n. The thickness of the bottom flange cleat (th_{bc}) can be classified as the second most influential input variable, especially in terms of maximum moment capacity M_n.

7. Concluding Remarks

Modeling the plastic response of different components in beam-to-column bolted connections and their interactions is a challenging issue in the structural engineering community. In this paper, an informational artificial neural network (ANN) model combined with the metaheuristic artificial bee colony (ABC) algorithm was developed to model the initial stiffness ($S_{j,ini}$) and maximum moment capacity (M_n) of top-seat angle connections with double web angles (TSACWs). Two different formulations of component-based mechanical models of the literature (as proposed by Eurocode 3 ("EC3") or by Kong and Kim in 2017 ("KK")) were also investigated in detail.

The discussed comparisons and results confirmed that the efficiency of the component-based mechanical model depends on the number and accuracy of the relationships of the constitutive components. By defining a sufficient number of components and subsequently idealizing the physical behavior in analytical equations, the reliable application of the mechanical model to different connection configurations is possible.

Nevertheless, idealization typically resulted in equations that excluded several important physical behavioral features of TSACWs, such as slippage. On the other hand, unlike the conventional mechanical modeling process that involves idealization from the observed behavior to the mathematical equations, in the informational base method, the information about essential behavior is extracted from available experimental test data and processed using an ANN. Nevertheless, the ANN model was limited to providing the global response only for bolted connections that include the contributions of all the constitutive components. Using this method, as shown, it is impossible to represent individual components and their actual contribution. Therefore, the model does not offer an insight into the underlying mechanics. Overall, the following conclusions can thus be derived:

- both the EC3 and KK component-based models failed to capture the underlying mechanism for estimating $S_{j,ini}$ and M_n parameters. As a result, these were either underestimated or overestimated for the reference specimens. On the other hand, the herein developed ABC-ANN model proved to offer a reliable prediction of required parameters, as also emphasized by the ratio of observational to computational values (R^2), and thus suggesting the high potential and accuracy of the proposal.
- The ANN model combined with the ABC algorithm established an excellent agreement with the available experimental database. The results highlighted that the ANN model may be a reliable alternative to a component-based mechanical model to estimate the mechanical behavior of bolted beam-to-column connections. Using the values of weights and biases between the different ANN layers, the two output parameters ($S_{j,ini}$ and M_n) can be accurately predicted.
- The sensitivity analysis confirmed that (apart from the yield strength f_y that necessarily depends on material properties) the thickness of the top flange (th_{tc}) has a significant influence, while the moment inertia ratio of column to beam (I_{col}/I_b) has the least effect on both the predicted output parameters, $S_{j,ini}$ and M_n.

Author Contributions: This research paper results from a joint collaboration of all the involved authors. All authors contributed to the paper drafting and review, I.F., M.N., R.P. and C.B. All authors have read and agreed to the published version of the manuscript.

Funding: This work was supported by funding from the Federal State Autonomous Educational Institution of Higher Education South Ural State University (National Research University). Further, MDPI is also acknowledged for providing waived fees (Guest Editor C.B.) in support of standard publication APCs.

Institutional Review Board Statement: The study did not require ethical approval.

Informed Consent Statement: Not applicable.

Data Availability Statement: Data sharing not applicable.

Conflicts of Interest: The authors declare no conflict of interest.

References

1. ANSI/AISC. *AISC 358-05 Prequalified Connections for Special and Intermediate Steel Moment Frames for Seismic Applications*; American Institute of Steel Construction Inc.: Chicago, IL, USA, 2005.
2. Madas, P.J.; Elnashai, A.S. A component-based model for beam-column connections. In Proceedings of the 10th World Conference of Earthquake Engineering, Madrid, Spain, 19–24 July 1992; pp. 4495–4499.
3. Chisala, M.L. Modelling M–φ curves for standard beam-to-column connections. *Eng. Struct.* **1999**, *21*, 1066–1075. [CrossRef]

4. Kim, J.; Ghaboussi, J.; Elnashai, A.S. Mechanical and informational modeling of steel beam-to-column connections. *Eng. Struct.* **2010**, *32*, 449–458. [CrossRef]

5. Pinheiro, L.; Silveira, R.A. Computational procedures for nonlinear analysis of frames with semi-rigid connections. *Lat. Am. J. Solids Struct.* **2005**, *2*, 339–367.

6. Hasan, M.J.; Ashraf, M.; Uy, B. Moment-rotation behaviour of top-seat angle bolted connections produced from austenitic stainless steel. *J. Constr. Steel Res.* **2017**, *136*, 149–161. [CrossRef]

7. Lacey, A.W.; Chen, W.; Hao, H.; Bi, K. Review of bolted inter-module connections in modular steel buildings. *J. Build. Eng.* **2019**, *23*, 207–219. [CrossRef]

8. Chen, Z.; Liu, J.; Yu, Y. Experimental study on interior connections in modular steel buildings. *Eng. Struct.* **2017**, *147*, 625–638. [CrossRef]

9. Chua, Y.; Liew, J.R.; Pang, S. Modelling of connections and lateral behavior of high-rise modular steel buildings. *J. Constr. Steel Res.* **2020**, *166*, 105901. [CrossRef]

10. Lacey, A.W.; Chen, W.; Hao, H.; Bi, K. New interlocking inter-module connection for modular steel buildings: Simplified structural behaviours. *Eng. Struct.* **2021**, *227*, 111409. [CrossRef]

11. Thai, H.-T.; Vo, T.P.; Nguyen, T.-K.; Pham, C.H. Explicit simulation of bolted endplate composite beam-to-CFST column connections. *Thin-Walled Struct.* **2017**, *119*, 749–759. [CrossRef]

12. Natesan, V.; Madhavan, M. Experimental study on beam-to-column clip angle bolted connection. *Thin-Walled Struct.* **2019**, *141*, 540–553. [CrossRef]

13. En, B. *1-8: 2005. Eurocode 3: Design of Steel Structures-Part 1-8: Design of Joints*; British Standards Institution: London, UK, 2005.

14. Kong, Z.; Kim, S.-E. Moment-rotation behavior of top-and seat-angle connections with double web angles. *J. Constr. Steel Res.* **2017**, *128*, 428–439. [CrossRef]

15. Calado, L.; De Matteis, G.; Landolfo, R. Experimental response of top and seat angle semi-rigid steel frame connections. *Mater. Struct.* **2000**, *33*, 499–510. [CrossRef]

16. Wang, T.; Song, G.; Liu, S.; Li, Y.; Xiao, H. Review of Bolted Connection Monitoring. *Int. J. Distrib. Sens. Netw.* **2013**, *9*, 871213. [CrossRef]

17. Jiang, L.; Rasmussen, K.; Zhang, H. Experimental investigation of full-range action-deformation behaviour of top-and-seat angle and web angle connections. In Proceedings of the 9th international conference on Advances in Steel Structures, Hong Kong, China, 5–7 December 2018.

18. Yan, S.; Jiang, L.; Rasmussen, K.J. Full-range behaviour of double web angle connections. *J. Constr. Steel Res.* **2020**, *166*, 105907. [CrossRef]

19. Kong, Z.; Kim, S.-E. Numerical Estimation for Initial Stiffness and Ultimate Moment of Top-Seat Angle Connections without Web Angle. *J. Struct. Eng.* **2017**, *143*, 04017138. [CrossRef]

20. Hasan, M.J.; Al-Deen, S.; Ashraf, M. Behaviour of top-seat double web angle connection produced from austenitic stainless steel. *J. Constr. Steel Res.* **2019**, *155*, 460–479. [CrossRef]

21. Kim, J.H. *Hybrid Mathematical and Informational Modeling of Beam-to-Column Connections*; University of Illinois at Urbana-Champaign: Champaign, IL, USA, 2010.

22. Moradi, S.; Alam, M.S. Finite-Element Simulation of Posttensioned Steel Connections with Bolted Angles under Cyclic Loading. *J. Struct. Eng.* **2016**, *142*, 04015075. [CrossRef]

23. Amadio, C.; Bedon, C.; Fasan, M.; Pecce, M.R. Refined numerical modelling for the structural assessment of steel-concrete composite beam-to-column joints under seismic loads. *Eng. Struct.* **2017**, *138*, 394–409. [CrossRef]

24. Málaga-Chuquitaype, C.; Elghazouli, A. Component-based mechanical models for blind-bolted angle connections. *Eng. Struct.* **2010**, *32*, 3048–3067. [CrossRef]

25. Azizinamini, A.; Radziminski, J.B. Static and Cyclic Performance of Semirigid Steel Beam-to-Column Connections. *J. Struct. Eng.* **1989**, *115*, 2979–2999. [CrossRef]

26. Pucinotti, R. Cyclic mechanical model of semirigid top and seat and double web angle connections. *Steel Compos. Struct.* **2006**, *6*, 139–157. [CrossRef]

27. Azizinamini, A.; Bradburn, J.; Radziminski, J. Initial stiffness of semi-rigid steel beam-to-column connections. *J. Constr. Steel Res.* **1987**, *8*, 71–90. [CrossRef]

28. Kishi, N. *Moment-Rotation Relation of Top-and Seat-Angle with Double Web-Angle Connections*; School of Civil Engineering, Purdue University: West Lafayette, IN, USA, 1987.

29. Kishi, N.; Chen, W. Moment-Rotation Relations of Semirigid Connections with Angles. *J. Struct. Eng.* **1990**, *116*, 1813–1834. [CrossRef]

30. Pucinotti, R. Top-and-seat and web angle connections: Prediction via mechanical model. *J. Constr. Steel Res.* **2001**, *57*, 663–696. [CrossRef]

31. Shen, J.; Astaneh-Asl, A. Hysteresis model of bolted-angle connections. *J. Constr. Steel Res.* **2000**, *54*, 317–343. [CrossRef]

32. De Stefano, M.; De Luca, A.; Astaneh-Asl, A. Modeling of Cyclic Moment-Rotation Response of Double-Angle Connections. *J. Struct. Eng.* **1994**, *120*, 212–229. [CrossRef]

33. Danesh, F.; Pirmoz, A.; Daryan, A.S. Effect of shear force on the initial stiffness of top and seat angle connections with double web angles. *J. Constr. Steel Res.* **2007**, *63*, 1208–1218. [CrossRef]

4. Pirmoz, A. Moment–rotation behavior of bolted top–seat angle connections. *J. Constr. Steel Res.* **2009**, *65*, 973–984. [CrossRef]
5. SALEM, A.H. *Behavior and Design of Steel I-beam-to-Column Flushed Rigid Bolted Connections*; The American University: Cairo, Egypt, 2011.
6. Weynand, K.; Huter, M.; Kirby, P.A.; Simoes da Silva, L.A.P. SERICON–Databank on Joints in Building Frames. In Proceedings of the 1st COST C1 Workshop—International Conference on Control of the Semi-Rigid Behaviour of Civil Engineering Structural Connections, Liège, Belgium, 17–19 September 1998; pp. 217–228.
7. Kishi, N.; Chen, W.-F. *Data Base of Steel Beam-to-Column Connections*; Structural Engineering Area, School of Civil Engineering, Purdue University: West Lafayette, IN, USA, 1986.
8. Astaneh, A. Demand and supply of ductility in steel shear connections. *J. Constr. Steel Res.* **1989**, *14*, 1–19. [CrossRef]
9. Construction, A.I.O.S. *Seismic Provisions for Structural Steel Buildings*; American Institute of Steel Construction: Chicago, IL, USA, 2002.
10. Komuro, M.; Kishi, N.; Chen, W. *Elasto-Plastic FE Analysis on Moment-Rotation Relations of Top-and Seat-Angle Connections*; Connections in Steel Structures V: Amsterdam, The Netherlands, 2004; pp. 111–120.
11. Faella, C.; Piluso, V.; Rizzano, G. *Prediction of the Flexural Resistance of Bolted Connections with Angles*; IABSE Reports; 1996; pp. 309–318. Available online: https://structurae.net/en/literature/conference-paper/prediction-of-the-flexural-resistance-of-bolted-connections-with-angles (accessed on 1 February 2021).
12. Flood, I.; Kartam, N. Neural Networks in Civil Engineering. I: Principles and Understanding. *J. Comput. Civ. Eng.* **1994**, *8*, 131–148. [CrossRef]
13. Kulkarni, P.; Londhe, S.; Deo, M. Artificial Neural Networks for Construction Management: A Review. *Soft Comput. Civ. Eng.* **2017**, *1*, 70–88.
14. Asteris, P.G.; Nikoo, M. Artificial bee colony-based neural network for the prediction of the fundamental period of infilled frame structures. *Neural Comput. Appl.* **2019**, *31*, 4837–4847. [CrossRef]
15. Hasanzade-Inallu, A.; Zarfam, P.; Nikoo, M. Modified imperialist competitive algorithm-based neural network to determine shear strength of concrete beams reinforced with FRP. *J. Central South Univ.* **2019**, *26*, 3156–3174. [CrossRef]
16. Haykin, S. *Neural Networks: A Comprehensive Foundation*; Prentice Hall: Upper Saddle River, NJ, USA, 1999.
17. Bishop, C.M. *Pattern Recognition and Machine Learning*; Springer: New York, NY, USA, 2006.
18. Najimi, M.; Ghafoori, N.; Nikoo, M. Modeling chloride penetration in self-consolidating concrete using artificial neural network combined with artificial bee colony algorithm. *J. Build. Eng.* **2019**, *22*, 216–226. [CrossRef]
19. Karaboga, D.; Basturk, B. On the performance of artificial bee colony (ABC) algorithm. *Appl. Soft Comput.* **2008**, *8*, 687–697. [CrossRef]
20. Karaboga, D.; Akay, B. A comparative study of Artificial Bee Colony algorithm. *Appl. Math. Comput.* **2009**, *214*, 108–132. [CrossRef]
21. Asteris, P.; Plevris, V. Neural Network Approximation of the Masonry Failure under Biaxial Compressive Stress. In Proceedings of the SEECCM III-3rd South-East European Conference on Computational Mechanics-an ECCOMAS and IACM Special Interest Conference, Kos Island, Greece, 12–14 June 2013.
22. Li, J.; Heap, A.D. *A Review of Spatial Interpolation Methods for Environmental Scientists*; Record 2008/23; Geoscience Australia: Canberra, Australia, 2008; p. 137.

applied
sciences

MDPI

Article

Influence of Additional Bracing Arms as Reinforcement Members in Wooden Timber Cross-Arms on Their Long-Term Creep Responses and Properties

Muhammad Rizal Muhammad Asyraf [1,*], Mohamad Ridzwan Ishak [1,2,3,*], Salit Mohd Sapuan [3,4] and Noorfaizal Yidris [1]

1   Department of Aerospace Engineering, Universiti Putra Malaysia, UPM, Serdang 43400, Selangor, Malaysia; nyidris@upm.edu.my
2   Aerospace Malaysia Research Centre (AMRC), Universiti Putra Malaysia, UPM, Serdang 43400, Selangor, Malaysia
3   Laboratory of Biocomposite Technology, Institute of Tropical Forestry and Forest Products (INTROP), Universiti Putra Malaysia, UPM, Serdang 43400, Selangor, Malaysia; sapuan@upm.edu.my
4   Advanced Engineering Materials and Composites Research Centre (AEMC), Department of Mechanical and Manufacturing Engineering, Universiti Putra Malaysia, UPM, Serdang 43400, Selangor, Malaysia
*   Correspondence: asyrafriz96@gmail.com (M.R.M.A.); mohdridzwan@upm.edu.my (M.R.I.)

Abstract: Previously, numerous creep studies on wood materials have been conducted in various coupon-scale tests. None had conducted research on creep properties of full-scale wooden cross-arms under actual environment and working load conditions. Hence, this research established findings on effect of braced arms on the creep behaviors of Virgin Balau (*Shorea dipterocarpaceae*) wood timber cross-arm in 132 kV latticed tower. In this research, creep properties of the main members of both current and braced wooden cross-arm designs were evaluated under actual working load conditions at 1000 h. The wooden cross-arm was assembled on a custom-made creep test rig at an outdoor area to simulate its long-term mechanical behaviours under actual environment of tropical climate conditions. Further creep numerical analyses were also performed by using Findley and Burger models in order to elaborate the transient creep, elastic and viscoelastic moduli of both wooden cross-arm configurations. The findings display that the reinforcement of braced arms in cross-arm structure significantly reduced its creep strain. The inclusion of bracing system in cross-arm structure enhanced transient creep and stress independent material exponent of the wooden structure. The addition of braced arms also improved elastic and viscoelastic moduli of wooden cross-arm structure. Thus, the outcomes suggested that the installation of bracing system in wooden cross-arm could extend the structure's service life. Subsequently, this effort would ease maintenance and reduce cost for long-term applications in transmission towers.

Keywords: balau wood; cross-arm; transmission tower; bracing system; creep; findley's power law model; burger model

Citation: Asyraf, M.R.M.; Ishak, M.R.; Sapuan, S.M.; Yidris, N. Influence of Additional Bracing Arms as Reinforcement Members in Wooden Timber Cross-Arms on Their Long-Term Creep Responses and Properties. *Appl. Sci.* **2021**, *11*, 2061. https://doi.org/10.3390/app11052061

Received: 9 December 2020
Accepted: 24 December 2020
Published: 26 February 2021

Publisher's Note: MDPI stays neutral with regard to jurisdictional claims in published maps and institutional affiliations.

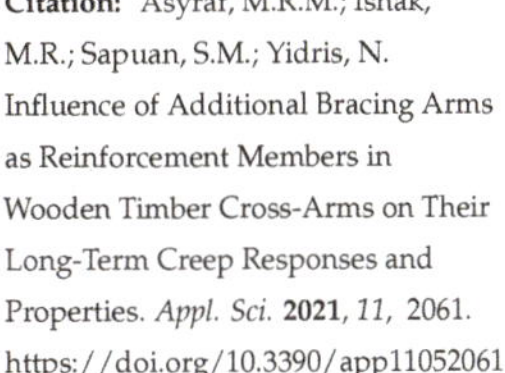

1. Introduction

Anisotropic materials has been widely used in many large structures such as bridges, buildings, pedestrian walkways and cross-arm in transmission tower. Wood and wood composites are those common anisotropic materials used to build many civil structures [1–4]. However, many studies reported that most wooden structures experienced premature failures after long period of service, especially when continuously exposed toward extreme weather [5–7]. These premature failures are also contributed by creep and natural defect such as wood ageing process, which subsequently may lead to structural collapse [8–13].

Creep is one of the major concerned by structural engineers in order to eliminate any possibilities of structural collapse during its service operation. Creep is a term referred to the tendency of a solid material to move slowly or deform permanently under the influence of persistent mechanical stresses. They are divided into several phases starting from instantaneous deformation followed by primary (transient), secondary (steady-state) and tertiary (accelerated). This mechanical phenomenon usually due to shear yielding, chain slippage, void formation, and also breakage of fibres [14,15]. To be specific, the creep response of wooden materials is dependent based on their level of stress, operation time, and surrounding temperature [16]. Numerous creep studies of wooden materials have been carried out by many researchers in many ways including development of creep test rigs [17–19], and creep numerical analyses [20,21] and coupon tests [22,23].

Currently, creep properties of wooden cross-arm in latticed transmission tower is still unexplored based on the previous literatures. Earlier, several research studies reported that the wooden cross-arm has shorter life span (less than 20 years of service) than its life expectancy due to its natural wood defects [24,25]. The wood defects occur due to natural fibre and wood defects as the wood is exposed to a constant loading for a prolonged time [26–28]. Hence, this issue has led engineers and researchers to find a solution in order to extend lifespan of the wooden cross-arms. One of the solution is the implementation of bracing system in the current cross-arm structure as suggested by Sharaf et al. [29]. Thus, this study evaluates the influence bracing system on long-term mechanical performance of wooden cross-arm.

Evaluating creep properties and responses of a full-scale wooden cross-arm used in 132 kV transmission tower could eliminate numerous exaggerated factors that happen in coupon scale test. The geometry and profile of the material could be neglected when the test is conducted in coupon scale, such as flexural, tensile, and compressive properties. Thus, more reliable data collection would be achieved when a full-scale size cross-arm is used in carrying out creep test to understand the mechanical behaviour during long-term loading condition. The investigation of long-term behaviour of main component member for the cross-arm structure could provide a more holistic perspective in order to evaluate the behaviour of the whole structure either with or without bracing.

Nowadays, many latticed transmission towers are still used the conventional wooden cross-arm to transmit electrical power. A literature survey revealed that no previous works have evaluated the creep properties of full-scale wooden cross-arms used for 132 kV towers [30,31]. Thus, this manuscript is expected to examine the effect of addition braced arms on the creep properties and responses of wooden cross-arms with its actual loading conditions. The study also intended to set a baseline for creep profiling of full-scale wooden cross-arms. At the end, the outputs from this study would create a practical perspective for engineers to understand the long-term mechanical performance of the conventional wooden structure.

2. Methodology

2.1. Materials

Balau timber wood or *Shorea Dipterocarpaceae* was used as a cross-arm's material to examine long-term creep behaviours. The cross-arm in 132 kV latticed transmission tower was composed of one tie member and two main members. All cross-arm members were fabricated from the same Balau tree trunk obtained at Bahau, Malaysia. The timber wood was cut individually to form continuous square section with the dimension of 102 mm × 102 mm. In term of size, the lengths of the main and tie members of the cross-arm was 3651 mm and 3472 mm, respectively. Each of the cross-arm members was joined together by using both bolts and nuts, as well as its mild steel fastener brackets. These members were assembled using fastener brackets in order to be incorporated on the creep test rig with constant loading.

For the braced arms, they connected with main and tie members by using custom-made fastener brackets. The braced arms were in square section beams with dimension of 50 mm × 50 mm. The arms are also made from Balau timber wood. In total, there are five braced arms interconnected with main and tie members. Those braced arms encompassed of two long members (connected in the middle of the tie member to the end of the main member) and three short members (joined at the middle to the middle of every member). The lengths of the long (tie-main), short (tie-main), and short (main-main) bracing members were 186, 50, and 40 cm, respectively.

2.2. Methods

The test was set up based on actual position of cross-arm in latticed transmission tower. The creep test was performed on the specialised cantilever beam creep test rig used for cross-arm testing. To evaluate the strain measurement, ten dial gauges were positioned 0.61 m in between five points under two main members. Figure 1 illustrates the positions and the length of between the dial gauges under the wooden cross-arm. The load was implemented at the bottom side of the joining parts of all members.

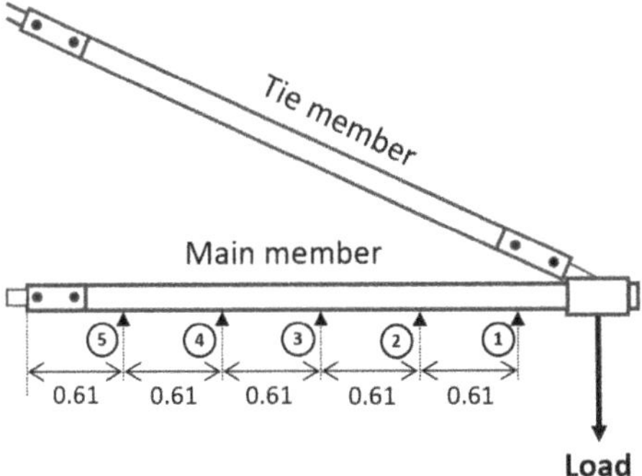

Figure 1. Positions of dial gauges under the cross-arm to measure creep strain pattern, in meter.

Detail drawing with dimensions of cantilever beam creep test rig with wooden cross-arm is displayed in Figure 2. The test rig was manufactured from mild steel square hollow section with dimension of 100 mm × 100 mm and 1.9 mm thickness. The rigidity and specifications of the test rig can be found in Table 1 [19]. The wooden cross-arm was installed on the test rig using forklift and it is manually fixed on the test rig's fastener brackets. A dead load was assembled at free end of the cross-arm to mimic the actual cross-arm conditions in the transmission tower as shown in Figure 3. In general practice, the wooden cross-arm is installed in suspension latticed transmission tower to carry the electric cables and insulators [32]. To be specific, the cross-arm was mounted on the test rig at height of 2100 mm from ground to hang the dead load for the creep test.

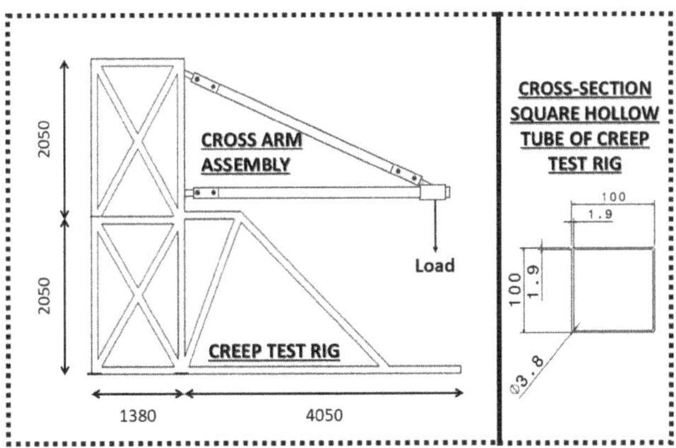

Figure 2. Schematic diagram of wooden cross-arm used in creep test rig.

Table 1. Specification of material of creep test rig [19].

Properties	Specification
Material	Mild steel
Tensile strength, MPa	766
Yield strength, MPa	572
Pipe shape	Square hollow section
Pipe size (width/height/thickness), mm	100/100/1.9
Total size (width/length/height), mm	1525/430/4100

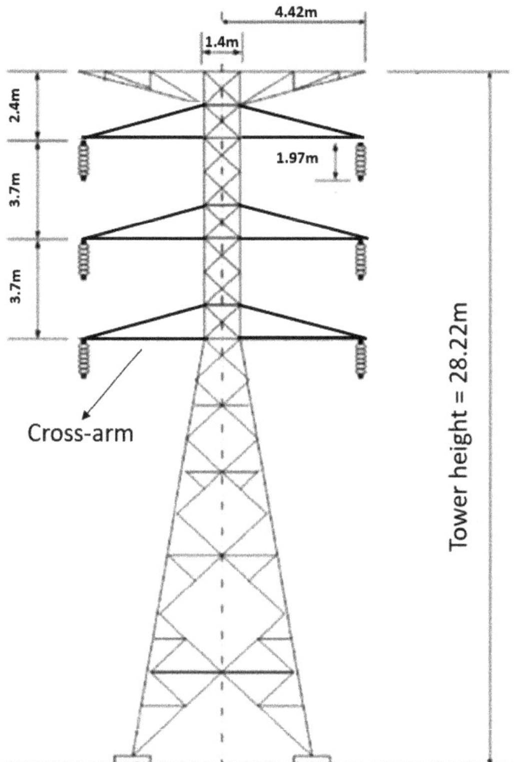

Figure 3. Dimension of latticed transmission tower. Adapted with permission from IEEE, 2021 [33].

The geometry and dimensions of latticed transmission tower is shown in Figure 3 [33]. The steel tower and wooden cross-arm was connected using specialised fittings. The assemblymen of wooden cross-arm to steel tower is manually fixed by bolt and nut. Based on Figure 4, the cross-arm was attached to the steel tower by sandwiching two steel plate fittings with end cross-arm member and they are manually connected by using bolt and nut.

As shown in Figure 5, the bracing members were connected with cross-arm's members via custom-made fastener brackets as aforementioned. The custom-made steel fastener bracket was designed and manufactured specifically to fit the shape of cross-arm. For the assembly process, the brackets were manually installed by using bolt and nut after the cross-arm was completely placed and fixed at the test rig. Figure 6 displays schematic diagram of two set configurations of cross-arm structure, which are current design (without braced arms) and braced design (with braced arms) configuration.

Figure 4. Connection of cross-arm to steel tower using steel plate fittings.

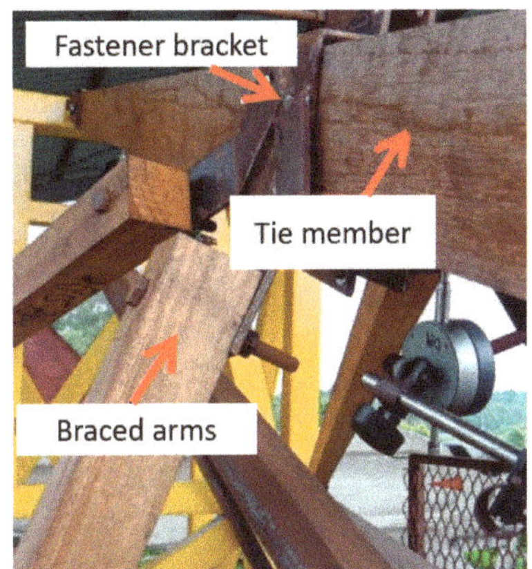

Figure 5. Mild steel fastener bracket joins the braced arms with tie members.

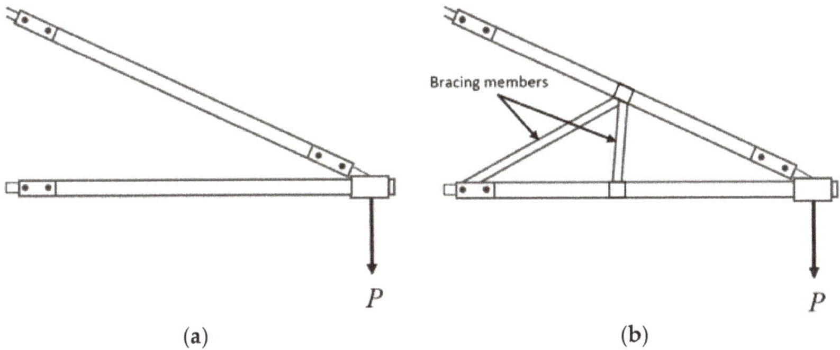

Figure 6. Cross-arm configurations: (**a**) current design configuration; and (**b**) braced design configuration.

In term of dead load, a hanging load was weighed based on actual working load of cross-arm (6.347 kN) before the experiment started. The test fulfilled the requirement time as in ASTM D2990, which evaluated the creep test at 1000 h of operation. Specifically, readings were taken at several specific time periods (0.1, 0.2, 0.5, 1, 2, 5, 20, 50, 100, 200, 500, 700 and 1000 h) to observe creep deformation. The condition of the experiment was set at open area that constantly exposed to actual tropical weather. At the end of experiment, the comparison of current (without bracing members) and braced (with bracing members) cross-arm designs was carried out in terms of long-term creep properties.

2.2.1. Creep Properties of Cross-Arms

A wooden cantilever beam structure usually experiences viscoelastic behaviour when a constant loading is continuously applied at the end of the beam. In common practice, the beam exhibits constant tension and compression actions on opposite sides of the beam, which induces a series of strain pattern depending on the applied load. When the displacement remains at certain positions along the time period, the viscoelastic beam usually expresses the stress response on the beam and gradually decreases. This shows that the viscoelastic beam responds to the material's viscous characteristic, which would decrease the total stress [34,35]. Similarly, a beam that is exposed to a constant load continues to deform as the material relaxes.

In general, the static elastic modulus (E_e) of the beam as shown in Figure 7 can generated based on Equation (1):

$$E_e = \frac{4PL^3}{ybh^3} \tag{1}$$

where, y is the deflection at the beam (m); E_e is the static elastic modulus (N/m^2); P is the force exerted on the beam (N); L is the total length (m) and b and h are the width and thickness of the beam (m), respectively.

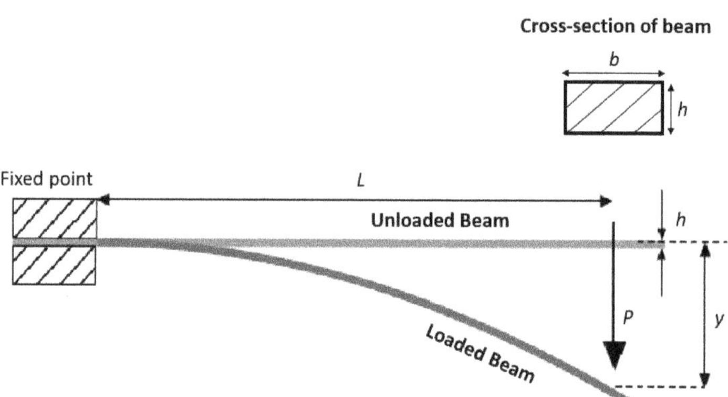

Figure 7. Schematic diagram of cross-arm structure when exposed applied force at the end of the cross-arm structure.

Given that the y deflection is already known, the maximum bending stress can be predicted by using Equation (1) [36]. In general, the maximum stress experienced by the cross-arm is usually located at fixed point $x = 0$, whereas the minimum stress is exhibited at the loading end $x = L$. The maximum and minimum stresses of the beam are formulated on the basis of Equation (2).

$$\sigma = \frac{P(L-x)\frac{h}{2}}{I} = \frac{6P(L-x)}{bh^3} \tag{2}$$

Equation (3), are formulated based on Hooke's law equation. The Equation (3) is functioned to calculate the creep strain at a specific time and specific location across the beam.

$$\varepsilon_t = \frac{\sigma_n}{E_e} \tag{3}$$

where, ε_t is the creep strain at a specific time and location point across the beam. σ_n is specific stress and E_e is the static elastic modulus at the specific point on the cross-arm.

2.2.2. Constitutive Creep Models

Findley power law model is an empirical mathematical model that simulates the creep properties of anisotropic material. The model is presented as in Equation (4) [37].

$$\varepsilon_t = At^n + \varepsilon_0 \tag{4}$$

where, A and n as transient creep strain and time exponent respectively, while ε_0 is the instantaneous strain after exerted the load.

To assess the time-dependent responses of the Balau wooden material on the basis of the flexural information, a reliable creep model has to be recognised. One of the models used in order to identify the relationship between the structure and creep behaviour was Burger model [5,34]. This model can be expressed in Equation (5).

$$\varepsilon_t = \varepsilon_e + \varepsilon_d + \varepsilon_v \tag{5}$$

The mathematical formulation in Equation (5) comprises ε_e, ε_d and ε_d, which are called the elastic strain, viscoelastic strain and viscous strain, respectively.

Equation (6) was derived on the basis of Equation (5) and the physical elements of Burger models, such as spring and dashpot elements.

$$\varepsilon_t = \frac{\sigma}{E_e} + \frac{\sigma}{E_d}[1 - exp(-t/\tau)] + \frac{\sigma}{\eta_v}t; \ \tau = \frac{E_d}{\eta_d} \tag{6}$$

At this point of view, these three strain components later were derived into stress, elastic modulus and viscoelastic modulus as shown Equation (6). Both elastic and viscoelastic moduli are essential behaviour of the material. Perez et al. [38] and Chandra and Sobral [39] established that the Burger model compromises of combination of three elements including a linear elastic spring, a dash pot, and a Kelvin-Voight element (dash-pot and combination of dash-pot and spring). Figure 8 visualise the long term behaviour of viscoelastic material under Burger model.

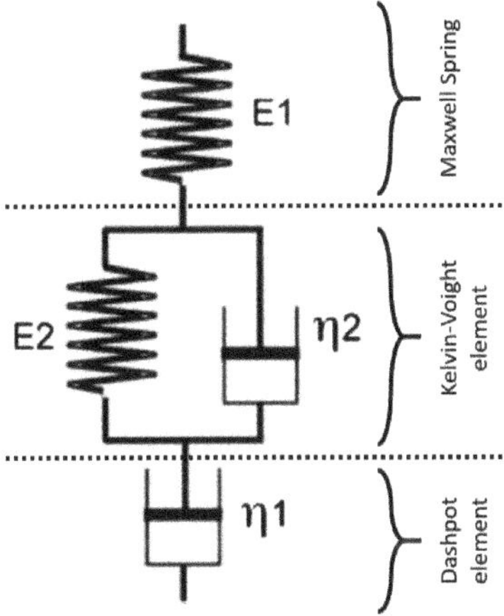

Figure 8. Schematic diagram of Burger model.

3. Results and Discussion

3.1. Strain-Time Curve

The creep strain-time graphs for both current and braced wooden cross-arm configurations at each point of main members are presented in Figure 9. Based on the curves, it can be seen that creep pattern are divided into three phases which are instantaneous deformation, primary and secondary creeps. As expected in the early experiment, the highest creep strain is located at y3, which at the centre of cross-arm's main member beams. This was probably due to the each of main members of the structure experienced compression from both ends when the force is exerted at free end of the cross-arm. As mentioned by Kanyilmaz (2017) [40], a simple interconnected members of a structure without bracing systems would induce inelastic behaviour, which tends to experience buckling at the centre of the structure arm. This established that the operational cross-arm would experience buckling action due to external forces from the dead weigh at the end of the structure.

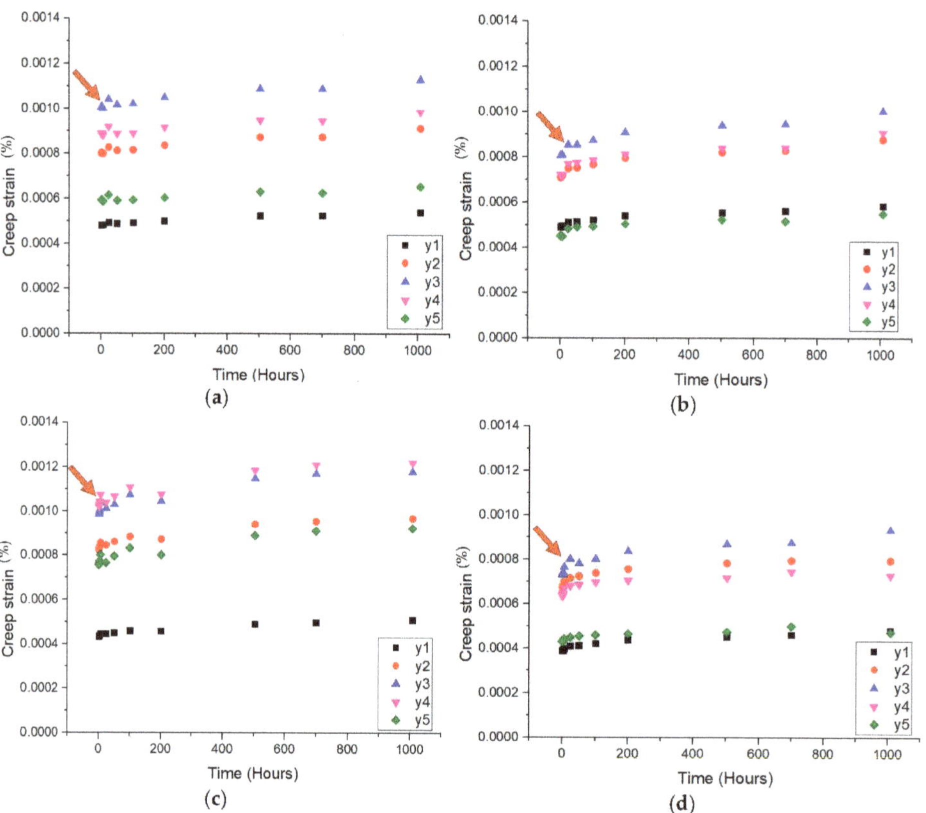

Figure 9. Creep strain-time curves for current wooden timber cross-arm for (**a**) left and (**c**) right; braced wooden timber cross-arm for left (**b**) and right (**d**) main member.

In the meantime, the findings also depicts the creep strain patterns have two distinct stages: elastic and viscoelastic stages. However, the transition period from the elastic period to the viscoelastic phase was extended for the current design cross-arm (red arrow's location shown in Figure 9). This established that the braced wooden cross-arm provide the structure become more stable in viscoelastic stage.

Based on Figure 9b,d, the curves in both left and right main members of braced design cross-arm exhibit a similar pattern. The similar creep strain pattern for both left and right wooden cross-arm is probably due to the addition of bracing system provide better structural integrity. Subsequently, a symmetrical shape and deformation pattern would permit during the cross-arm service to grasp the power cables and insulators, which would reduce any potential sudden failure of the structure after years of service.

Besides that, the installation of the brace arms reduces the creep strain along the operation time. This can be observed where the instantaneous deformation of the braced design cross-arm are noticeably less than current design cross-arm at any points along the individual member. For instance, at the middle of cross-arm (y3), the finding displays that the creep strain value for the current wooden cross-arm was higher than the braced wooden cross-arm. At this point of view, the inclusion of bracing system in the structure would significantly provide higher creep resistant performance by reduced the creep strain approximately 31%. This outcomes is tally with a research conducted by Patil et al. [41]. They mentioned that the improvement of mechanical response of a structure is significant enhanced as the addition of bracing members could resist lateral forces which subsequently avoid from buckling.

3.2. Findley Power Law Model

Table 2 tabulates the values of *A* parameter and stress-independent material exponent, *n*. *A* and *n* parameters were discovered based on Equation (4) by using Origin Pro 2016.

Table 2. Average parameters obtained from Findley power law for current and braced wooden cross-arms.

Main Member Arm	Location	*A*		*n*		Adj. R^2	
		Current Cross-Arm	Braced Cross-Arm	Current Cross-Arm	Braced Cross-Arm	Current Cross-Arm	Braced Cross-Arm
Right	1	5.818×10^{-7}	6.086×10^{-6}	0.669	0.395	0.973	0.989
	2	6.977×10^{-7}	1.005×10^{-5}	0.726	0.402	0.957	0.981
	3	1.128×10^{-6}	1.255×10^{-5}	0.673	0.396	0.936	0.977
	4	1.927×10^{-7}	1.350×10^{-5}	0.895	0.372	0.888	0.951
	5	4.224×10^{-8}	1.429×10^{-5}	1.045	0.278	0.833	0.938
Left	1	1.647×10^{-6}	6.331×10^{-6}	0.551	0.389	0.981	0.991
	2	5.076×10^{-6}	2.969×10^{-5}	0.486	0.234	0.958	0.992
	3	6.377×10^{-6}	1.952×10^{-5}	0.496	0.331	0.933	0.960
	4	5.313×10^{-6}	6.858×10^{-5}	0.528	0.123	0.920	0.958
	5	4.278×10^{-6}	2.682×10^{-5}	0.533	0.155	0.917	0.928

The obtained values of *A* parameter are resumed on Figure 10. From Figure 10, it is apparent that *A* parameter is increased for braced wooden cross-arm. This result acknowledged that the implementing of bracing system could enhanced the transient creep strain of the structure. On the other hand, the left main member had higher *A* parameter value compared to the right main member in both cross-arms. This indicated that the left main member had grain orientation in either radial, tangential, or longitudinal direction which differed in terms of their creep resistance performance [42,43].

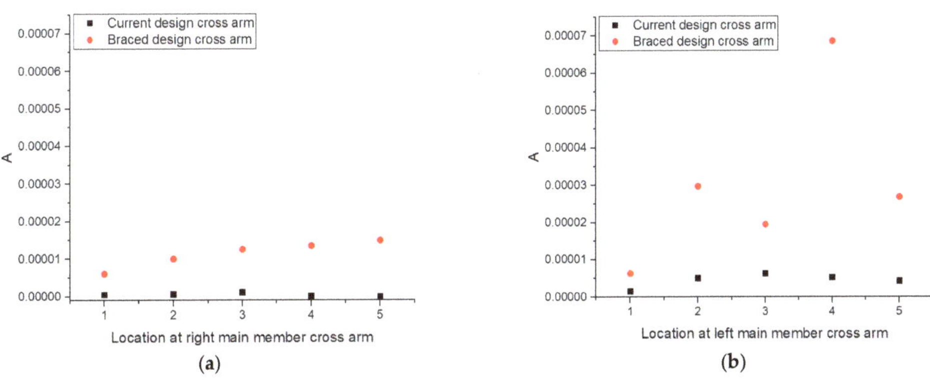

Figure 10. *A* parameter for current and braced wooden cross-arms: (**a**) right; (**b**) left main members

In terms of stress-independent material exponent, the findings showed that the average value for the existing right cross-arm (0.8016) was higher than the existing left cross-arm (0.5188) as shown in Table 3. This was probably due to both members of cross-arms having different grain direction and orientation, which would affect the n exponent. According to Hill (2006) [44], the cell wall swelled significantly more either in radial or tangential directions rather than longitudinal orientation during water absorption process. This was affected by winding angle of microfibrils within the wood fibre layer. On the other hand, the manufacturer might cut each beam separately from different tree trunks or different heights. According to Machado et al. [45] and Van Duong and Matsumura [46], a significant decrease of wood density would affect their bending properties when the height of timber varied. Apart from that, different heights of tree may contribute to different chemical compositions, such as cellulose, hemicellulose, lignin, and ash, which affect their physico-mechanical properties [47]. These factors might contribute to different values of stress-independent material exponent, n, for left and right main members of the same cross-arm. However, when the cross-arm was incorporated and added with bracing arms, it seemed that the right and left main members of the cross-arm had almost similar value of n exponent, which are 0.3686 and 0.2464, respectively (Table 3). This attribute was due to the addition of brace arms in wooden cross-arm structure which would improve the distribution of stress across the cross-arm members.

Table 3. Stress independent material exponent, n of both current and braced wooden cross-arms.

Cross-Arm Configuration	Current		Braced	
	Right	Left	Right	Left
Stress independent material exponent, n	0.8016	0.5188	0.3686	0.2464

Lastly, the adjusted regression (Adj. R^2) forms for both existing designs (WOB) and braced (WB) wooden cross-arm had high value, narratively close to 1. This showed that the Findley's power law model fit the experimental data very well. Moreover, it explained that the creep of both cross-arms experienced two levels of creep, which are primary and secondary stages. This model forecasted the secondary stage well, but it cannot forecast the tertiary creep, which can determine the time of failure. On the other hand, the braced wooden cross-arm (0.989–0.928) exhibited higher Adj. R^2 value compared to the existing design (0.981–0.833). This explained that the braced wooden cross-arm followed the creep principles involving primary and secondary creep stages, and the creep data produced was less exaggerated due to better structural integrity.

3.3. Burger Model

Experimental graphs were fitted by means of the Burger model. A computational software, OriginPro 7.5, was implemented to plot a non-linear curve fit to identify four parameters (E_e, E_d, η_d, η_k) as shown in Table 4. The Burger model is usually executed in creep data evaluation due to its elaborate elastic and viscoelastic properties of anisotropic beam based on the working load within the creep period. In general, the elastic material demonstrated no residual deformation when the stress was detached from the structure. On the other hand, the viscoelastic property displayed a stress relaxation condition over time. The basis of the model was a combination of Burger's elements within the working load condition. However, the structure system would differently respond in terms of creep strain depending on the types of elements in the model [21].

Figure 11 depicts the values of E_e at different locations in wooden cross-arm for both current and additional bracing arms. The elasticity parameter or E_e was obtained from the data of instantaneous creep strain after execution of stress. From the results, it showed that the elastic performance of the cross-arm's beam decreased from the fixed point to free end. This happened due to stress along the cantilever beam which would decrease linearly from fixed to free end [34]. Apart from that, the finding displayed that braced cross-arm had higher E_e value compared to the existing cross-arm. This could be due to the retrofitting of braced arms which provided a sufficient restraining force to ease the residual deflection. Moreover, the increasing elastic modulus, E_e, indicated the enhancement of the tensile modulus [48]. This would increase their stiffness to resist plastic deformation to maintain their working load during their long-term service period [49].

Table 4. Average parameters obtained from Burger model for both current and braced wooden cross-arms.

Main Member Arm	Location	E_e		η_k		Adj R^2	
		Current Cross-Arm	Braced Cross-Arm	Current Cross-Arm	Braced Cross-Arm	Current Cross-Arm	Braced Cross-Arm
Right	1	4.62×10^{10}	4.53×10^{10}	4.11×10^{14}	3.72×10^{14}	0.970	0.991
	2	5.52×10^{10}	6.22×10^{10}	4.53×10^{14}	4.07×10^{14}	0.961	0.987
	3	6.54×10^{10}	8.20×10^{10}	6.15×10^{14}	5.37×10^{14}	0.938	0.983
	4	9.91×10^{10}	12.2×10^{10}	9.84×10^{14}	7.60×10^{14}	0.888	0.973
	5	18.5×10^{10}	24.4×10^{10}	19.2×10^{14}	20.9×10^{14}	0.824	0.969
Left	1	5.11×10^{10}	5.59×10^{10}	3.41×10^{14}	2.38×10^{14}	0.973	0.851
	2	5.33×10^{10}	6.35×10^{10}	3.67×10^{14}	3.43×10^{14}	0.941	0.695
	3	6.70×10^{10}	8.74×10^{10}	3.86×10^{14}	3.50×10^{14}	0.922	0.802
	4	8.53×10^{10}	13.4×10^{10}	5.51×10^{14}	8.74×10^{14}	0.901	0.559
	5	14.5×10^{10}	24.9×10^{10}	7.26×10^{14}	19.9×10^{14}	0.910	0.518

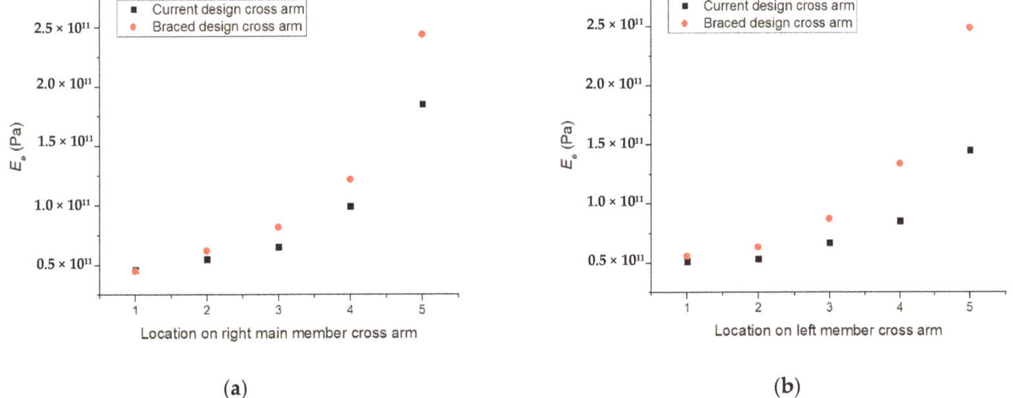

Figure 11. E_e parameter for current and braced wooden cross-arms: (**a**) right; (**b**) left main members.

In this study, another parameter was being evaluated, η_k, which represented the relaxation coefficient for the viscoelastic property. Moreover, the viscoelastic parameter, also known as irrecoverable creep strain, demonstrated the relaxation response over time [23]. Figure 12 displays the viscoelastic properties for both cross-arms (braced and existing cross-arms), showing relatively the same values of viscoelastic modulus. This illustrated that the bracing system did improve the viscoelastic properties of the cross-arm's structure in terms of relaxation time especially at y5 location. This showed that the bracing arms increased the relaxation of the cross-arm under long-term constant load. Apart from that, both cross-arms exhibited linear viscoelastic property along the beam length since the set working load was below the critical value of applied stress. If a critical stress was achieved during the creep cycle, the creep rate grew disproportionately faster [50].

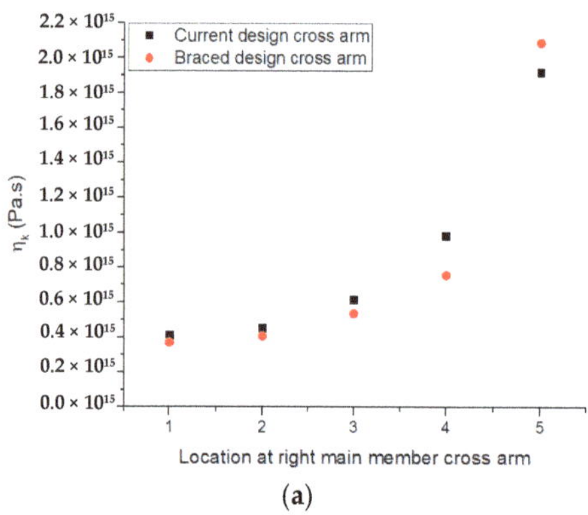

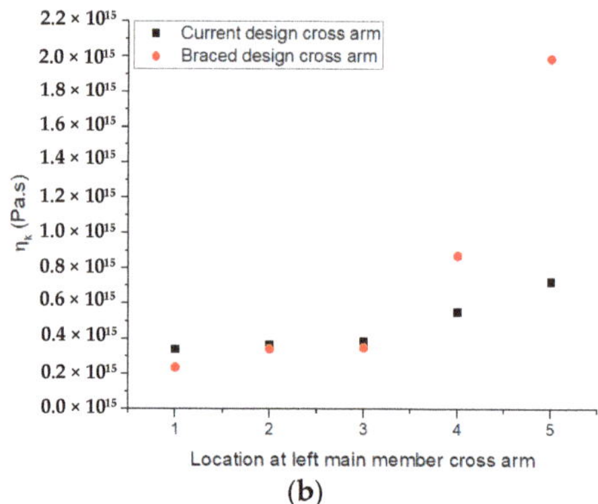

Figure 12. η_k parameter for current and braced wooden cross-arms: (**a**) right; (**b**) left main members.

3.4. Creep Models Accuracy and Validation

In terms of accuracy of data plotted in creep strain-time analysis, the adjusted regression (Adj. R^2) or coefficient of determination was used to fit these numerical models with experimental outputs. These Adj. R^2 values are tabulated in Tables 2 and 3 for the Findley and Burger models, respectively. From these tables, it was found that the adj. R^2 for Findley's power law model exhibited higher values, ranging from 0.833 to 0.989 rather than the Burger model, ranging from 0.518 to 0.913. The Burger model forecasted a relationship of viscosity and time linearly, which was affected during the numerical model fitting with experimental data [51]. Thus, from the adjusted regression evaluation, the best numerical model suitable to analyse the finding of creep for wooden cross-arm was Findley's power law model. This observation showed that the wooden cross-arm experienced a steady-state creep in a long-term period and did not permit any sign of tertiary creep phase to suddenly fail [52,53] during long-term period and not permit any sign of tertiary creep phase to be suddenly failed. However, the project also required the use of the Burger model in order to examine the effect of addition of bracing arms on elastic and viscoelastic properties of wooden cross-arm. Thus, both studies were required to achieve a holistic view and analysis on creep behaviours for the existing and braced wooden cross-arms.

These creep models (Findley's power law model and Burger model) were validated by comparing their instantaneous strain with experimental results. Fundamentally, the instantaneous strains were principally proportional to the applied stress according to Hooke's law. Table 5 summarises the evaluation of instantaneous strain between the experimental outcomes with two numerical models (Findley and Burger models) for both existing and braced cross-arms at y3 location. Since the y3 location exhibited the most highly severe in terms of creep strain, thus, the creep strain results were compared with the numerical outputs. As seen in Table 5, all percentage errors recorded were below than 5%. This showed that all numerical models including the Findley and Burger models fitted the experimental data accurately. According to Zhang et al. [54], the acceptable value for percentage error when comparing the experimental outputs with numerical values should be less than 20%. When the percentage error below 20%, it displayed that the plotted experimental data were not severely deviated, and consistent within the principles proposed from the exact numerical model. In this study, the experimental data plotted to elaborate the creep properties of the wooden cross-arms were verified with precise and consistent values.

Table 5. Comparison of instantaneous strain value between experimental outputs and numerical models located at y3 for current and braced wooden cross-arms.

Configuration	Model	Inst. Strain	Located at y3 at Main Member			
			Right	Percentage Error (%)	Left	Percentage Error (%)
Current cross-arm	Experimental data	$\varepsilon \ (10^{-3})$	1.006	-	0.988	-
	Findley model	$\varepsilon_0 \ (10^{-3})$	1.010	0.398	0.994	0.604
	Burger model	$\varepsilon_0 \ (10^{-3})$	1.010	0.398	0.986	0.202
Braced cross-arm	Experimental data	$\varepsilon \ (10^{-3})$	0.806	-	0.731	-
	Findley model	$\varepsilon_0 \ (10^{-3})$	0.798	0.993	0.722	1.231
	Burger model	$\varepsilon_0 \ (10^{-3})$	0.806	0.000	0.756	3.420

4. Conclusions

The creep properties of Balau wood timber cross-arms reinforced with additional braced arms was significantly reduced as compared to existing design wooden cross-arms. Thus, the implementation of bracing system in cross-arm structures display a good potential for application existing wooden cross-arms in latticed transmission tower. Many previous studies conducted creep experiment of wooden specimens in laboratory with controlled environment. This approach is typical for intended baseline characterizations. However, no study has been carried out on full-scale size of wooden cross-arm in actual environment of transmission tower. This study is narrowed to compare the braced and current design of wooden cross-arms with actual working load and environment conditions. The comparison depicts that the creep strain of the main member for braced wooden cross-arm had reduced about 15–21% compared to existing design of wooden cross-arm. In addition, the addition of braced arms in cross-arm structures would effectively enhance the stability of the viscoelastic stage, which would reduce the failure probability. Additional creep analyses were carried out using Findley and Burger models discovered that braced wooden cross-arm has greater elastic modulus. This indicates that incorporation of extra braced arms contribute better flexure property for overall structure. The braced systems increased the viscoelastic modulus of the cross-arm, thus enhancing relaxation during creep. Moreover, the results also displays the braced wooden cross-arm permit more stable stress independent material exponent between right and left main members. This shows that bracing system in cross-arm would induce better dimensional stability of the structure. As a conclusion, the implementation of additional braced arms would be highly advantageous during construction of latticed transmission tower which could prolong the cross-arm's service life.

It is suggested that the connection of the braced arms with cross-arm's members could possibly affect the overall long-term mechanical properties of the structure. In the case of timber structures, the flexibility of wood in connection, and the flexibility of the connectors themselves have a very significant impact on the deformation of the entire structure. This aspect is vital for the creep study of a complex structure in order to further elaborate the potential failures and problems in the future. Thus, it is highly suggested that further study could be conducted to examine the effect of braced arms connection on creep behaviours of the wooden cross-arm.

Author Contributions: Conceptualization, M.R.I. and M.R.M.A.; methodology, M.R.M.A.; software, M.R.M.A.; validation, M.R.I., S.M.S. and N.Y.; investigation, M.R.M.A.; resources, M.R.I.; writing—original draft preparation, M.R.M.A.; writing—review and editing, M.R.M.A., and M.R.I.; visualization, M.R.M.A.; supervision, M.R.I., S.M.S. and N.Y.; project administration, M.R.I.; funding acquisition, M.R.I. All authors have read and agreed to the published version of the manuscript.

Funding: This research work was funded by Ministry of Higher Education, Malaysia for financial support through the enumerator service under Fundamental Research Grant Scheme (FRGS): FRGS/1/2019/TK05/UPM/02/11 (5540205) and Geran Putra: (9634000) by Universiti Putra Malaysia to carry out all research activities.

Institutional Review Board Statement: Not applicable.

Informed Consent Statement: Not applicable.

Data Availability Statement: The data used to support the findings of this study are included within the article.

Acknowledgments: This research work was funded by Ministry of Higher Education, Malaysia for financial support through the enumerator service under Fundamental Research Grant Scheme (FRGS): FRGS/1/2019/TK05/UPM/02/11 (5540205) and Geran Putra: (9634000) by Universiti Putra Malaysia to carry out all research activities. The authors are also very thankful to Department of Aerospace Engineering, Faculty of Engineering, UPM for providing space and facilities for the project. Moreover, all authors are very appreciate and thankful to Jabatan Perkhidmatan Awam (JPA) and Kursi Rahmah Nawawi for providing scholarship award and financial aids to the first author to carry out this research project.

Conflicts of Interest: The authors declare no conflict of interest.

References

1. Jaafar, C.N.A.; Rizal, M.A.M.; Zainol, I. Effect of kenaf alkalization treatment on morphological and mechanical properties of epoxy/silica/kenaf composite. *Int. J. Eng. Technol.* **2018**, *7*, 258–263. [CrossRef]
2. Jaafar, C.N.A.; Zainol, I.; Rizal, M.A.M. Preparation and characterisation of epoxy / silica / kenaf composite using hand lay-up method. In Proceedings of the 27th Scientific Conference of the Microscopy Society Malaysia (27th SCMSM 2018), Melaka, Malaysia, 3–4 December 2018; Universiti Putra Malaysia: Serdang, Malaysia; pp. 2–6.
3. Shahroze, R.M.; Chandrasekar, M.; Senthilkumar, K.; Senthilmuthukumar, T.; Ishak, M.R.; Asyraf, M.R.M. A review on the various fibre treatment techniques used for the fibre surface modification of the sugar palm fibres. In Proceedings of the Seminar Enau Kebangsaan, Bahau, Negeri Sembilan, Malaysia, 1 April 2019; Universiti Putra Malaysia: Serdang, Malaysia; pp. 48–52.
4. Asyraf, M.R.M.; Rafidah, M.; Ishak, M.R.; Sapuan, S.M.; Yidris, N.; Ilyas, R.A.; Razman, M.R. Integration of TRIZ, Morphological Chart and ANP method for development of FRP composite portable fire extinguisher. *Polym. Compos.* **2020**, *41*, 2917–2932. [CrossRef]
5. Asyraf, M.R.M.; Ishak, M.R.; Sapuan, S.M.; Yidris, N.; Ilyas, R.A. Woods and composites cantilever beam: A comprehensive review of experimental and numerical creep methodologies. *J. Mater. Res. Technol.* **2020**, *9*, 6759–6776. [CrossRef]
6. Rawi, I.M.; Rahman, M.S.A.; Ab Kadir, M.Z.A.; Izadi, M. Wood and fiberglass crossarm performance against lightning strikes on transmission towers. In Proceedings of the International Conference on Power Systems Transient (IPST), Seoul, Korea, 26–29 June 2017; Sung-Kyun-Kwan University: Seoul, Korea; pp. 1–6.
7. Sapuan, S.M. *Tropical Natural Fibre Composites: Properties, Manufacture and Applications*; Springer Science+Business Media Singapore: Singapore, 2014; ISBN 9780857095244.
8. Ma, X.; Jiang, Z.; Tong, L.; Wmang, G.; Cheng, H. Development of creep models for glued laminated bamboo using the time-temperature superposition principle. *Wood Fiber Sci.* **2015**, *47*, 1–6.
9. Liu, T. Creep of wood under a large span of loads in constant and varying environments. Part 1: Experimental observations and analysis. *Holz Roh. Werkst.* **1993**, *51*, 400–405. [CrossRef]
10. Ilyas, R.A.; Sapuan, S.M.; Asyraf, M.R.M.; Atikah, M.S.N.; Ibrahim, R.; Dele-Afolabia, T.T. Introduction to biofiller reinforced degradable polymer composites. In *Biofiller Reinforced Biodegradable Polymer Composites*; Sapuan, S.M., Jumaidin, R., Hanafi, I., Eds.; CRC Press: Boca Raton, FL, USA, 2020; pp. 1–23.
11. Ilyas, R.A.; Sapuan, S.M.; Atiqah, A.; Ibrahim, R.; Abral, H.; Ishak, M.R.; Zainudin, E.S.; Nurazzi, N.M.; Atikah, M.S.N.; Ansari, M.N.M.; et al. Sugar palm (*Arenga pinnata* [Wurmb.] Merr) starch films containing sugar palm nanofibrillated cellulose as reinforcement: Water barrier properties. *Polym. Compos.* **2020**, *41*, 459–467. [CrossRef]
12. Ilyas, R.; Sapuan, S.; Atikah, M.; Asyraf, M.; Rafiqah, S.A.; Aisyah, H.; Nurazzi, N.M.; Norrrahim, M. Effect of hydrolysis time on the morphological, physical, chemical, and thermal behavior of sugar palm nanocrystalline cellulose (*Arenga pinnata* (Wurmb.) Merr). *Text. Res. J.* **2020**, 004051752093239. [CrossRef]
13. Ilyas, R.A.; Sapuan, M.S.; Norizan, M.N.; Norrrahim, M.N.F.; Ibrahim, R.; Atikah, M.S.N.; Huzaifah, M.R.M.; Radzi, A.M.; Izwan, S.; Azammi, A.M.N.; et al. Macro to nanoscale natural fiber composites for automotive components: Research, development, and application. In *Biocomposite and Synthetic Composites for Automotive Applications*; Sapuan, M.S., Ilyas, R.A., Eds.; Woodhead Publishing Series: Amsterdam, The Netherland, 2020.
14. Asyraf, M.R.M.; Ishak, M.R.; Sapuan, S.M.; Yidris, N.; Shahroze, R.M.; Johari, A.N.; Rafidah, M.; Ilyas, R.A. Creep test rig for cantilever beam: Fundamentals, prospects and present views. *J. Mech. Eng. Sci.* **2020**, *14*, 6869–6887. [CrossRef]
15. Johari, A.N.; Ishak, M.R.; Leman, Z.; Yusoff, M.Z.M.; Asyraf, M.R.M. Influence of CaCO₃ in pultruded glass fibre/unsaturated polyester composite on flexural creep behaviour using conventional and TTSP methods. *Polimery* **2020**, *65*, 46–54. [CrossRef]
16. Asyraf, M.R.M.; Ishak, M.R.; Razman, M.R.; Chandrasekar, M. Fundamentals of creep, testing methods and development of test rig for the full-scale crossarm: A review. *J. Teknol.* **2019**, *81*, 155–164. [CrossRef]

7. Asyraf, M.R.M.; Ishak, M.R.; Sapuan, S.M.; Yidris, N. Conceptual design of multi-operation outdoor flexural creep test rig using hybrid concurrent engineering approach. *J. Mater. Res. Technol.* **2020**, *9*, 2357–2368. [CrossRef]
8. Asyraf, M.R.M.; Ishak, M.R.; Sapuan, S.M.; Yidris, N. Conceptual design of creep testing rig for full-scale cross arm using TRIZ-Morphological chart-analytic network process technique. *J. Mater. Res. Technol.* **2019**, *8*, 5647–5658. [CrossRef]
9. Asyraf, M.R.M.; Ishak, M.R.; Sapuan, S.M.; Yidris, N.; Ilyas, R.A.; Rafidah, M.; Razman, M.R. Evaluation of design and simulation of creep test rig for full-scale cross arm structure. *Adv. Civ. Eng.* **2020**, 6980918. [CrossRef]
10. Gamalath, S.S. Long Term Creep Modelling of Wood Using Time Temperature Superposition Principle. Ph.D. Thesis, Virginia Polytechnic Institute and State University, Blacksburg, VA, USA, 1991.
11. Moutee, M.; Fafard, M.; Fortin, Y.; Laghdir, A. Modeling the creep behavior of wood cantilever loaded at free end during drying. *Wood Fiber Sci.* **2005**, *37*, 521–534.
12. Kaboorani, A.; Blanchet, P.; Laghdir, A. A rapid method to assess viscoelastic and mechanosorptive creep in wood. *Wood Fiber Sci.* **2013**, *45*, 1–13.
13. Segovia, F.; Blanchet, P.; Laghdir, A.; Cloutier, A. Mechanical behaviour of sugar maple in cantilever bending under constant and variable relative humidity conditions. *Int. Wood Prod. J.* **2013**, *4*, 225–231. [CrossRef]
14. Rawi, I.M.; Ab Kadir, M.Z.A. Investigation on the 132kV overhead lines lightning-related flashovers in Malaysia. In Proceedings of the International Symposium on Lightning Protection (XIII SIPDA), Balneario Camboriu, Brazil, 28 September–2 October 2015; Institute of Electrical and Electronics Engineers (IEEE): New York, NY, USA, 2015; pp. 239–243.
15. Engineering Department of TNB Transmission Division. *Investigation Report on Wooden Crossarm Failure at 132kV KKSRPPAN L2*; Engineering Department of TNB Transmission Division: Selangor, Malaysia, 2013.
16. Xu, Y.; Lee, S.Y.; Wu, Q. Creep analysis of bamboo high-density polyethylene composites: Effect of interfacial treatment and fiber loading level. *Polym. Compos.* **2011**, *32*, 692–699. [CrossRef]
17. Asyraf, M.R.M.; Ishak, M.R.; Sapuan, S.M.; Yidris, N.; Ilyas, R.A.; Rafidah, M.; Razman, M.R. Potential application of green composites for cross arm component in transmission tower: A brief review. *Int. J. Polym. Sci.* **2020**, 8878300. [CrossRef]
18. Asyraf, M.R.M.; Ishak, M.R.; Sapuan, S.M.; Yidris, N.; Johari, A.N.; Ashraf, W.; Sharaf, H.K.; Chandrasekar, M.; Mazlan, R. Creep test rig for full-scale composite crossarm: Simulation modelling and analysis. In Proceedings of the Seminar Enau Kebangsaan, Bahau, Negeri Sembilan, Malaysia, 1 April 2019; Universiti Putra Malaysia: Serdang, Malaysia, 2019; pp. 34–38.
19. Sharaf, H.K.; Ishak, M.R.; Sapuan, S.M.; Yidris, N. Conceptual design of the cross-arm for the application in the transmission towers by using TRIZ–morphological chart–ANP methods. *J. Mater. Res. Technol.* **2020**, *9*, 9182–9188. [CrossRef]
20. Johari, A.N.; Ishak, M.R.; Leman, Z.; Yusoff, M.Z.M.; Asyraf, M.R.M. Creep Behaviour Monitoring of Short-term Duration for Fiber-glass Reinforced Composite Cross-arms with Unsaturated Polyester Resin Samples Using Conventional Analysis. *J. Mech. Eng. Sci.* **2020**, *14*, 7361–7368. [CrossRef]
21. Beddu, S.; Syamsir, A.; Arifin, Z.; Ishak, M. Creep behavior of glass fibre reinforced polymer structures in crossarms transmission line towers. *AIP Conf. Proc.* **2018**, *2031*, 20039. [CrossRef]
22. Bakar, M.S.A.; Mohamad, D.; Ishak, Z.A.M.; Yusof, Z.M.; Salwi, N. Durability control of moisture degradation in GFRP cross arm transmission line towers. *AIP Conf. Proc.* **2018**, *2031*, 20027. [CrossRef]
23. Hassan, N.H.N.; Bakar, A.H.A.; Mokhlis, H.; Illias, H.A. Analysis of arrester energy for 132kV overhead transmission line due to back flashover and shielding failure. In Proceedings of the PECon 2012 IEEE International Conference on Power and Energy, Kota Kinabalu, Malaysia, 2–5 December 2012; pp. 683–688.
24. Hunt, J.F.; Zhang, H.; Huang, Y. Analysis of cantilever-beam bending stress relaxation properties of thin wood composites. *BioResources* **2015**, *10*, 3131–3145. [CrossRef]
25. Feng, S.H.; Zhao, Y.K. The summary of wood stress relaxation properties and its influencing factors. *Wood Proc. Mach.* **2010**, *25*, 39–40.
26. Liu, H.W. *Mechanics of Materials*; China Machine Press: Beijing, China, 2004.
27. Loni, S.; Stefanou, I.; Valvo, P.S. Experimental study on the creep behaviour of GFRP pultruded beams. In *Proceedings of the AIMETA 2013–XXI Congresso Nazionale dell'Associazione Italiana di Meccanica Teorica e Applicata, Torino, Italy, 17–20 September 2013*; Politecnico di Torino: Torino, Italy, 2013; pp. 1–10.
28. Pérez, C.J.; Alvarez, V.A.; Vázquez, A. Creep behaviour of layered silicate/starch-polycaprolactone blends nanocomposites. *Mater. Sci. Eng. A* **2008**, *480*, 259–265. [CrossRef]
29. Chandra, P.K.; Sobral, P.J.d.A. Calculation of viscoelastic properties of edible films: Application of three models. *Ciência e Tecnol. Aliment.* **2006**, *20*, 250–256. [CrossRef]
30. Kanyilmaz, A. Role of compression diagonals in concentrically braced frames in moderate seismicity: A full scale experimental study. *J. Constr. Steel Res.* **2017**, *133*, 1–18. [CrossRef]
31. Patil, D.M.; Sangle, K.K. Seismic behaviour of different bracing systems in high rise 2-D steel buildings. *Structures* **2015**, *3*, 282–305. [CrossRef]
32. Ozyhar, T.; Hering, S.; Niemz, P. Viscoelastic characterization of wood: Time dependence of the orthotropic compliance in tension and compression. *J. Rheol.* **2013**, *57*, 699–717. [CrossRef]
33. Jiang, J.; Erik Valentine, B.; Lu, J.; Niemz, P. Time dependence of the orthotropic compression Young's moduli and Poisson's ratios of Chinese fir wood. *Holzforschung* **2016**, *70*, 1093–1101. [CrossRef]

44. Hill, C.A.S. *Wood Modification: Chemical, Thermal and Other Processes*, 1st ed.; John Wiley & Sons: Chichester, UK, 2006; ISBN 9780470021729.
45. Machado, J.S.; Louzada, J.L.; Santos, A.J.A.; Nunes, L.; Anjos, O.; Rodrigues, J.; Simões, R.M.S.; Pereira, H. Variation of wood density and mechanical properties of blackwood (*Acacia melanoxylon* R. Br.). *Mater. Des.* **2014**, *56*, 975–980. [CrossRef]
46. Van Duong, D.; Matsumura, J. Within-stem variations in mechanical properties of Melia azedarach planted in northern Vietnam. *J. Wood Sci.* **2018**, *64*, 329–337. [CrossRef]
47. Ishak, M.R.; Sapuan, S.M.; Leman, Z.; Rahman, M.Z.A.; Anwar, U.M.K. Characterization of sugar palm (Arenga pinnata) fibres: Tensile and thermal properties. *J. Therm. Anal. Calorim.* **2012**, *109*, 981–989. [CrossRef]
48. Ranade, A.; Nayak, K.; Fairbrother, D.; D'Souza, N.A. Maleated and non-maleated polyethylene-montmorillonite layered silicate blown films: Creep, dispersion and crystallinity. *Polymer* **2005**, *46*, 7323–7333. [CrossRef]
49. Harries, K.A. Enhancing the stability of structural steel components using fibre-reinforced polymer (FRP) composites. In *Rehabilitation of Metallic Civil Infrastructure Using Fiber Reinforced Polymer (FRP) Composites: Types Properties and Testing Methods*; Karbhari, V.M., Ed.; Woodhead Publishing: Cambridge, UK, 2014; pp. 117–139, ISBN 9780857096531.
50. Younes, M.F.; Abdel Rahman, M.A. Tensile relaxation behaviour for multi layes fiberglass fabric/epoxy composite. *Eur. J Mater. Sci.* **2016**, *3*, 1–13.
51. Yang, J.L.; Zhang, Z.; Schlarb, A.K.; Friedrich, K. On the characterization of tensile creep resistance of polyamide 66 nanocomposites. Part II: Modeling and prediction of long-term performance. *Polymer* **2006**, *47*, 6745–6758. [CrossRef]
52. Siengchin, S. Dynamic mechanic and creep behaviors of polyoxymethylene/boehmite alumina nanocomposites produced by water-mediated compounding: Effect of particle size. *J. Thermoplast. Compos. Mater.* **2013**, *26*, 863–877. [CrossRef]
53. Siengchin, S.; Karger-Kocsis, J. Structure and creep response of toughened and nanoreinforced polyamides produced via the latex route: Effect of nanofiller type. *Compos. Sci. Technol.* **2009**, *69*, 677–683. [CrossRef]
54. Zhang, Z.; Zhang, W.; Zhai, Z.J.; Chen, Q.Y. Evaluation of various turbulence models in predicting airflow and turbulence in enclosed environments by CFD: Part 2—comparison with experimental data from literature. *HVAC R Res.* **2007**, *13*, 871–886. [CrossRef]

Article

Numerical Analysis-Based Blast Resistance Performance Assessment of Cable-Stayed Bridge Components Subjected to Blast Loads

Jungwhee Lee, Keunki Choi * and Chulhun Chung**

Department of Civil & Environmental Engineering, Dankook University, Gyeonggi-do 16890, Korea;
jwhee2@dankook.ac.kr (J.L.); chchung5@dankook.ac.kr (C.C.)
* Correspondence: ckk0204@dankook.ac.kr; Tel.: +82-31-8005-3493

Received: 19 October 2020; Accepted: 26 November 2020; Published: 28 November 2020

Abstract: Cable-stayed bridges are infrastructure facilities of a highly public nature; therefore, it is essential to ensure operational safety and prompt response in the event of a collapse or damage, which are caused by natural and social disasters. Among social disasters, blast accidents can occur in cable-stayed bridges as a result of explosions produced by vehicle collisions or terrorist attacks; this can lead to the degradation in their structural performances and subsequent collapse. In this research, a procedure to assess structural blast-resistance performance is suggested based on a numerical analysis approach, and the feasibility of the procedure is demonstrated by performing an example assessment. The suggested procedure includes (1) selection of major structural components that severely affect the global structural behavior, (2) set-up blast hazard scenarios consisting of various blast levels and locations, and (3) assessment of the components using numerical blast simulation. By performing an example assessment, the critical blast level for each component could be determined and the blast location that affects the considering components the most severely could be found as well. The scenario-based assessment process employed in this study is expected to facilitate the evaluation of bridge structures under blasts in both existing bridges and future designs.

Keywords: cable-stayed bridge; social disaster; blast scenario; blast analysis; LS-DYNA

---

1. Introduction

Various important infrastructure facilities are currently under construction, both domestically and internationally. The 9/11 terrorist attacks in the United States aroused social awareness of the safety of important infrastructure facilities such as cable-stayed bridges. Furthermore, the preparedness of infrastructure facilities against disasters has emerged as a major issue, and the need for an engineering review of the safety of infrastructure facilities has been recognized; at present, many countries are making continuous budget investments in maintenance and related fields to prepare for such disasters. In particular, disaster-induced accidents on long-span bridges (e.g., cable-stayed bridges) can result in enormous losses of life and significant damage to the economy, because these infrastructure facilities are of a highly public nature. Disasters can be classified into natural disasters (e.g., earthquakes, typhoons, and floods) and social disasters caused by accidents (e.g., fire, collisions, explosions) and terrorist activities. Among the latter category, blast accidents in cable-stayed bridges can occur as a result of explosions caused by vehicle collisions and terrorist attacks; these can lead to significant deterioration in their structural performance, which may cause the bridge to collapse.

Over the last seven decades, more than six hundred terrorist attacks on bridges and other infrastructure facilities have been recorded [1]. In addition, the Seohae Bridge in South Korea was exposed to a blast risk from a large vehicle accident in 2006. In 2007, the Al-Sarafiya Bridge in Bagdad,

Iraq, collapsed due to an attack by a suicide bomber. Furthermore, vehicle blast accidents have occurred on the Riyadh Bridge, Saudi Arabia (2012) and the Sanmenxia Bridge, China (2013). Despite these bridge accidents, no criteria or method has been established to evaluate entire bridge structures prior to construction. Several studies have conducted full-scale experiments to evaluate the performances of piers and decks (structural components of general road bridges) in the event of blasts [2–7]. Based on previous experiments, Williamson et al. [8] suggested three blast design categories, and Williams and Williamson [9] developed a numerical model to describe the spalling of concrete piers using the simulation software package LS-DYNA. Furthermore, several studies have conducted blast analyses for general road bridges using 2D and 3D models [10–13]. However, previous studies had limited capacity to directly or indirectly evaluate the overall performance of cable-stayed bridges subjected to blasts; thus, many researchers have recently begun to numerically simulate blasts in such bridges. Deng and Jin [14] examined stress distributions in a bridge using a 3D cable-stayed bridge model in the simulation software ANSYS AUTODYN; they used a blast pressure derived from 1D analysis, without distinct blast scenario settings. Tang and Hao [15,16] verified the response of a cable-stayed bridge to displacement and stress by numerically analyzing the case of a truck blast occurring 0.5 m from the pier and 1.0 m above the deck. Son and Lee [17] conducted fluid-structure interaction analyses for two types of pylons (hollow steel box and concrete-filled composite pylons) by applying a compressive force to the pylon top. Bojanowski and Balcerzak [18] introduced the procedure and results of a numerical analysis method for evaluating the performance of a complete cable-stayed bridge system under a blast load, and they specified and applied the blast load cases of cars and trucks. Hashemi et al. [19,20] conducted blast analysis for steel cable-stayed bridges, by setting bridge blast scenarios with three load levels (small, medium, and large) and examining the damage and responses of components. Pan et al. [21] defined several scenarios of human-installed explosives and blast accidents caused by trucks, and they analyzed blast load responses for a slab-on-girder, box-girder, and long-span cable-stayed bridges. Farahmand-Tabar et al. [22] evaluated the risks of a progressive collapse and performance degradation in suspension bridges by using a 2D suspension bridge model.

To evaluate the performance of cable-stayed bridges under different blast loads, the aforementioned studies analyzed the overall behaviors and damage characteristics of bridges by modeling either the deck and pylon only or the entire structure [14–22]. Furthermore, when setting the blast loads, they often chose randomized blast loads, rather than setting them according to specific scenarios [14–17,22]. Most of the existing studies have focused on the behaviors of entire bridges, and very few cases have analyzed the damage types or levels of major components (e.g., the cables) according to different scenarios. Therefore, this study conducts blast analysis by defining a detailed finite element model for the major components of the target bridges and setting appropriate blast scenarios, as shown in the procedure in Figure 1. Furthermore, the effects of the blast load are examined by analyzing the component responses and damage types.

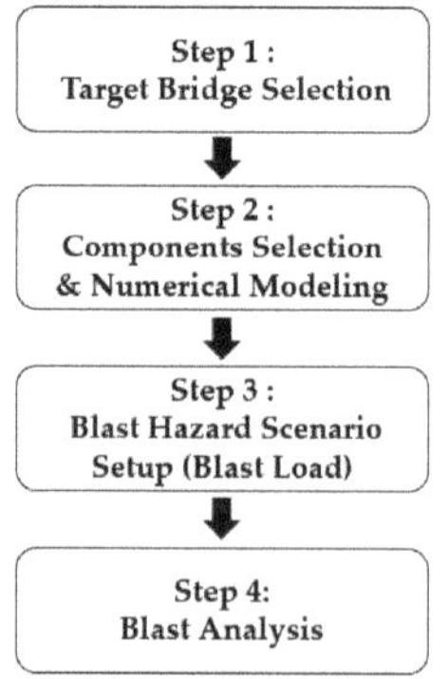

Figure 1. Blast analysis process.

2. Target Structure and Numerical Modeling

2.1. Target Structure

To analyze the performances of the major components of a cable-stayed bridge under a blast load, we selected the target bridge as shown in Figure 2; it is a cable-stayed bridge featuring three spans, with a main span length of 470 m and a side span length of 200 m. It is a composite superstructure formed of precast reinforced concrete (RC) slabs and steel girders (edge girder, center stringer, and crossbeam). The deck is 310 mm thick near the pylon and 260 mm thick in other sections, and it is made of 68 standard segments (length (L) = 12.3 m). In addition, the bridge features 144 cables of varying areas and lengths and two RC pylons (PY1 and PY2), which contain hollow sections. To evaluate each component of the cable-stayed bridge, a single 260-mm thick segment at the center of the span was selected from the superstructure. We selected the following cables for analysis: Cable no. 37 (area (A) = 0.012450 m^2, L = 252 m), which is the longest cable, has the largest cross-sectional area, and extends to the mid-point of the superstructure; Cable no. 53 (A = 0.005185 m^2, L = 81 m), which has the smallest cross-sectional area; and Cable no. 68 (A = 0.009385 m^2, L = 206 m), which is attached to the side span, where the largest stress is applied. We selected PY2 as the pylon; its base segment (CB1 in the figure) is 2.37 m taller than that of PY1.

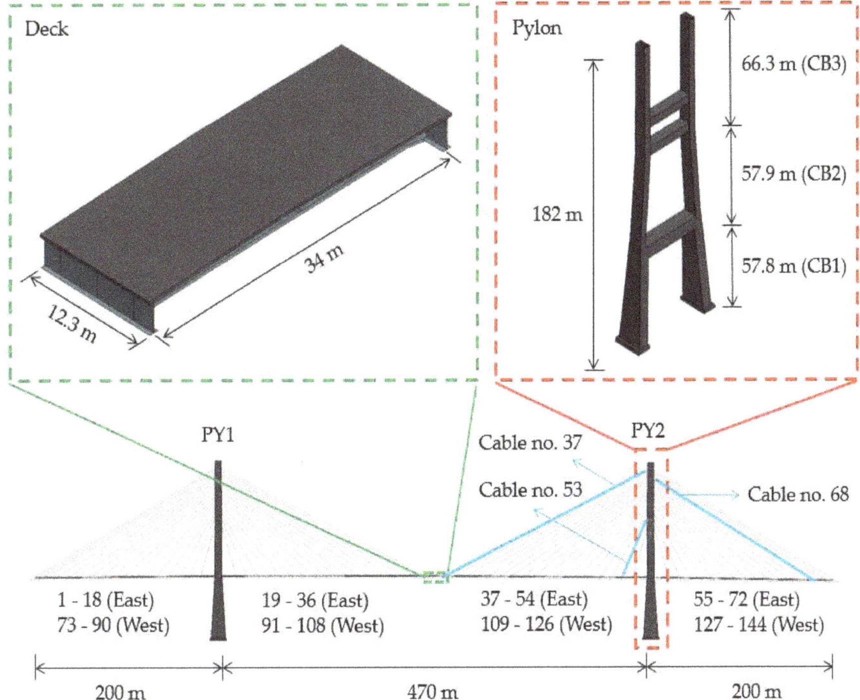

Figure 2. Target bridge components.

Tables 1 and 2 summarize the specifications of the steel girders and rebars comprising each component; these values were taken from [23].

Table 1. Steel girder details.

	Part	Thickness (mm)
	Upper flange	50
Edge Girder	Lower flange	50
	Web	22
	Upper flange	30
Center Stringer	Lower flange	20
	Web	12
	Upper flange	30
Crossbeam	Lower flange	30
	Web	14

Table 2. Rebar details.

Part	Vertical Rebar Diameter and Spacing (mm)	Lateral Rebar Diameter and Spacing (mm)
Deck	H19@150	H16@150
Pylon (CB1)	H35@125	H22@100, H19@250
Pylon (CB2, CB3)	H29@125	H22@100

2.2. Finite Element Model Development

Finite element modeling was performed using the general-purpose finite element program LS-DYNA [24]. For the deck model, 8-node solid elements, 4-node shell elements, and 2-node beam elements are used for the concrete slab, steel girder, and reinforcing bars, respectively. The deck model consists of 288,332 nodes and 198,550 elements, and the edge lengths range from 0.05 to 0.15 m. The concrete slab of the model is divided into 5 layers of elements. Stay cables are modeled as solid round bar shapes with 8-node solid elements. The numerical models of the cable #37, 53, and 68 consist of 87,362, 39,668, 73,205 nodes and 77,686, 35,316, 65,232 elements, respectively. The overall edge length ranges from 0.10 to 0.50 m, and the finer mesh with 0.02 to 0.05 m of edge length is generated at the region of the application of the blast load. In the pylon model, there is a concrete part and a reinforcing bar part. The concrete part and the reinforcing bar parts consist of 8-node solid elements and 2-node beam elements, respectively, and the pylon model has 1,185,084 nodes and 917,438 elements. The section submitted to the blast load was modeled with smaller element sizes of 0.05 to 0.20 m. The remaining sections were configured using element sizes of approximately 0.5 m. For the rebars, the specifications in Table 2 were applied. The composite conditions for the concrete and rebars were implemented using the *CONSTRAINED_LAGRANGE_IN_SOLID command. The contact condition between the steel girders was applied considering the case of a large, load-induced deformation, using the *AUTOMATIC_SINGLE_SURFACE command. Figures 3–5 illustrate the finite element models of the superstructure, cables, and pylon.

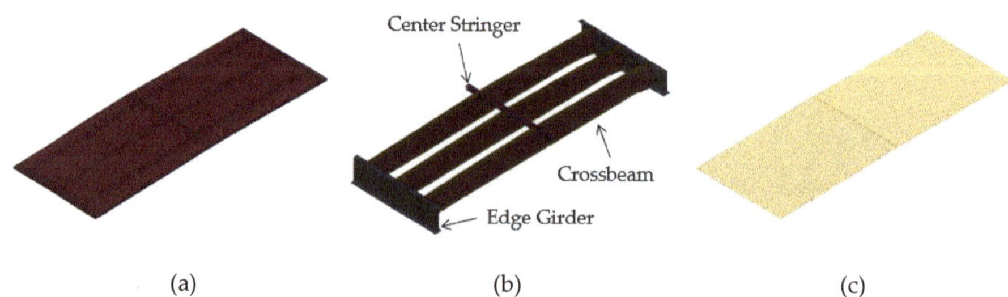

Figure 3. Deck model: (**a**) slab, (**b**) steel girder, and (**c**) rebar.

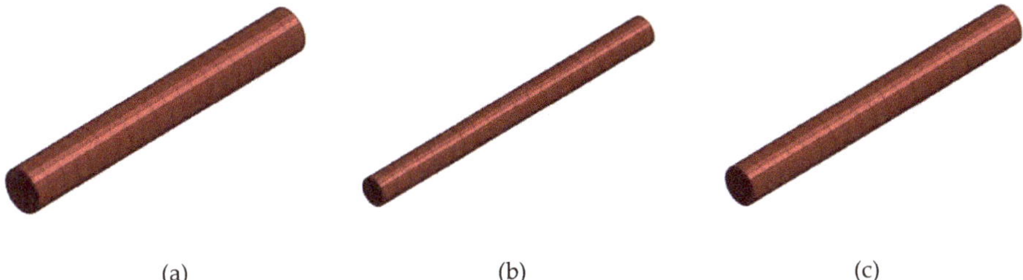

(a)   (b)   (c)

Figure 4. Cable model: (**a**) Cable no. 37, (**b**) Cable no. 53, and (**c**) Cable no. 68.

(**a**)   (**b**)

Figure 5. Pylon model: (**a**) concrete part and (**b**) rebar.

2.3. Material Model

When selecting a material model for blast analysis, various material properties—in particular, the strength increasing effect due to the strain rate—must be considered. Therefore, we selected the following material model from LS-DYNA:

For the concrete, the *MAT_CSCM_CONCRETE material model was applied; this captures the nonlinear material behaviors of concrete, as well as its stiffness degradation arising from damage, erosion, and the strain rate effect [25,26]. A compressive strength (f_{ck}) of concrete was selected as 39.23 MPa, and 1.05 was used as the ERODE parameter, to simulate erosion of the concrete element in the blast analysis.

To model the steel plates in the girders, rebars, and cables, the *MAT_PLASTIC_KINEMATIC material model was applied. This model considers the strain rate effect, isotropy, and kinematic hardening of the material, and it assumes a bi-linear material nonlinearity. The material constants C and p (used in the Cowper–Symonds model, which describes the strain rate effect) were determined using the following equation, developed for high-strength steel [27]. Usually, the C value of 40 is applied to normal structural steel, but this value is known to overestimate the strain rate effect of high-strength steel. Therefore, the experimentally validated Equation (1) is used to calculate the C value in order to prevent overestimation of the strain rate effect.

$$C = \begin{cases} 92000 \cdot \exp\left(\frac{\sigma_0}{364}\right) - 194000, & \sigma_0 > 270 \text{ MPa}, \\ 40, & \sigma_0 \leq 270 \text{ MPa}, \end{cases} \quad p = 5. \tag{1}$$

where, σ_0 is the yield stress (MPa). In addition, the strengths selected for the steel plates in the girders were varied according to the thickness. The input parameters used for each material model are listed in Table 3.

Table 3. Material model input parameters.

Part	Category	Input Parameter
Concrete	Deck, Pylon	f_{ck} = 39.23 MPa, E = 33.107 GPa, ρ = 2,300 kg/m^3, ERODE = 1.05
Steel plate	Girder (Thickness 0–16 mm)	f_y = 365 MPa, f_u = 520 MPa, E = 210 GPa, ρ = 7850 kg/m^3, ε_{fail} = 0.15, C = 5.67 × 10^4 s^{-1}, p = 5
	Girder (Thickness 17–40 mm)	f_y = 355 MPa, f_u = 520 MPa, E = 210 GPa, ρ = 7850 kg/m^3, ε_{fail} = 0.19, C = 4.99 × 10^4 s^{-1}, p = 5
	Girder (Thickness 41–75 mm)	f_y = 335 MPa, f_u = 520 MPa, E = 210 GPa, ρ = 7850 kg/m^3, ε_{fail} = 0.21, C = 3.69 × 10^4 s^{-1}, p = 5
Rebar	Deck, Pylon	f_y = 400 MPa, f_u = 560 MPa, E = 200 GPa, ρ = 7850 kg/m^3, ε_{fail} = 0.16, C = 8.21 × 10^4 s^{-1}, p = 5
Cable (BS-5896 Super Grade)	Stay Cable	f_y = 1569 MPa, f_u = 1765 MPa, E = 200 GPa, ρ = 7850 kg/m^3, ε_{fail} = 0.035, C = 6.65 × 10^6 s^{-1}, p = 5

2.4. Initial Load and Boundary Conditions

To assess the structural performance of each major bridge component such as pylon, deck, and cables, initial compressive forces and tensile forces of the components should be considered. During the application of tension to the stay cables, compressive forces result in the pylon and deck while tensile forces are generated in the cables as shown in Figure 6a. The distribution of compressive forces along the bridge's length is depicted in Figure 6b. Since the compressive forces which are transferred from each stay cable are accumulated, compression of the deck section gradually increases as it approaches the pylon. In another point of view, a very small amount of compression results on the deck section of the center of the mid-span. The deck segment of the current numerical analysis is located at the center of the mid-span, and the design compression of the section is approximately 1.8% of the maximum compression of the deck, according to the design documents [23,28]. Therefore, the compressive force of the considered segment is regarded as negligible, and only the self-weight of the segment is considered as the initial load.

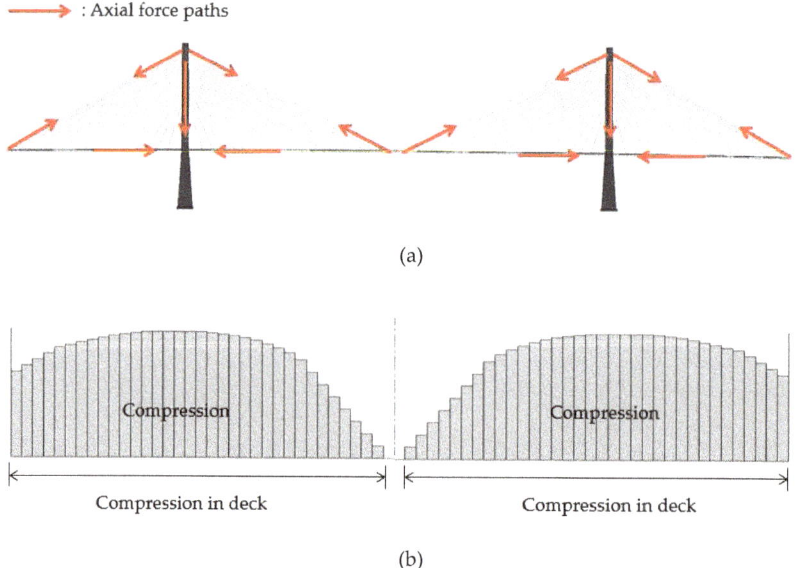

Figure 6. Axial force in cable-stayed bridge: (**a**) axial force paths and (**b**) axial force in deck.

Initial loads and boundary conditions of each numerical model of the bridge components are shown in Figure 7. As previously mentioned, only self-weight is considered as the initial load for the deck model. As a boundary condition of the deck model, the longitudinal symmetry condition is considered, and the z-directional translation is constrained at the location of cable anchorage.

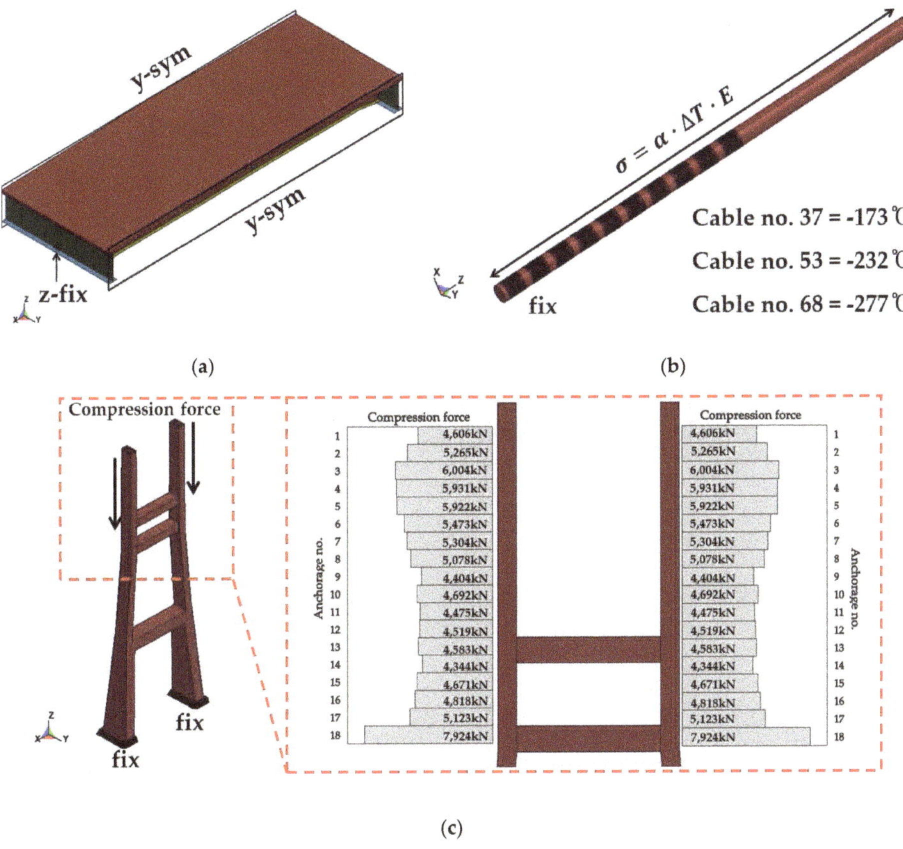

Figure 7. Initial load and boundary conditions: (**a**) deck, (**b**) cable, and (**c**) pylon.

For the cable models, translational DOFs of surface nodes at each end region are constrained considering the anchored length. Thermal load is utilized to apply the initial stresses of each cable model of nos. 37, 53, and 68, which are 415, 556, and 665 MPa, respectively. For thermal load analysis, the temperature change is calculated using the thermal stress Equation (2) with the general thermal expansion coefficient of steel, $1.2 \times 10^{-5}/°C$. The calculated temperature drops are -173, -232, and $-277 °C$ for Cable nos. 37, 53, and 68, respectively.

$$\sigma = \alpha \cdot \Delta T \cdot E \tag{2}$$

where, σ_T is thermal stress (MPa), α is coefficient of thermal expansion (/°C), ΔT is temperature change (°C), and E is elastic modulus (MPa). The LS-DYNA commands, *LOAD_THERMAL_VARIABLE and *MAT_ADD_THERMAL_EXPANSION, are used for thermal analysis.

For the pylon model, a fixed boundary condition is applied on the bottom surface of each pylon leg. Also, the compressive forces transferred from the stay cables are applied on the nodes at cable anchorages as initial loads. The amounts of compressive forces are shown in Figure 7c, and the total compression is approximately 186,000 kN. Self-weight is also considered as initial load, and the LS-DYNA command, *CONTROL_DYNAMIC_RELAXATION, is used for initial load analysis.

3. Blast Scenario and Load Cases

3.1. Blast Scenario

Selecting an appropriate scenario from among the various possible blast events and setting the size of the blast load are essential to the performance evaluation and risk analysis. Blasts that occur on bridges are unpredictable and—on both domestic and overseas bridges—have been caused by explosions from terrorist attacks, vehicle accidents, and other events (e.g., wars). When conducting blast analysis for bridges, most previous researchers have used random blast loads, rather than setting separate blast or vehicle explosion scenarios [14–17,22]. This study defined blast accident scenarios in terms of vehicle-based terrorist attacks, considering the current likelihood of war-related blast accidents to be low. A virtual scenario describing a terrorist attack conducted through explosive-laden vehicles was assumed. Thus, the quantity of explosive (TNT) and height of the explosion above the road were set according to the maximum load and ground clearance of general-purpose vehicles driven in South Korea. A TNT load of 1000–1500 kg was considered, which is classified as a very large load considering the explosive definitions introduced in [29] or the quantities of explosive used in the American Society of Mechanical Engineers-American Concrete Institute explosion tests [30]. Table 4 presents the blast scenario specifications employed in this study.

Table 4. Blast scenarios.

Blast Scenario.	Vehicle Type	Ground Clearance (mm)	Maximum Load (kg)	TNT Load (kg)
Scenario A	Passenger car	700	226.8	226.8
Scenario B	Truck	800	1500	500
Scenario C	Truck	800	1500	1000
Scenario D	Truck	800	1500	1500

3.2. Blast Load Cases

The blast load cases defined in this study are illustrated in Figure 8; they were set by considering the positions of the target components and lanes. For the superstructure case, it was assumed that the vehicle explodes in the second lane, either between crossbeams or immediately above one. For the pylon and cable case, explosions were assumed to occur on the shoulder, where the vehicle is closest to the target component. In blast load modeling, the load was applied after calculating the explosion pressure using the scaled distance, which was calculated from the TNT quantity and the distance from the explosion location to the target structure. This method allows the load of CONWEP to be applied indirectly using the *LOAD_BLAST_ENHANCED and *LOAD_SEGMENT_SET commands provided in LS-DYNA. Therefore, we applied the blast load using these commands. Because the blast load is reflected from the deck at the moment of detonation, the shock wave was assumed to propagate hemispherically. Table 5 summarizes the blast load cases in terms of the target components and scaled distances.

Table 5. Blast load cases.

Blast Scenario	Component	TNT (kg)	Z (Scaled Distance, m/kg$^{1/3}$)
Scenario A		226.8	0.115
Scenario B	Deck	500	0.101
Scenario C		1000	0.080
Scenario D		1500	0.070

Table 5. *Cont.*

Blast Scenario	Component	TNT (kg)	Z (Scaled Distance, m/kg$^{1/3}$)
Scenario A		226.8	0.525
Scenario B	Cable	500	0.403
Scenario C		1000	0.320
Scenario D		1500	0.279
Scenario A		226.8	1.181
Scenario B	Pylon	500	0.907
Scenario C		1000	0.720
Scenario D		1500	0.629

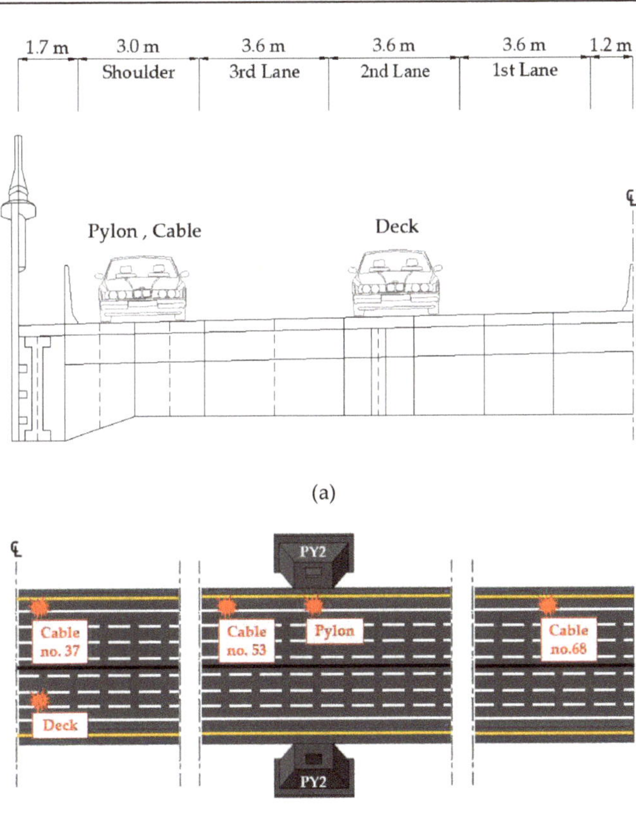

Figure 8. Blast load position: (**a**) cross-sectional view and (**b**) plan view.

4. Blast Analysis Results

Numerical blast simulations were performed to evaluate the explosion resistance performances of major components of the cable-stayed bridge. The stress, displacements, and damage levels of the components were examined. The numerical analysis results show that the response sizes and damage ranges of all components increase with the blast load. The blast analysis results of the major components are described below for different scenarios. In addition, Table 6 summarizes the time step and CPU time utilized for the analysis of each numerical model. The numerical analysis of this study was all performed with a single CPU.

Table 6. Summary of utilized time step size and CPU time.

Model	Number of Elements	Time Step Size (10^{-6} s)	CPU Time (s)
Deck	198,550	4.826–4.943	49,670–50,568
Cable no. 37	77,868	0.861–0.868	54,115–56,275
Cable no. 53	35,316	0.545–0.554	37,015–40,345
Cable no. 68	65,232	0.735–0.745	52,948–54,090
Pylon	917,438	7.257–11.092	87,009–95,547

4.1. Deck

Table 7 summarizes the damage types and maximum stresses of the steel girder in each scenario, for a blast load applied between and above the crossbeams of the superstructure. The slab's displacement and response to rebar stress were not examined, because the analysis results included the occurrence of perforation. In blast analysis between the superstructure crossbeams, the damage range of the slab was found to increase with the blast load; however, the main damage occurred only between the crossbeams. In the damage assessments for each scenario, scabbing of the slab was observed in Scenario A, and slab perforations were observed in Scenarios B–D. When the effective stress and degree of damage of the steel girder were examined, the effective stress was found to increase with the blast load. Furthermore, the maximum stresses in Scenarios A–D were 434, 452, 479, and 509 MPa, respectively, and deformation occurred in the upper flange of the crossbeam. The ultimate strength (f_u) of the steel girder was 520 MPa; hence, it was concluded that the steel girder plastically deforms but does not fracture when a blast load is applied between the crossbeams of the superstructure.

Table 7. Summary of blast analysis results of the deck.

Blast Scenario	Blast Position	Damage Type of Slab	Steel Girder	
			Maximum Stress	Damage Type
Scenario A	2nd lane/ Between crossbeams	Scabbing	434 MPa	Plastic deformation
Scenario B		Perforation	452 MPa	
Scenario C			479 MPa	
Scenario D			509 MPa	
Scenario A	2nd lane/ Above crossbeam	Scabbing	602 MPa	Plastic deformation
Scenario B		Perforation	-	Ruptured
Scenario C			-	
Scenario D			-	

In the analysis of a blast above the superstructure crossbeam, the damage range of the slab was found to increase with the blast load and was wider than that seen for the blast between the crossbeams. The damage types of each scenario were identical to those of the previous case. When the effective stress of the steel girder was examined, a stress of 602 MPa (exceeding the ultimate strength (f_u) of the steel girder (520 MPa)) occurred momentarily in the upper flange and web of the crossbeam in Scenario A; however, it appears that no damage occurred owing to the strain rate effect of the material. In Scenarios B–D, the crossbeam was damaged, and the damage range increased with the blast load. This suggests that in the case of the superstructure, a blast load applied to the crossbeam causes greater damage (and has a larger effect on the structural behavior of the cable-stayed bridge) than the blast load applied between the crossbeams.

Figures 9–12 show the damage and effective stress contours generated by the blast load. The fringe range of the effective stress contour was set to 0–520 MPa, considering the ultimate strength of the steel girder; the resulting output time was 0.05 s, at which point the analysis was terminated.

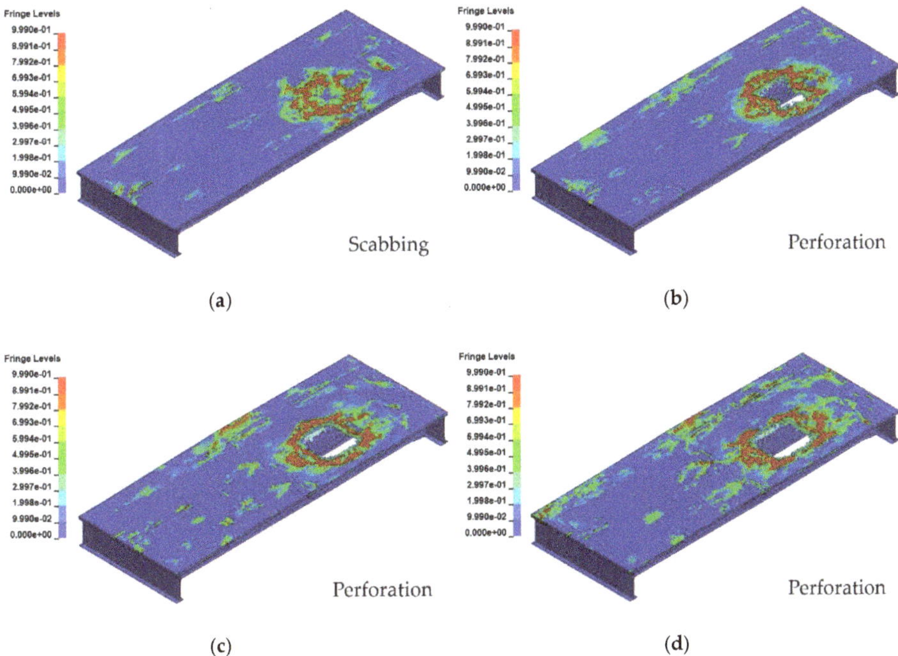

Figure 9. Damage contours of the deck (between the crossbeams): (**a**) Scenario A, (**b**) Scenario B, (**c**) Scenario C, and (**d**) Scenario D.

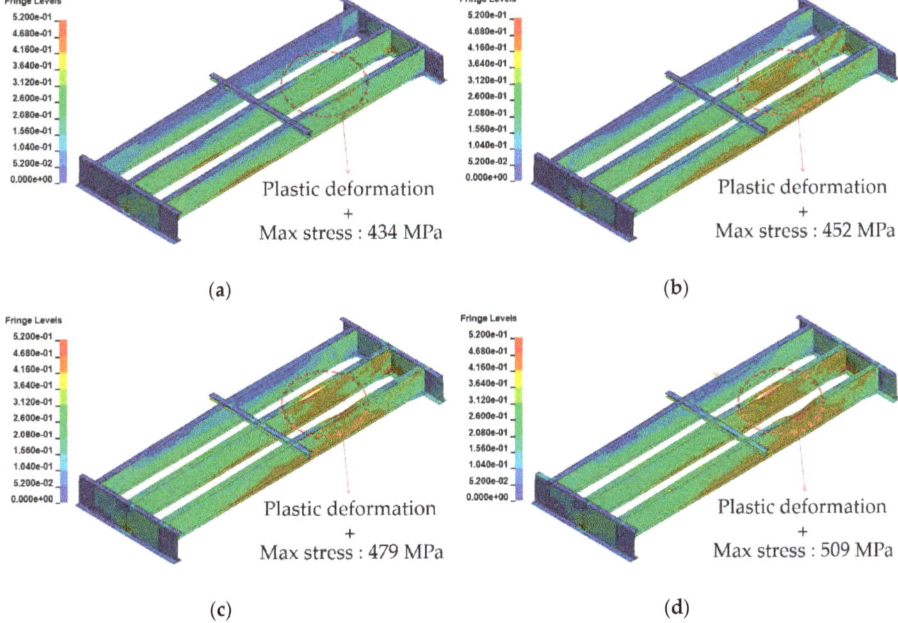

Figure 10. Effective stress contours of the steel girder (between the crossbeams): (**a**) Scenario A, (**b**) Scenario B, (**c**) Scenario C, and (**d**) Scenario D (fringe level unit: MPa).

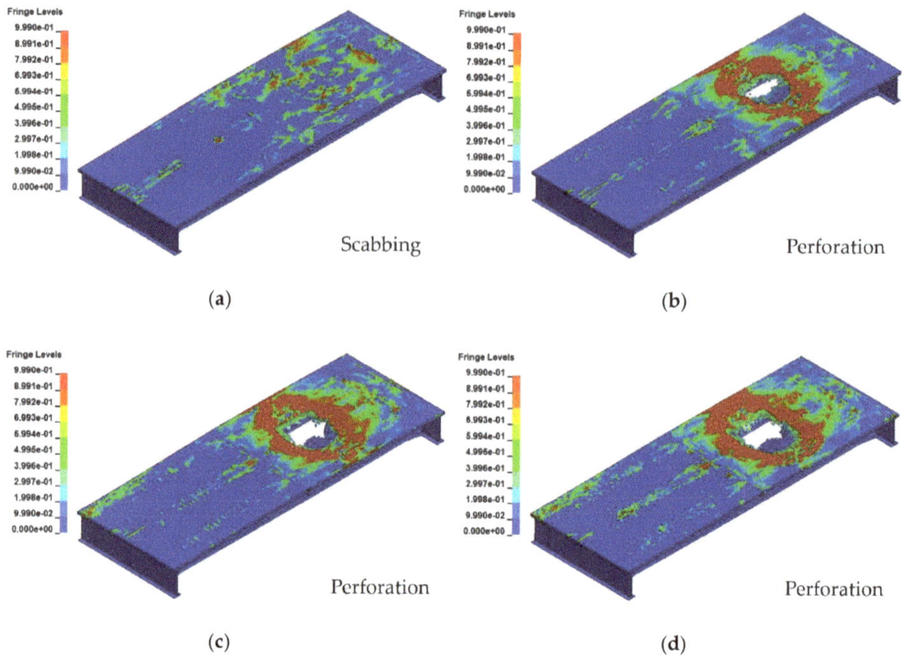

Figure 11. Damage contours of the deck (above the crossbeam): (**a**) Scenario A, (**b**) Scenario B, (**c**) Scenario C, and (**d**) Scenario D.

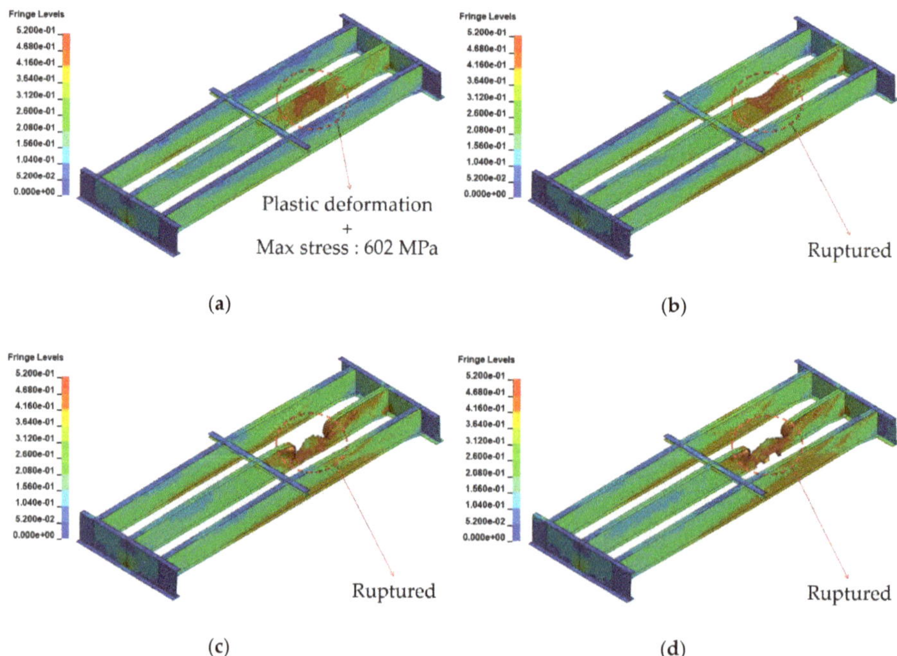

Figure 12. Effective stress contours of the steel girder (above the crossbeam): (**a**) Scenario A, (**b**) Scenario B, (**c**) Scenario C, and (**d**) Scenario D (fringe level unit: MPa).

4.2. Cable

Table 8 presents the maximum displacement and maximum effective stress of each scenario when a blast load was applied to each cable selected in this study (see Figure 2). The displacements and effective stresses on the cables increased with the blast load. In scenario D of Cable no. 53, a momentary stress was found to exceed the static tensile strength (f_{pu}) of the cable (1765 MPa); however, no damage or breakage was observed, owing to the strain rate effect of the material. Furthermore, when the initial stress was excluded, the stresses resulting from the blast load only were found to be approximately 531–1292 MPa for Cable no. 37, 824–1258 MPa for Cable no. 53 and 646–1095 MPa for Cable no. 68 depending on the scenario. Cable no. 37 is observed to have the largest stress variation compared to the initial stress of the design. The reason for this is that Cable no. 37 has a maximum diameter, or maximum blast load reception area. In addition, the maximum displacement and the maximum effective stress occurred on Cable no. 53, which has a minimum cable diameter. These results suggest that the diameter of the cable has a greater influence on the behavior of stay-cables under blast load, or a cable with a small diameter is more vulnerable to blast load. Figures 13 and 14 show the displacement and effective stress contours resulting from the blast load. The fringe range of the displacement contour was set as 0–400 mm, and that of the effective stress contour was set to 0–1765 MPa, considering the tensile strength of the cable. The result output time was 0.18 s and 0.05 s, at which point the maximum displacement and effective stress have occurred.

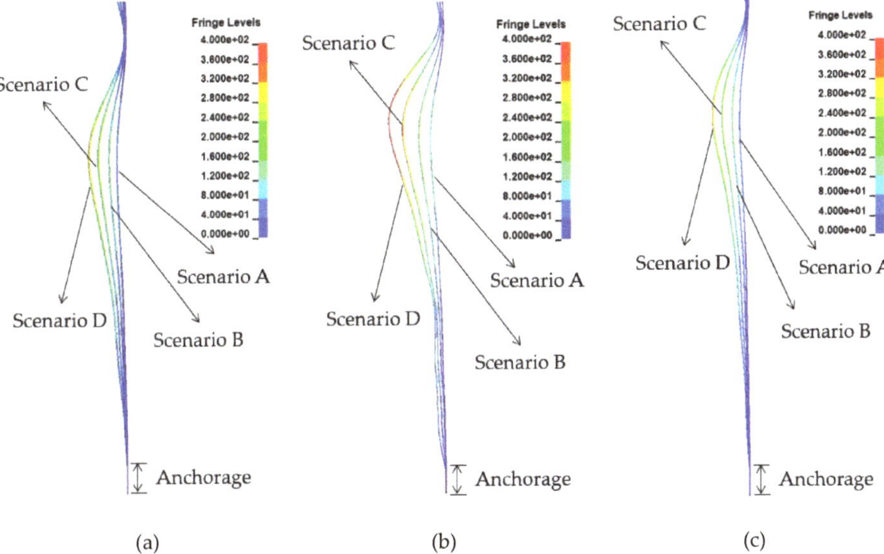

Figure 13. Displacement contours of cables: (**a**) Cable no. 37, (**b**) Cable no. 53, and (**c**) Cable no. 68 (fringe level unit: mm).

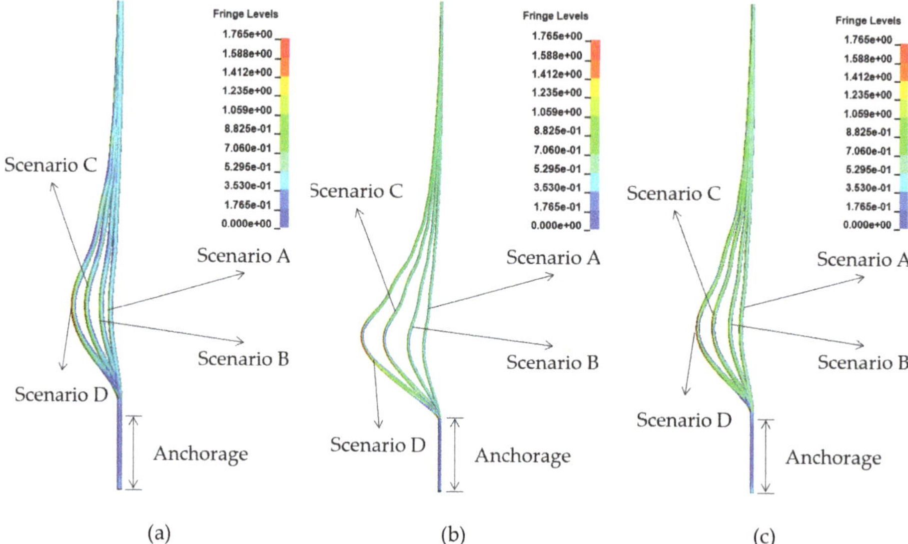

Figure 14. Effective stress contours of cables: (**a**) Cable no. 37, (**b**) Cable no. 53, and (**c**) Cable no. 68 (fringe level unit: GPa).

Table 8. Summary of blast analysis results for the cables.

Blast Scenario	Blast Position	Maximum Displacement	Maximum Stress
Scenario A		74.65 mm	946 MPa
Scenario B	Shoulder/	133.81 mm	1395 MPa
Scenario C	Cable no. 37	216.94 mm	1663 MPa
Scenario D		289.89 mm	1707 MPa
Scenario A		111.33 mm	1380 MPa
Scenario B	Shoulder/	199.62 mm	1656 MPa
Scenario C	Cable no. 53	321.73 mm	1730 MPa
Scenario D		428.15 mm	1814 MPa
Scenario A		74.94 mm	1311 MPa
Scenario B	Shoulder/	134.96 mm	1641 MPa
Scenario C	Cable no. 68	220.02 mm	1703 MPa
Scenario D		294.46 mm	1760 MPa

4.3. Pylon

Table 9 presents the damage type, rebar stress, and transverse displacement of the pylon under the different blast load scenarios. The damage range of the pylon column was found to increase with the blast load, and the uppermost crossbeam of the pylon was also considerably damaged by the continuous compression. In Scenarios A and B, the concrete element was not eroded, though damage was observed. In Scenario C, partial spalling of the pylon column occurred near the blast load application point and on the underside of the pylon crossbeam. In Scenario D, concrete failure occurred in the portion of the pylon column level with the explosion; at that location the rebar was exposed. The rebar stress was examined based on a yield strength (f_y) of 400 MPa, and the rebars in which the stress exceeded the yield strength are marked in red. In Scenarios A and B, very few rebars yielded at the position of the largest blast load. In Scenario C, numerous rebars began to yield and the damage range was increased. The rebars were fractured in Scenario D. It is believed that the rebar in the uppermost crossbeam of the pylon received a relatively small impact from the blast load; however, it yielded owing to the continuous compression of the pylon. The maximum transverse displacements of the pylon are as follows: in Scenario A, local displacements of 9.2 and 18.7 mm occurred at the top of the pylon and at the height of the deck surface, respectively; in Scenarios B and C, local displacements of 19.1 and 59.1 mm at the pylon top and 32.8 and 78.6 mm at the height of the deck surface were observed, respectively; and in Scenario D, displacements of 636.6 and 649.7 mm occurred at the pylon top and the height of the deck surface, respectively, along with a partial fracture of the pylon. Thus, the blast affected the local behavior of the pylon in Scenarios A–C; however, in Scenario D, it affected the overall behavior of the pylon and was even predicted to result in its collapse. Figures 15–17 show the damage contours, axial stress contours, and displacement contours caused by the blast load. The measurement range of the axial stress contours was set to 0–400 MPa considering the yield strength of the rebar, and the measurement range of the displacement contours was set separately, considering the displacement for each scenario. The resulting output time was 1 s, at which point the analysis was terminated.

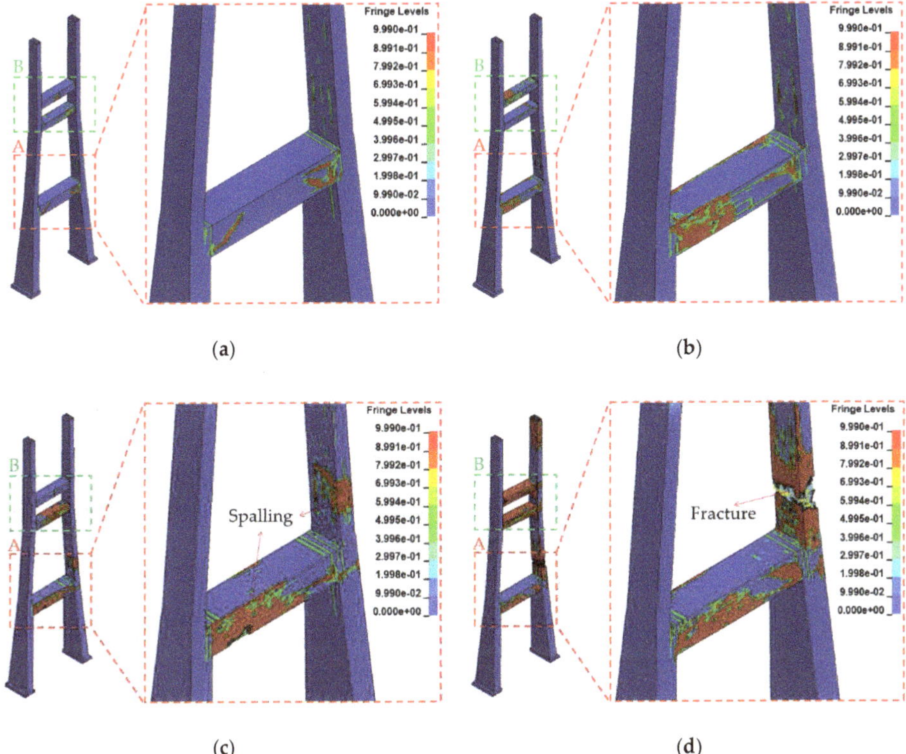

(a)

(b)

(c)

(d)

Figure 15. Damage contours of the pylon: (**a**) Scenario A, (**b**) Scenario B, (**c**) Scenario C, and (**d**) Scenario D.

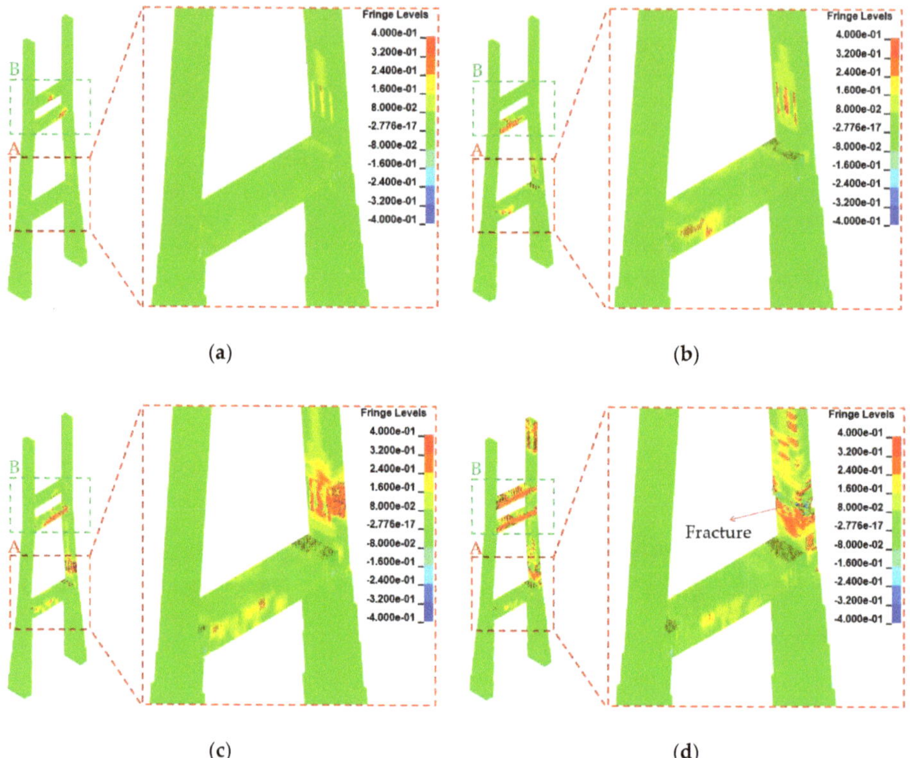

Figure 16. Axial stress contours of the pylon rebars: (**a**) Scenario A, (**b**) Scenario B, (**c**) Scenario C, and (**d**) Scenario D (fringe level unit: GPa).

Table 9. Summary of blast analysis results for the pylon.

Blast Scenario	Blast Position	Damage Type			Maximum Transverse Displacement	
		Concrete	Rebar		Upper Part	Road Surface
			A Zone	B Zone		
Scenario A		-	Yield	Yield	9.2 mm (←)	18.7 mm (→)
Scenario B	Shoulder	-	Yield	Yield	19.1 mm (←)	32.8 mm (→)
Scenario C		Spalling	Yield	Yield	59.1 mm (←)	78.6 mm (→)
Scenario D		Fracture	Fracture	Yield	636.6 mm (→)	649.7 mm (←)

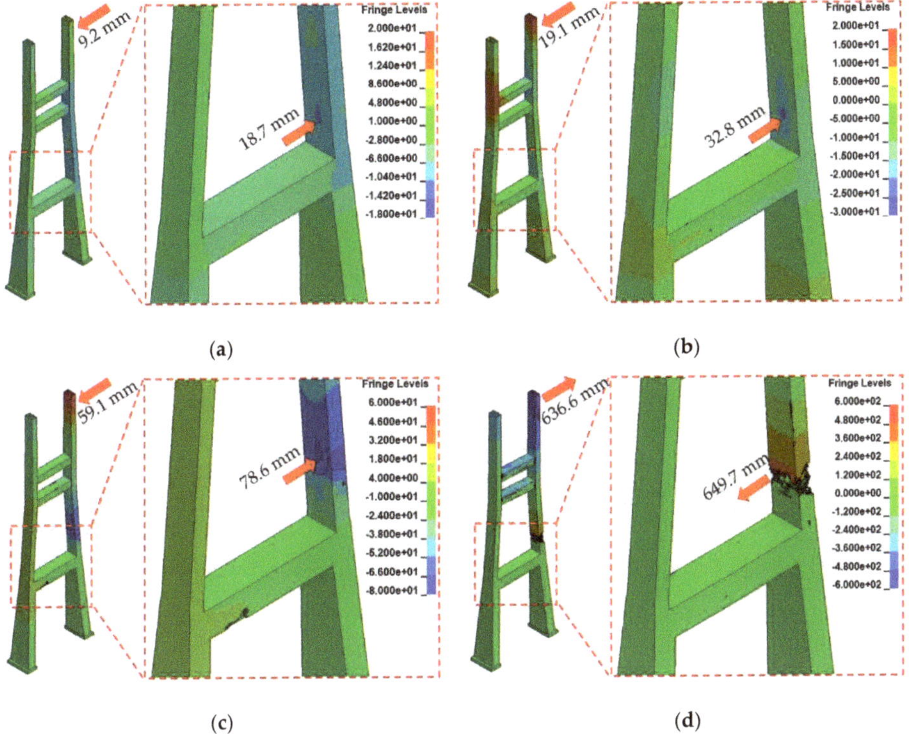

Figure 17. Displacement contours of the pylon: (**a**) Scenario A, (**b**) Scenario B, (**c**) Scenario C, and (**d**) Scenario D (fringe level unit: mm).

5. Conclusions

The performances of key components in a cable-stayed bridge (deck, cable, and pylon) were evaluated via numerical analysis, and all responses for the components and damage type were examined for a range of strategically targeted blast events. The main findings of this study are as follows:

- The study introduced an appropriate blast scenario setting method to facilitate the performance evaluations of the main components of a cable-stayed bridge under a blast load. Furthermore, a method for modeling these components was proposed, which was sensitive to the structural characteristics of such bridges.
- A superstructure blast analysis was conducted in each scenario according to the blast position, and the results were examined. The results showed that the damage range of the components increased with the blast load. The damage types observed included the scabbing and perforation of the slab, as well as the deformation and fracture of the steel girder. Furthermore, the superstructure was more vulnerable to a blast above the crossbeams than between them.
- A blast analysis was conducted in each scenario for cables of different diameters, lengths, and design stresses. The results showed that the maximum displacement and effective stress in each cable increased with the blast load, and no cable fractures occurred in all scenarios. The increase in effective stress (relative to the initial stress of the cable) generated by the blast load was approximately 531–1292 MPa for Cable no. 37 (which had the maximum diameter), 824–1258 MPa for Cable no. 53 (which had the minimum diameter) and 646–1095 MPa for Cable no. 68 (which had the maximum design stress). As such, the maximum stress variation was

observed on Cable no. 37. However, the maximum displacement and effective stress occurred on Cable no. 53, which had a minimum cable diameter. Therefore, in the case of cables, a blast near the cable with a small diameter tends to have a large impact on the behavior of the cable-stayed bridge.

- For evaluating pylon characteristics, the case wherein a blast occurred on the hard shoulder was examined. The results show that the damage range of the pylon increased with the blast load, and the concrete and rebars were fractured in some scenarios. The damage and rebar yield occurred primarily in the pylon column and the upper cross beams, wherein the applied blast load was largest. The significant damage in the upper crossbeams was considered to be due to the continuous compression of the pylon. Furthermore, the maximum transverse displacement caused by the blast load was examined, and it was found to have an impact on the local behavior of the pylon for Scenarios A–C; however, for Scenario D, it was found to affect the overall behavior of the pylon, with displacements of 636.6 and 649.7 mm being observed at the top of the pylon and at the deck height in the column, respectively. Therefore, the pylon may be severely damaged under Scenario D.

These results indicate that the blast analysis method introduced in this study will be useful for evaluating the blast load performances of the individual components of cable-stayed bridges. If a performance evaluation of the entire bridge system can be conducted alongside this method, it will become possible to conduct a detailed review of the structural performances of bridges during operation.

Author Contributions: Conceptualization, methodology, writing—review and editing, and supervision, J.L.; methodology, formal analysis, investigation, and writing—original draft preparation, K.C.; conceptualization and supervision, C.C. All authors have read and agreed to the published version of the manuscript.

Funding: This work was supported by a Korea Agency for Infrastructure Technology Advancement (KAIA) grant, funded by the Ministry of Land, Infrastructure, and Transport (Grant No.20SCIP-B119963-05); it was also supported by the Basic Science Research Program through the National Research Foundation of Korea (NRF-2018R1A6A1A07025819).

Conflicts of Interest: The authors declare no conflict of interest.

References

1. Jenkins, B.M. *Protecting Surface Transportation System and Patrons from Terrorist Activities*; Mineta Transportation Institute, San José State University: San Jose, CA, USA, 1997.
2. Ray, J.C. Validation of numerical modeling and analysis of steel bridge towers subjected to blast loadings. In Proceedings of the Structures Congress: Structural Engineering and Public Safety, St. Louis, MO, USA, 18–21 May 2006.
3. Fujikura, S.; Bruneau, M.; Lopez-Garcia, D. Experimental Investigation of multihazard resistant bridge piers having concrete-filled steel tube under blast loading. *J. Bridge Eng.* **2008**, *13*, 586–594. [CrossRef]
4. Fujikura, S.; Bruneau, M. Experimental Investigation of seismically resistant bridge piers under blast loading. *J. Bridge Eng.* **2011**, *16*, 63–71. [CrossRef]
5. Williams, E.B.; Bayrak, O.; Davis, C.; Williams, G.D. Performance of bridge columns subjected to blast loads. I: Experimental program. *J. Bridge Eng.* **2011**, *16*, 693–702. [CrossRef]
6. Foglar, M.; Kovar, M. Conclusions from experimental testing of blast resistance of FRC and RC bridge decks. *Int. J. Impact Eng.* **2013**, *59*, 18–28. [CrossRef]
7. Hajek, R.; Fladr, J.; Pachman, J.; Stoller, J.; Foglar, M. An experimental evaluation of the blast resistance of heterogeneous concrete-based composite bridge decks. *Eng. Struct.* **2019**, *179*, 204–210. [CrossRef]
8. Williams, E.B.; Bayrak, O.; Davis, C.; Williams, G.D. Performance of bridge columns subjected to blast loads. II: Results and recommendations. *J. Bridge Eng.* **2011**, *16*, 703–710. [CrossRef]
9. Williams, G.D.; Williams, E.B. Response of reinforced concrete bridge columns subjected to blast loads. *J. Struct. Eng.* **2011**, *137*, 903–913. [CrossRef]
10. Winget, D.G.; Marchand, K.A.; Williamson, E.B. Analysis and design of critical bridges subjected to blast loads. *J. Struct. Eng.* **2005**, *131*, 1243–1255. [CrossRef]
11. Anwarul Islam, A.K.M.; Yazdani, N. Performance of AASHTO girder bridges under blast loading. *Eng. Struct.* **2008**, *30*, 1922–1937. [CrossRef]

12. Pan, Y.; Chan, B.Y.B.; Cheung, M.M.S. Blast loading effects on an RC slab-on-girder bridge superstructure using the multi-euler domain method. *J. Bridge Eng.* **2013**, *18*, 1152–1163. [CrossRef]

13. Andreou, M.; Kotsoglou, A.; Pantazopoulou, S. Modeling blast effects on a reinforced concrete bridge. *Adv. Civ. Eng.* **2016**, *4167329*, 1–11.

14. Deng, R.-B.; Jin, X.-L. Numerical simulation of bridge damage under blast loads. *WSEAS Trans. Comput.* **2009**, *8*, 1564–1574.

15. Tang, E.K.C.; Hao, H. Numerical simulation of a cable-stayed bridge response to blast loads, part I: Model development and response calculations. *Eng. Struct.* **2010**, *32*, 3180–3192. [CrossRef]

16. Hao, H.; Tang, E.K.C. Numerical simulation of a cable-stayed bridge response to blast loads, part II: Damage prediction and FRP strengthening. *Eng. Struct.* **2010**, *32*, 3193–3205. [CrossRef]

17. Son, J.; Lee, H.-J. Performance of cable-stayed bridge pylons subjected to blast loading. *Eng. Struct.* **2011**, *33*, 1133–1148. [CrossRef]

18. Bojanowski, C.; Balcerzak, M. Response of a large span stay cable bridge to blast loading. In Proceedings of the 13th International LS-DYNA Users Conference, Dearborn, MI, USA, 8–10 June 2014.

19. Hashemi, S.K.; Bradford, M.A.; Valipour, H.R. Dynamic response of cable-stayed bridge under blast load. *Eng. Struct.* **2016**, *127*, 719–736. [CrossRef]

20. Hashemi, S.K.; Bradford, M.A.; Valipour, H.R. Dynamic response and performance of cable-stayed bridge under blast load: Effect of pylon geometry. *Eng. Struct.* **2017**, *137*, 50–66. [CrossRef]

21. Pan, Y.; Ventura, C.E.; Cheung, M.M.S. Performance of highway bridges subjected to blast loads. *Eng. Struct.* **2017**, *151*, 788–801. [CrossRef]

22. Farahmand-Tabar, S.; Barghian, M.; Vahabzadeh, M. Investigation of the progressive collapse in a suspension bridge under the explosive load. *Int. J. Steel Struct.* **2019**, *19*, 2039–2050. [CrossRef]

23. Korea Expressway Corporation. *Practical applications of Long-termly Measured Data on Cable-supported Bridges*; Korea Expressway Corporation: Gimcheon, Korea, 2007.

24. LSTC (Livermore Software Technology Corporation). *LS-DYNA Keyword User's Manual*; Livermore Software Technology Corporation: Livermore, CA, USA, 2018.

25. Murray, Y.D. *User Manual for LS-DYNA Concrete Material Model 159*; Report No, FHWA-HRT-05-062; Federal Highway Administration: Olympia, WA, USA, 2007.

26. Murray, Y.D. *User Manual for LS-DYNA Concrete Material Model 159*; Report No, FHWA-HRT-05-063; Federal Highway Administration: Olympia, WA, USA, 2007.

27. Choung, J.M.; Shim, C.S.; Kim, K.S. Plasticity and fracture behavior of marine structural steel, part I: Theoretical background of strain hardening and rate hardening. *J. Ocean. Eng. Technol.* **2011**, *25*, 134–144. [CrossRef]

28. Gimsing, N.J.; Georgakis, C.T. *Cable Supported Bridges: Concept and Design*, 3rd ed.; John Wiley & Sons: Hoboken, NY, USA, 2012.

29. Krauthammer, T. *Modern Protective Structures*; CRC Press: Boca Raton, FL, USA, 2008.

30. Orbovic, N.; Grimes, J.; El-Domiaty, K.; Florek, J. ASME blast test on pre-stressed concrete slabs. In Proceedings of the 24th Conference on Structural Mechanics in Reactor Technology, Busan, Korea, 20–25 August 2017; BEXCO: Busan, Korea, 2017.

Publisher's Note: MDPI stays neutral with regard to jurisdictional claims in published maps and institutional affiliations.

Article

Numerical Investigation of the Collapse of a Steel Truss Roof and a Probable Reason of Failure

Mertol Tüfekci [1,*], **Ekrem Tüfekci** [2] **and Adnan Dikicioğlu** [2]

[1] Department of Mechanical Engineering Imperial College London, Exhibition Road, London SW7 2AZ, UK
[2] Faculty of Mechanical Engineering, Istanbul Technical University, Gumussuyu-Istanbul 34437, Turkey; tufekcie@itu.edu.tr (E.T.); dikicioglu@itu.edu.tr (A.D.)
* Correspondence: tufekcime@itu.edu.tr or m.tufekci17@imperial.ac.uk

Received: 6 October 2020; Accepted: 1 November 2020; Published: 3 November 2020

Abstract: This study investigated the failure of the roof, with steel truss construction, of a factory building in Tekirdag in the northwestern part of Turkey. The failure occurred under hefty weather conditions including lightning strikes, heavy rain, and fierce winds. In order to interpret the reason for the failure, the effects of different combinations of factors on the design and dimensioning of the roof were studied. Finite element analysis, using the commercial software Abaqus (Dassault Systèmes, Vélizy-Villacoublay, France), was performed several times under different assumptions and considering different factors with the aim of determining the dominant factors that were responsible for the failure. Each loading condition gives out a characteristic form of failure. The scenario with the most similar form of failure to the real collapse is considered as the most likely scenario of failure. In addition, the factors included in this scenario are expected to be the responsible factors for the partial collapse of the steel truss structure.

Keywords: steel truss; roof structure; partial collapse; finite element analysis; lightning strike

---

1. Introduction

Engineers aim to make human life easier and to enhance life quality. Using mathematical calculations and experimentation, engineers try to predict the behaviour of a system and design it accordingly [1]. However, there have been cases that ended in structural failures, and some of these failures caused financial loss or even cost lives. Hadipriono studied nearly 150 major failures of structures around the world and determined that the major failures were due to lateral impact forces [2]. Moreover, Klasson published a survey covering failures of slender roofs [3]. Even the simplest structures, which have the most predictable behaviours, fail under unexpected conditions that exceed the designated safety margins [4,5].

Trusses are one of the most widely used and easy-to-design light structures [6]. They are able to carry very large loads relative to their own weights over very large spans. This is one of the main reasons that truss-type structures are preferred for building roofs and bridges. The possible loads are standardised to help engineers design structures in a very simple and straightforward way [7]. The standardised loads can be multiple times greater than the structures' own weight. In the case of unexpectedly excessive loads of accumulated snow or rainwater, failure of the designed structure may be unavoidable [8,9]. Geis et al. studied more than 1000 snow-induced building failure incidents all over the world [10]. An accumulated mass may cause failure in various ways. Excessive weight loads acting on the structures lead to a different load distribution than that of the designed distribution due to some members entering the plastic region and/or buckling [11–13].

In addition, the loads caused by dynamic effects such as earthquakes could lead to failure [14–16]. Structures are more vulnerable to dynamic loads than they are to static loads [17–19]. A way to improve the performance of structures against dynamic loads is to add damping to the structure [20]. Earthquakes are not the only sources that cause dynamic loads on structures. One of the main dynamic effects that may lead to sudden or progressive collapses is wind [21–24]. The wind is a very significant factor that leads to lateral loads to which building structures are relatively more sensitive than they are to vertical loads [25].

A crucial dynamic effect that is a very complicated phenomenon is lightning strike [26]. A lightning strike can cause partial damage or complete failure of structures [27,28]. Specifically, there are cases where the main reason reported for structural failure is lightning strike [29–32]. In Australia, it is reported that 21% of insurance claims for damage to buildings are caused by lightning strikes [33]. A lightning strike can affect a structure in many different ways, with the two most dangerous processes being blast (local and rapid pressure oscillations) and heating [34]. Previous research proved the heating effects of the lightning strike phenomenon [35,36]. It is known that a lightning strike increases the temperature rapidly in a very short period of time [29,37]. The electric current that is caused by the lighting strike heats up the structure so that it can start fires [38]. Besides, fire alone can also be the main factor that leads to structural failure [39–42].

Lightning is one of the key points of this study. Of course, structures are not affected just by one of the consequences of the lightning strike phenomenon. Instead, the combination of these enumerated parameters creates a synergy effect, which increases the resultant influence of the factors so that the combination becomes greater than a superposition of these contributing load components [43]. The effects of a lightning strike can be even more dramatic on the structures when there is accumulated water on top of the structure, since the local blasts are amplified by the presence of the water and thermal changes influence the structure [34,44].

There are different types of failure mechanisms as well as different types of reasons for the collapse. Thus, it is possible to find possible reasons for the failure just from the mode/form of the failure. One of the most important types of failure mechanisms is fatigue [45]. Fatigue may lead to the progressive collapse of structures [46,47]. Progressive collapse may happen instantly following the changed load distribution of members due to buckling and/or failure of some members [48,49]. However, it does not have to happen at the time of the initial failure of individual members but can happen after a certain number of members have failed successively [50,51].

There has been experimental research focusing on the mechanical behaviour of some specific components and/or structures that are used in trusses [15,52–55]. In past studies by various researchers, real roof structures were put through experimental testing to analyse their mechanical behaviour [15,56,57]. However, experimentation is not always possible and/or feasible. Therefore, numerical studies are conducted to predict the mechanical behaviour as well as to predict plausible failures. For truss-type structures, the finite element method is usually employed since it is one of the most powerful tools in engineering [14,22,57,58]. Pieraccini et al. studied the collapse of a spatial truss roof of a gym building in North Italy during a moderate snowfall by using the finite element method [59]. They reported that the elements and connections made by ductile or brittle materials influenced the bearing capacity of the roof structures and created geometric imperfections.

On the other hand, not all the failures occur slowly. As mentioned earlier, some collapses occur suddenly under excessive, impulsive loadings. This study aimed to study a case in which a sudden partial failure of a steel truss roof of a factory built in July 2011 happened during heavy weather conditions on 22 October 2012 in Cerkezkoy Tekirdag in the northwestern part of Turkey [8]. The main focus was on developing a plausible theory for the failure by employing numerical analyses performed using the finite element method, which covers various loading conditions and their combinations. Not only the truss roof but also the combination of the truss roof with the column supports were analysed. Thus, the results of numerical analyses were used to compare the obtained failure modes to the real failure, considering different criteria for the failure. The extreme loads due to the ponding of rainfall or snow accumulation and the temperature shock due to the thunderstorm were considered in the analyses. Even under these potential mechanical overloads, the observed damage was different from the actual damage: The damage caused by these excessive loads occurred in different regions than those in which the actual damage occurred. While the studies in the literature mostly describe failures due to vertical loads caused by water/snow accumulation or horizontal loads caused by wind, this study showed that the critical regions created by temperature change are similar to the actual damaged zones.

2. Theory and Case Study

2.1. Overview of the Structure and the Case

The building consists of six partitions with different sizes and an administration office. The design and all the relevant details of the structure were provided by the company that undertook the design and the construction of the building. The steel truss roof structure, which is supported on steel-reinforced concrete columns, covers a total area of 30,180 m^2.

The slope of the roof is 1% in each direction. The parapets of 15 and 25 cm heights surround the roof, which is equipped with a siphonic drainage system. The layout of the structure is presented schematically in Figure 1; the damaged parts are marked in white. The entire structure is made of seven partitions, including an administrative office and six halls where the industrial manufacturing took place.

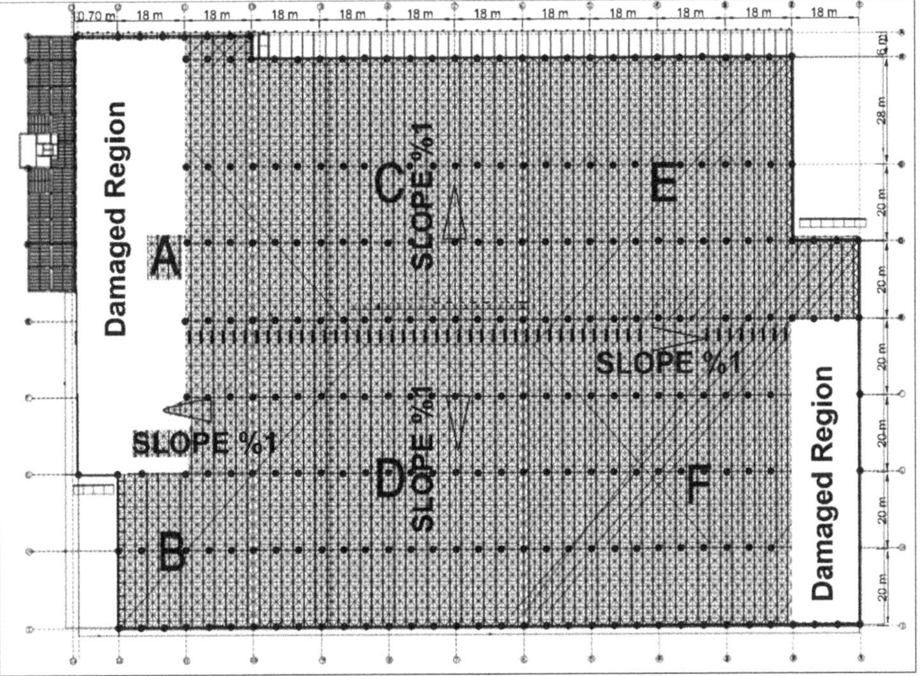

Figure 1. Layout and damaged regions of the roof structure.

The main dimensions of the unit truss system are presented in Figure 2. The steel carrier system consists of a double-layer truss with a depth of 2 m. The bars of this system have circular tube cross-sections with different wall thicknesses and diameters depending on their predicted loads. At both ends of the bars, screw threads are located to enable connecting those members to the mero system. The mero system includes steel spherical parts that have screw holes. It is noted that no eccentricity at the nodes is observed after the assembly of the individual members. The blocks are divided from each other with expansion joints.

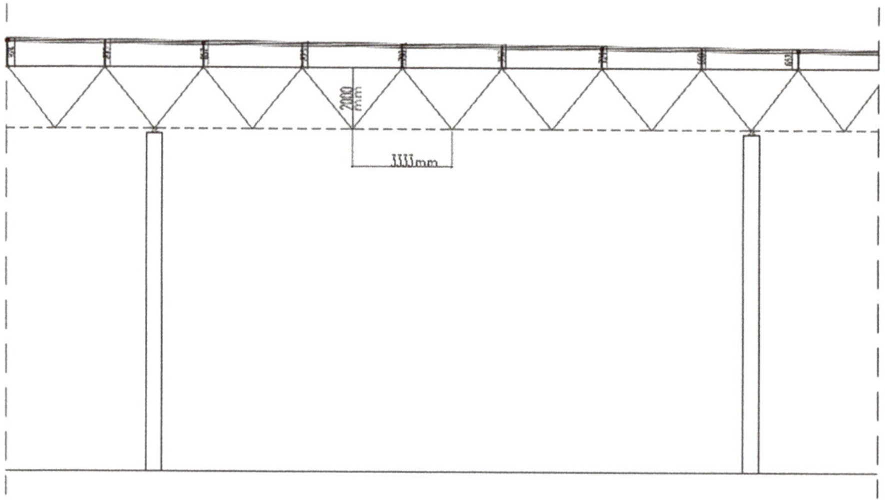

Figure 2. Unit truss structure.

The steel space truss roof structures are mounted on reinforced concrete columns that are 11 m tall with an 80 cm by 80 cm cross-section. The number of columns and their configuration can be seen in the plan view of the building, which is given in Figure 1. No failure, no deformation, and no cracks were observed on these columns during the investigations. Only some broken wings were observed on their top section, which was assumed to have happen during the collapse of the roof. Thus, it was concluded that the columns did not play a significant role in the collapse mechanism of this building. On top of the columns, the spherical joints are welded to the bearing plates. Figure 3 represents an overview of the real building structure, and Figure 4 shows the damaged roof of the space frame system on the night of the incident.

Figure 3. Overview of the roof structure.

Figure 4. Damage to the roof's space truss system.

On the day of the incident, more than 500 lightning events were detected in the region by the British Meteorological Office's (The Met Office) ATDnet (Arrival Time Difference Network) system and the Vaisala Global Lightning Dataset GLD360. It was determined that lightning struck the building and the lightning rods worked. Meteorological data indicate that the thunderstorms were rainy and the wind blew from 40 to 60 km/h from the north and northeast directions and up to 70 to 80 km/h locally. The rainfall was 22.6 mm during the collapse, although equally heavy local rainfalls had occurred several times in recent years. Around the roof, there are several siphonics at 5 cm height. The heights of the parapets on the roof are 15 cm and 25 cm. Although no scuppers are required to be used on such a shallow parapet, there are several scuppers of 15 cm height on the roof structure.

2.2. Structural Members

Members with steel tube sections of different diameters and thicknesses are used in the truss roof system. The pipe elements with screwed cone ends are connected to the hot forged steel spherical joints. These joints ensure that no eccentricity occurs and the only axial forces are developed in the bars of the truss system. The yield strength of the material (S235JR) of the bars and the cone parts at the ends of them (DIN 2458) is 235 MPa, ultimate tensile strength is 510 MPa, and the allowable stress is 144 MPa. The bolts used for mounting the bars to the spherical joints are made of 10.9 material quality. The tensile strength of the bolts is 1000 MPa, yield strength 900 MPa, and the allowable strength is 360 MPa, in which the safety factor is 2.5. The spherical joints used to connect the rods to each other and the supports are made of hot forged steel with a yield stress of 330 MPa and a tensile strength of 590 MPa. The strength values and the chemical decompositions of the materials used in the truss roof system were validated by Piroglu et al. [8].

The support spheres are fixed onto the concrete columns by using square or circular plates bolted to the columns. The supports are made of EN C45 steel, and Teflon plates are placed under the sliding supports to reduce friction. The details of the joints and supports are given in Figure 5.

Figure 5. View of the spherical joints and the supports.

3. Analysis

Two main groups of analyses were performed within the scope of this study. The first group was called the global analysis, where all the load-carrying members of the building were considered as beams and bars, while the second group considered the individual columns and supports.

The first group included multiple case scenarios with different combinations of loads. Only two major cases are presented in this study for the sake of simplicity.

The second group consisted of two stages. The first stage investigated the effective material properties of the concrete column reinforced with steel rods under uniaxial tensile strain conditions. Thus, the column was handled as a composite material. The second stage investigated the individual column support connection under the highest forces determined by the global model acting on an individual support.

All the simulations were conducted using the commercial software Abaqus (Dassault Systèmes, Vélizy-Villacoublay, France) under static loading assumptions.

The values of the three most important material properties, namely Young's modulus (E), Poisson's ratio (ν), and the thermal expansion coefficient (α) are shown in Table 1.

Table 1. The material properties considered in the analyses.

Property	Steel	Concrete
Young's modulus (E)	210 GPa	32 GPa
Poisson's ratio (ν)	0.28	0.20
Thermal expansion coefficient (α)	13 μm/(m K)	11.5 μm/(m K)

3.1. Global Analysis

The given design of the structure was recreated in a compatible computer aided design (CAD) format and was imported into the commercial finite element software mentioned above. The finite element model was used to simulate some assumptions and certain loading conditions with a consideration of geometric nonlinearities. The boundary conditions were defined as appropriate to the actual structure and the loads applied to the truss roof system were defined as given in Table 2, which are greater than the loads that are assumed for the design. So, the main purpose of this global analysis was to generate a better understanding of the mechanical behaviour of the roof and to predict the failure modes under different loads and assumptions. It is worthy of note that all the analyses were run in a static manner. Since the equivalent static loads are determined using the relevant Turkish standards, it is appropriate to use this simplification [8].

Table 2. The loads considered in the analyses.

Load Description	Value
Weight of the space truss system	$140 \ \text{N/m}^2$
Dead loads	$350 \ \text{N/m}^2$ ($240 \ \text{N/m}^2$ given in design calculations)
Snow load	$2000 \ \text{N/m}^2$ ($1000 \ \text{N/m}^2$ given in design calculations.)
Vertical wind loads	$960 \ \text{N/m}^2$
Equivalent earthquake vertical load	$950 \ \text{N/m}^2$
Equivalent earthquake horizontal load	$200 \ \text{N/m}^2$ ($180 \ \text{N/m}^2$ given in design calculations)

One of the most important aspects here is that the presented strength and elastic properties of the materials and members used were put through tests and experimental processes and were found to be consistent in general with the standards [8]. So, in the analyses performed for this study, the values that were presented in the reports of the company were mainly used.

Different dead, live, snow, temperature, and earthquake loads were considered in the design. Rain load caused by an excessive accumulation of water, especially on low-slope or flat roofs where the parapets are mounted around the roof, can cause partial or total destruction if they are not considered in the design. However,, rainfall ponding on the roof is considered in the design. The design code used gives wind loads for buildings of 20–90 m tall as $0.8 \ \text{kN/m}^2$ (pressure) on the windward and $-0.4 \ \text{kN/m}^2$ (suction) on the leeward sides. For the seismic loads, the earthquake acceleration is taken as 0.3 g, where g is the gravitational acceleration, so the earthquake loads are overestimated [8].

The loads acting on the space truss system are to be transmitted to the bars over the spherical steel elements, called nodes. The weight of the space truss roof system is taken as $140 \ \text{N/m}^2$, the dead load from coatings and the installation is given in the design documents as $240 \ \text{N/m}^2$, but it was assumed to be $350 \ \text{N/m}^2$ in the analysis. The snow load is given in the design calculations as $1000 \ \text{N/m}^2$ and was taken as $2000 \ \text{N/m}^2$ in the analysis. The equivalent earthquake horizontal load is given as $180 \ \text{N/m}^2$ in the design calculations and was considered here as $200 \ \text{N/m}^2$. The equivalent earthquake vertical load was $950 \ \text{N/m}^2$ and the vertical wind load was $960 \ \text{N/m}^2$. Table 2 gives the loads used in the analyses.

The boundary conditions of the space truss system and the columns were arranged as the original design of the structure.

The bars of the space truss were modelled based on the standards of DIN S235 JR for quality linear elastic materials, and columns were modelled as C35 grade homogeneous and linear elastic steel-reinforced concrete with the elasticity modulus, which was determined using the homogenisation of the composite column analysis. The bar elements in the space truss system and the beam elements in the columns were assembled as a whole system of 160,133 elements and 461,847 nodes. Figure 6 displays the model for global analysis in a commercial finite element software. With this model, a static global analysis of the space truss system was performed with the above-mentioned loads being quite a bit larger than the design loads.

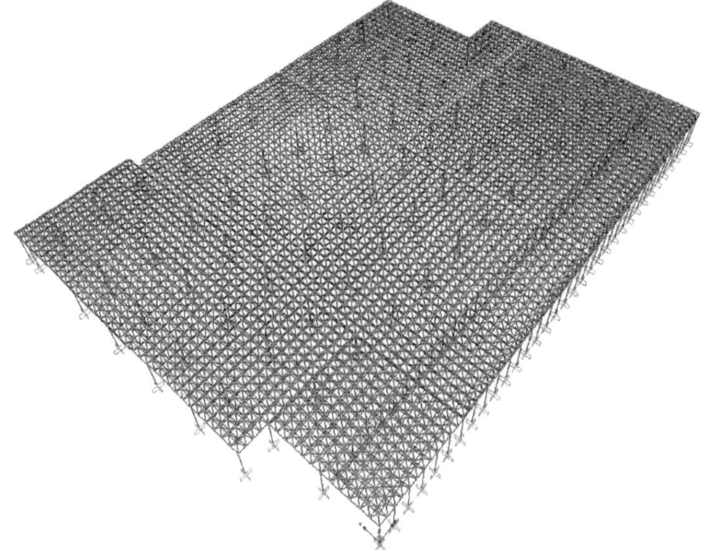

Figure 6. Three-dimensional finite element model prepared for finite element (FE) analysis.

3.2. Analysis of the Columns and Supports

This section covers the analyses performed to gain more insights into the mechanical behaviour of the columns and the supports, which either were not very accurately represented in the global model or to increase the accuracy of the global model. The first stage of these analyses aimed to homogenise the composite column and the second stage investigates the mechanical behaviour of the column and support behaviour focusing on the stress/strength of the structure.

3.2.1. Analysis of the Steel-Reinforced Columns

The main purpose of this analysis was to understand the effective material properties of the steel-reinforced concrete columns by using three-dimensional finite elements. Here, the column was assumed to be a composite beam and its material properties were calculated with homogenisation. Thus, the aim was to obtain a more realistic global model. For this purpose, a model was created considering the steel reinforcement of the concrete columns. In this model, the column material was considered as a composite material with concrete and reinforcing steel rods. The steel rods were placed in the concrete as shown in Figure 7. The rods and the concrete were rigidly bonded to each other. According to the general theory of mechanics of composite materials, lateral reinforcement steel binders, which are predicted to have no significant effect on axial stiffness, were not included in the analysis for the sake of simplicity. The steel reinforcing rods were taken as given in the project and as described in the previous section.

To represent the behaviour, a cross-sectional face was subjected to axial displacement restrictions whereas the opposite face, which is the other cross-sectional face, was subjected to a uniform axial displacement. Using the results of the analysis, the effective Poisson's ratio and Young's modulus were determined. A similar procedure was also performed for the effective thermal expansion coefficient. The results of this analysis were used in the global analysis.

Figure 7. CAD of a column with steel-reinforced concrete.

3.2.2. Analysis of the Spherical Support with Steel-Reinforced Columns

This part of the study was interested in the mechanical behaviour and safety of the column and spherical support structures. This study used the highest mechanical loads acting on a single truss-column connection node calculated in the global analysis. It is also important to note that this analysis considered only the mechanical loads and did not consider any thermal effects. The analysis of the column support connection was performed with a three-dimensional finite element model.

The column support connection was built such that the sphere was mounted on the support plate, which is embedded in the steel-reinforced concrete with a bolt connection. The structure of this model is given in Figure 8.

Figure 8. CAD of the column support connection with steel-reinforced concrete.

The contact interfaces between the steel rods and concrete were rigidly bonded and so too were the contact surfaces between the spherical support and the steel plate. The bolts were attached to the model with a standard preload and connected to the steel plate and concrete with friction.

The bottom cross-sectional face was connected to a grounded spring with 6 degrees of freedom and with the stiffness values of the rest of the column, to save computational time.

4. Results and Discussion

In this section, the results of the analyses are presented and discussed. Based on the interpretations of the results that match the actual failure form of the structure, a scenario, which is judged the most plausible, is described and discussed.

4.1. Analysis of the Steel-Reinforced Columns

The homogenisation study was conducted and the results are presented here. Figure 9 shows the equivalent stress distribution in the case of axial loading of the composite structure and Figure 10 shows displacement distribution. As a result of this analysis, the elasticity modulus and the resultant thermal expansion coefficient of the steel-reinforced concrete were calculated as 48 GPa and 12.8 μm/(m K), respectively. This axial elasticity modulus value was 50 GPa in the global analysis.

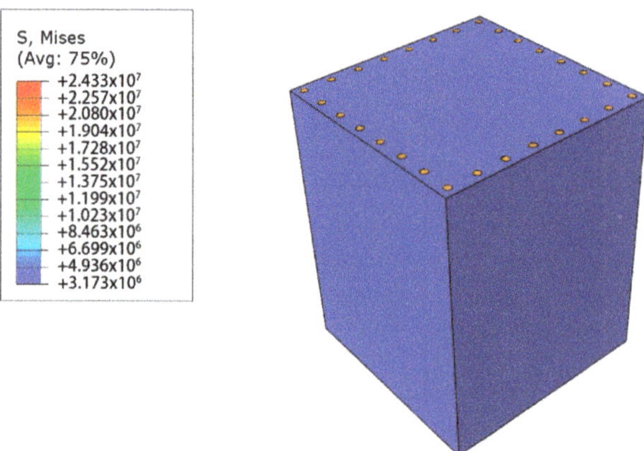

Figure 9. Stress distribution in the case of axial stress on the steel-reinforced concrete column.

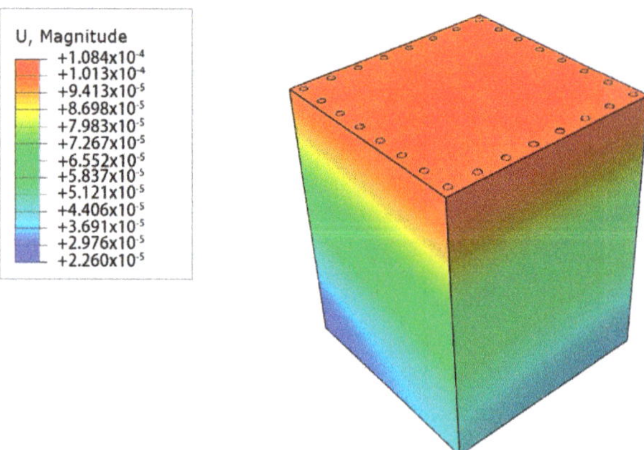

Figure 10. Displacement distribution of the steel-reinforced concrete column in the case of axial stress.

4.2. Global Analysis with Only Mechanical Loads

The displacements and the stresses occurring in the truss roof system were calculated by using the finite element analysis software. Figure 11 shows the vertical displacement (in meters) under the considered loads. The largest displacements of about 4.7 cm occurred in blocks A, C, and E, where the gap between the columns was the largest (28 m), as was expected. It should be noted here that due to the positive direction of the axis, which is upwards, blue colours are around zero and red is the largest in terms of absolute values. The region of the larger resultant displacements is not where the real damage happened. The deformations of the column structures are so small that they do not suggest any safety issues. The analyses demonstrate clearly that the system is safe and that the results are consistent with the results obtained by the engineering bureau.

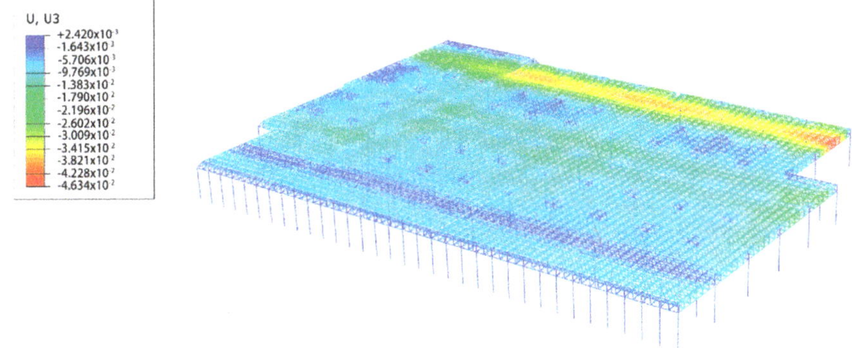

Figure 11. Vertical displacement of the steel-reinforced concrete column system.

Considering the stress analysis, the equivalent stresses using the von Mises criteria were calculated for all the structures. Figure 12 gives the von Mises stresses in the roof structure (in Pa). It can be seen that all stresses of the roof structure together with the columns are within the safety limits. The collapsed regions are also within the safety limits. As well as the displacements, the stresses occurring in the column structures are well below the safety limits. Although exaggerated loads are applied on the roof structure, the stresses and displacements in the beams and rods show that the structure would not fail under these combinations of excessive loads.

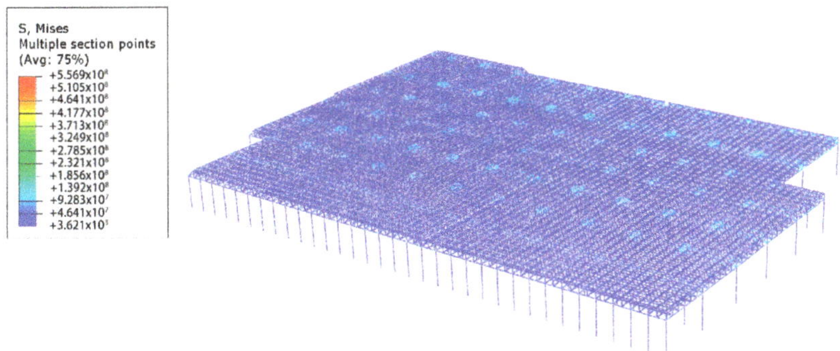

Figure 12. Equivalent von Mises stress distribution of the steel-reinforced concrete columns.

The greatest stresses occur at these "U" shaped connections shown in Figure 13; however, the displacements are not excessive. It is important to note that this connection structure does not exist in the real application. These connections are used in the CAD/FE model in order not to distort the dimensions of the truss structure. In reality, there are walls with certain thicknesses whereas the FE model used beam elements. To connect the trusses with the supports, such a "dummy" structure was used. Therefore, the stress values that were read at these connections were disregarded.

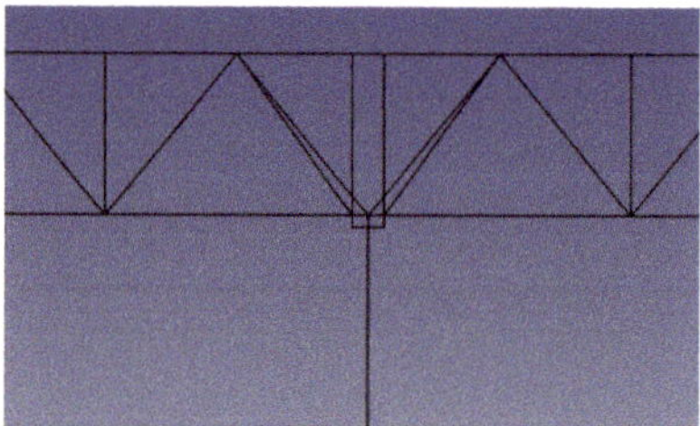

Figure 13. Connections with the supports and steel truss at the block interfaces.

4.3. Analysis of the Spherical Support with Steel-Reinforced Columns

After concluding that the global structure is safe, the next step was to analyse an individual column support by using the finite element method. The results of the global analysis were used to determine the support reactions. At the most dangerous column support connection, the horizontal force was 95 kN and vertical force was 940 kN. The forces were taken as 120 kN in horizontal and 1000 kN in vertical directions.

The equivalent stress distribution obtained in this case is shown in Figure 14. As can be clearly seen from this figure, the stress levels are below the yield stress, even though extreme loads are applied. Figure 15 shows a post-damage image of one of the columns to which the space frame roof system was attached.

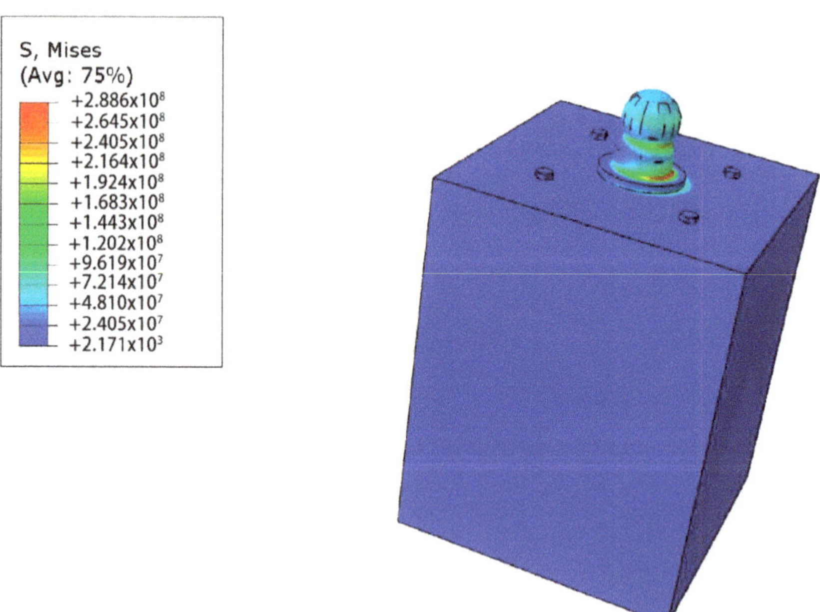

Figure 14. Equivalent stress distribution of the column support connection.

Figure 15. Damage on the support to which the space truss roof system was attached.

Figure 16 shows the distribution of the resultant displacement that occurs in the column support connection. Here, it can be seen that the maximum value of the displacement compounds of the spherical support was around 0.3 mm.

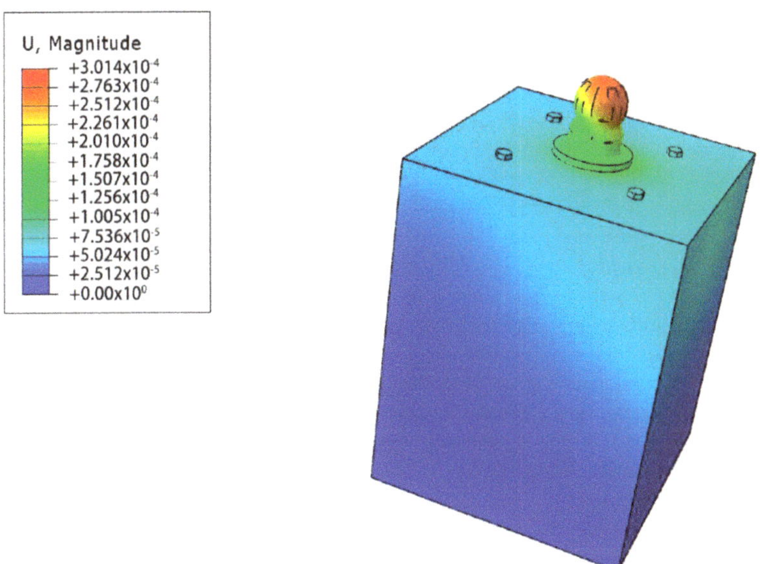

Figure 16. Resultant displacement distribution of the column support connection.

4.4. Global Analysis with Mechanical Loads and Thermal Effects

With these results obtained from the global analysis it was found that increasing horizontal forces may cause damage, various damage scenarios were investigated by changing the loads applied to the space frame system. As the first case, the vertical load was increased considerably and no damage was predicted in the real damaged zones A and F, as given in Figure 1. As the next case, the structure was tested by increasing the horizontal load. Although the horizontal load value was increased to a level that cannot be reached by wind and similar factors, the damage that occurred cannot be explained with these combinations of loads. However, the structure was found to be more sensitive and vulnerable to the horizontal loads. Finally, it was understood that the space frame system with the columns is safe against vertical loads (snow, water, etc.) and horizontal loads (wind, etc.).

It is obvious that different factors must have been present in order to explain the real damage in the structure. Therefore, considering the natural events occurring on the night of the event, their effects on the structure were considered. It was assumed that reasons such as heavy rain that leads to ponding on the roof with a slope of 1% and a plausibly clogged water discharging drainage system would not cause damage similar to that which occurred. Besides, around the rooftop surface there are several scuppers with 15 cm height. Even with the assumption of full ponding of the roof with 15 cm height, it can be confidently claimed that the chosen loads for the analysis were on the cautious side and led to conservative results, which do not point out any damage predictions similar to the real damage.

On the other hand, a large number of lightning strikes were detected on the night of the incident. The effects of lightning depend on the energy of the lightning itself. The energy of lightning, which is impossible to fully determine, depends on many parameters, such as the altitude of the cloud where the lightning originates. It is well known that this discharged energy heats up a considerable amount of air surrounding the core of the lightning. The temperature rise is so rapid and significant that it can start fires near the location where lightning strikes [37]. Furthermore, when lightning strikes a building, the electric current passes through the structure, which leads to a sudden heat generation and a rapid increase of temperature [29,37,38]. Besides, quite large shock forces occur due to the interaction of large masses of hot and cold air around it, and also due to the humidity of the air and the ground, which is called the blast effect. In addition, it is known that the presence of water amplifies the blast effect and interacts with the structure under electric current [34]. Therefore, it is thought that these great shock forces can cause instant damage.

Even though it is judged that the structure is heated, it is impossible to predict or evaluate the exact or even approximate rise of temperature. For the purpose of understanding the underlying factors that led to the damage shown in Figure 1, the effects of temperature change on the structure were investigated using a simplified approach. The space frame system was given a temperature increase of 50 °C. Lightning is most likely to cause a local thermal shock and not heat up the entire structure. However, it is impossible to know the number of lightning strikes, the generated heat, the temperature rise, generated blast, etc.

Based on these assumptions, a 50 °C temperature increase was applied to the global model alongside the external loads as described in the previous global analysis. A new static model was built and the results are given in Figures 17–21. Figure 17 shows the equivalent stresses of von Mises occurring in the space truss systems and columns. It can be seen that the stresses coming out of this figure are quite safe. However, it has been found that the stress values in the support joints reach values exceeding the safety limits, and the supports reaching these values are concentrated in the areas where damage occurred. The reason for the great stress values in this global model is that the model is simple and gives a general idea. It would be appropriate to consider separately the situations in the supports with large stresses.

It is also to be noted that the "U" shaped connections between the steel truss structures and the supports on the walls display the greatest stresses, which is not realistic, as these connections do not exist in the real structures. Therefore, those stress values were disregarded.

Figure 18 shows the resultant displacements that occurred throughout the system. From this figure, it can be seen that displacements are suitable and larger displacements occurred in the regions where the damage occurred.

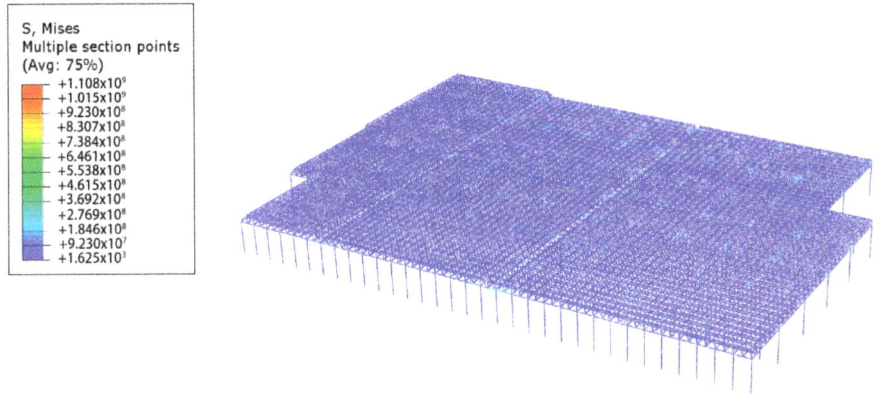

Figure 17. Stresses due to load and thermal effects in the space frame and steel-reinforced concrete column system.

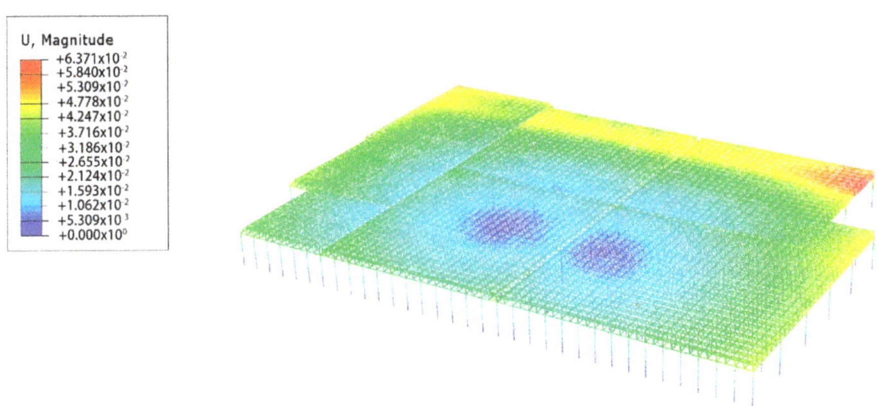

Figure 18. Resultant displacements due to load and thermal effects in the space frame and steel-reinforced concrete column system.

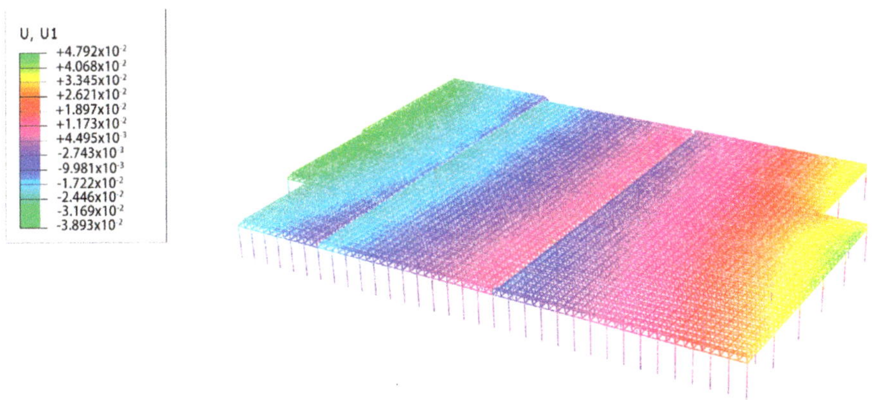

Figure 19. Displacements in the long-edge direction caused by load and thermal effects in the space frame and steel-reinforced concrete column system.

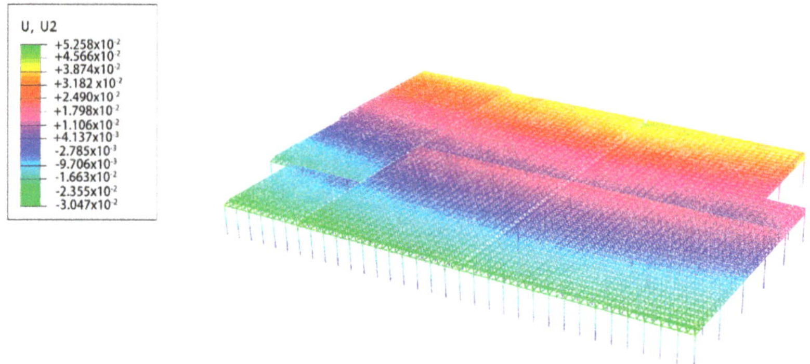

Figure 20. Displacements in the short-edge direction caused by load and thermal effects in the space frame and steel-reinforced concrete column system.

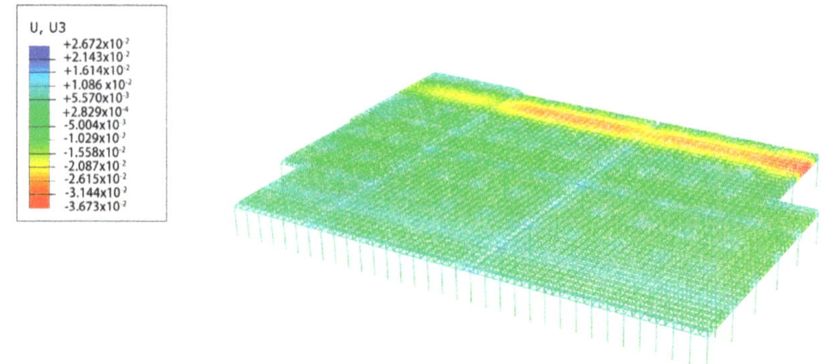

Figure 21. Displacements in the vertical direction caused by the loads in the space frame and steel-reinforced concrete column system.

Furthermore, Figure 19 shows the displacements in the long-edge direction, x-direction. This figure clearly shows that there were large displacements in the areas where the damage occurred. Figure 20 shows the displacements in the direction of the short edge of the building, the so-called y-direction. It is to be noted that the larger displacements were observed in the direction along the shorter edge, whereas the larger displacements due to thermal expansion were expected in the direction of the long edge. The reason behind this can be explained with different stiffness levels of the structure in different directions and boundary conditions of the supports. As can be seen from Figure 6, the number of columns in the first two rows of the halls B, D, and F are much more intensively positioned than that seen in the rest of the building. Moreover, the intensity of the columns in the x-direction is greater than the intensity of columns in the y-direction. Therefore, the stiffness in the x-direction is expected to be greater than the stiffness in the y-direction. The sliding supports and the stiffness of the structure allow the structure to deform in the y-direction more than it can in the x-direction, even though the tendency to deform is more in the x-direction than in the y-direction. This leads to the storage of a larger amount of energy in the x-direction. Additionally, since, there was no damage in the spherical supports themselves, as can be seen in Figure 15, the excessive displacements caused larger stress values in the bolt connections between the spherical parts and the members. Therefore, it was found plausible that this mechanism caused the actual failure.

The displacements of the structure in the vertical direction are given in Figure 20. Comparing the vertical displacements of the previous global analysis and this one (Figures 11 and 21), both cases display very similar distribution trends but the new one exhibits larger displacements. This is because there is a synergy effect based on the complex loading, and the thermal expansion leads to extra loading and elongation in horizontal directions.

5. Conclusions

The space truss roof structure, the columns, and the spherical steel supports of the factory building were put through several numerical analyses, and their behaviour was evaluated using the commercial software Abaqus (Dassault Systèmes, Vélizy-Villacoublay, France). These static stress analyses were performed using the finite element method and considering geometric nonlinearities. Based on the results of these simulations, the following conclusions about the real failure of the structure were drawn:

- According to the global analysis, the structure is quite strong and insensitive to the vertical dead loads (rain, snow, etc.) even at some higher values, which are judged to be extreme and unrealistic. The most dangerous region of the space frame system in terms of vertical loads is the corridor with the largest column span in the A, C, and E blocks. However, this region is still safe, and the damage occurred somewhere else.

- Using the results of the global analysis, the most critical support connection was individually investigated and found safe, although it is more sensitive to horizontal forces. However, in the analysis, all the effects that create lateral forces caused by the wind and similar homogeneous distributed loads could not make the structure exceed the safety limits and cause any damage in the locations where the real damage occurred.

- It is judged that the thermal expansion of structural members increases the horizontal forces at the supports and the stresses at the members. The temperature change was assumed to be caused by lightning strikes. To investigate the effects of the temperature change, a simplified analysis with a uniform temperature rise was conducted and produced results that point out critical regions that are the same as the real damaged zones.

- The significant change of temperature of the air surrounding the lightning channel and the electric current that passes through the structure may cause heating of the roof system rapidly and locally. The lightning at the scene may also cause a sudden pressure change that can be amplified by the presence of water on top of the structural surface. In addition to the thermal expansion, it is judged possible that the blast effect may have a significant role in the failure of the structure.

- Testing the idea of temperature change, the damage prediction was found to be consistent with the real failure mode. It was concluded that the bolts and the bolted members attached to the supports in the real damaged regions of the space truss roof system were overloaded and damaged, as it can be seen in Figure 15.

Author Contributions: Conceptualization, E.T. and A.D.; methodology, E.T. and A.D.; software, M.T.; validation, E.T., A.D., and M.T.; formal analysis, M.T., E.T., and A.D.; investigation, M.T.; resources, A.D.; data curation, M.T.; writing—original draft preparation, M.T., E.T., and A.D.; writing—review and editing, M.T., E.T., and A.D.; visualization, M.T.; supervision, E.T. and A.D.; project administration, A.D.; All authors have read and agreed to the published version of the manuscript.

Funding: This research received no external funding.

Acknowledgments: The authors would like to thank İnci Pir for her support in visualizing the results.

Conflicts of Interest: The authors declare no conflict of interest.

References

1. Wu, Z.; Rong, J.; Liu, C.; Liu, Z.; Shi, W.; Xin, P.; Li, W. Dynamic analysis of spatial truss structures including sliding joint based on the geometrically exact beam theory and isogeometric analysis. *Appl. Sci.* **2020**, *10*, 1231. [CrossRef]
2. Hadipriono, F.C. Analysis of Events in Recent Structural Failures. *J. Struct. Eng.* **1985**, *111*, 1468–1481. [CrossRef]
3. Klasson, A.; Björnsson, I.; Crocetti, R.; Hansson, E.F. Slender Roof Structures—Failure Reviews and a Qualitative Survey of Experienced Structural Engineers. *Structures* **2018**, *15*, 174–183. [CrossRef]

4. Deshpande, V.S.; Fleck, N.A. Collapse of truss core sandwich beams in 3-point bending. *Int. J. Solids Struct.* **2001**, *38*, 6275–6305. [CrossRef]

5. Ballarini, R.; La Mendola, L.; Le, J.L.; Monaco, A. Computational study of failure of hybrid steel trussed concrete beams. *J. Struct. Eng.* **2017**, *143*, 1–13. [CrossRef]

6. Wu, Y.; Xiao, Y. Steel and glubam hybrid space truss. *Eng. Struct.* **2018**, *171*, 140–153. [CrossRef]

7. Sun, W.; Zhang, Q. Universal equivalent static wind loads of fluctuating wind loads on large-span roofs based on compensation of structural frequencies and modes. *Structures* **2020**, *26*, 92–104. [CrossRef]

8. Piroglu, F.; Ozakgul, K.; Iskender, H.; Trabzon, L.; Kahya, C. Site investigation of damages occurred in a steel space truss roof structure due to ponding. *Eng. Fail. Anal.* **2014**, *36*, 301–313. [CrossRef]

9. Goto, Y.; Kawanishi, N.; Honda, I. Dynamic stress amplification caused by sudden failure of tension members in steel truss bridges. *J. Struct. Eng.* **2011**, *137*, 850–861. [CrossRef]

10. Geis, J.; Strobel, K.; Liel, A. Snow-Induced Building Failures. *J. Perform. Constr. Facil.* **2012**, *26*, 377–388. [CrossRef]

11. Altunişik, A.C.; Ateş, Ş.; Hüsem, M. Lateral buckling failure of steel cantilever roof of a tribune due to snow loads. *Eng. Fail. Anal.* **2017**, *72*, 67–78. [CrossRef]

12. Thai, H.T.; Kim, S.E. Nonlinear inelastic time-history analysis of truss structures. *J. Constr. Steel Res.* **2011**, *67*, 1966–1972. [CrossRef]

13. Shivarudrappa, R.; Nielson, B.G.; Asce, M. Sensitivity of load distribution in light-framed wood roof systems due to typical modeling parameters. *J. Perform. Constr. Facil.* **2013**, *27*, 222–234. [CrossRef]

14. Malla, R.B.; Nalluri, B.B. Dynamic effects of member failure on response of truss-type space structures. *J. Spacecr. Rockets* **1995**, *32*, 545–551. [CrossRef]

15. Parisi, M.A.; Piazza, M. Seismic behavior and retrofitting of joints in traditional timber roof structures. *Soil Dyn. Earthq. Eng.* **2002**, *22*, 1183–1191. [CrossRef]

16. Pollino, M.; Bruneau, M. Seismic testing of a bridge steel truss pier designed for controlled rocking. *J. Struct. Eng.* **2010**, *136*, 1523–1532. [CrossRef]

17. Sosorburam, P.; Yamaguchi, E. Seismic retrofit of steel truss bridge using buckling restrained damper. *Appl. Sci.* **2019**, *9*, 2791. [CrossRef]

18. Chou, C.C.; Chen, J.H. Seismic tests of post-tensioned self-centering building frames with column and slab restraints. *Front. Archit. Civ. Eng. China* **2011**, *5*, 323–334. [CrossRef]

19. Zhao, H.; Ding, Y.; Nagarajaiah, S.; Li, A. Longitudinal displacement behavior and girder end reliability of a jointless steel-truss arch railway bridge during operation. *Appl. Sci.* **2019**, *9*, 2222. [CrossRef]

20. Rezaei, S.; Akbari Hamed, A.; Basim, M.C. Seismic performance evaluation of steel structures equipped with dissipative columns. *J. Build. Eng.* **2020**, *29*. [CrossRef]

21. Qu, W.L.; Chen, Z.H.; Xu, Y.L. Dynamic Analysis of Wind Excited Truss Tower With Friction Dampers. *Comput. Struct.* **2001**, *79*, 2817–2831. [CrossRef]

22. Spyrakos, C.C.; Raftoyiannis, I.G.; Ermopoulos, J.C. Condition assessment and retrofit of a historic steel-truss railway bridge. *J. Constr. Steel Res.* **2004**, *60*, 1213–1225. [CrossRef]

23. Zheng, H.D.; Fan, J. Analysis of the progressive collapse of space truss structures during earthquakes based on a physical theory hysteretic model. *Thin-Walled Struct.* **2018**, *123*, 70–81. [CrossRef]

24. Maraveas, C.; Tsavdaridis, K.D. Assessment and retrofitting of an existing steel structure subjected to wind-induced failure analysis. *J. Build. Eng.* **2019**, *23*, 53–67. [CrossRef]

25. Foraboschi, P. Lateral load-carrying capacity of steel columns with fixed-roller end supports. *J. Build. Eng.* **2019**, *26*. [CrossRef]

26. Rakov, V.A. The physics of lightning. *Surv. Geophys.* **2013**, *34*, 701–729. [CrossRef]

27. Hartono, Z.A.; Ibrahim, R. A database of lightning damage caused by bypasses of air terminals on buildings in Kuala Lumpur, Malaysia. In Proceedings of the VI International Symposium on Lightning Protection, Santos, Brazil, 19–23 November 2001.

28. Bothma, J.G. Transmission line tower collapse investigation: A case study. In Proceedings of the IEEE Power and Energy Society Conference and Exposition in Africa: Intelligent Grid Integration of Renewable Energy Resources (PowerAfrica), Johannesburg, South Africa, 9–13 July 2012. [CrossRef]

29. Mills, B.; Unrau, D.; Pentelow, L.; Spring, K. Assessment of lightning-related damage and disruption in Canada. *Nat. Hazards* **2010**, *52*, 481–499. [CrossRef]

30. Zhang, W.; Meng, Q.; Ma, M.; Zhang, Y. Lightning casualties and damages in China from 1997 to 2009. *Nat. Hazards* **2011**, *57*, 465–476. [CrossRef]

31. Krausmann, E.; Renni, E.; Campedel, M.; Cozzani, V. Industrial accidents triggered by earthquakes, floods and lightning: Lessons learned from a database analysis. *Nat. Hazards* **2011**, *59*, 285–300. [CrossRef]

32. Patel, K. Effect of Lightning on Building and Its Protection Measures. *Int. J. Eng. Adv. Technol.* **2013**, *2*, 182.

33. Blong, R. Residential building damage and natural perils: Australian examples and issues. *Build. Res. Inf.* **2004**, *32*, 379–390. [CrossRef]

34. Shivalli, S. Lightning Phenomenon, Effects and Protection of Structures from Lightning Sanketa Shivalli. *IOSR J. Electr. Electron. Eng.* **2016**, *11*, 44–50.

35. Zhivlyuk, Y.; Mandel'shtam, S. On the temperature of lightning and force of thunder. *Sov. Phys. JETP* **1961**, *13*, 338–340.

36. Mu, Y.; Yuan, P.; Wang, X.; Dong, C. Temperature distribution and evolution characteristic in lightning return stroke channel. *J. Atmos. Solar-Terrestrial Phys.* **2016**, *145*, 98–105. [CrossRef]

37. Elsom, D.M.; Webb, J.D.C. Lightning Impacts in the United Kingdom and Ireland. In *Extreme Weather*; Wiley-Blackwell: Hoboken, NJ, USA, 2016; pp. 195–207.

38. Holle, R.L.; López, R.E.; Arnold, L.J.; Endres, J. Insured Lightning-Caused Property Damage in Three Western States. *J. Appl. Meteorol.* **1996**, *35*, 1344–1351. [CrossRef]

39. Kmet, S.; Tomko, M.; Demjan, I.; Pesek, L.; Priganc, S. Analysis of a damaged industrial hall subjected to the effects of fire. *Struct. Eng. Mech.* **2016**, *58*, 757–781. [CrossRef]

40. Usmani, A.S. Stability of the world trade center twin towers structural frame in multiple floor fires. *J. Eng. Mech.* **2005**, *131*, 654–657. [CrossRef]

41. Behnam, B. Fire Structural Response of the Plasco Building: A Preliminary Investigation Report. *Int. J. Civ. Eng.* **2019**, *17*, 563–580. [CrossRef]

42. Mwangi, S. Why Broadgate Phase 8 composite floor did not fail under fire: Numerical investigation using ANSYS® FEA code. *J. Struct. Fire Eng.* **2017**, *8*, 238–257. [CrossRef]

43. Usmani, A.S.; Chung, Y.C.; Torero, J.L. How did the WTC towers collapse: A new theory. *Fire Saf. J.* **2003**, *38*, 501–533. [CrossRef]

44. Meppelink, J. The Impact of a lightning Stroke on a Flat Roof When the Building is Filled with Water. In Proceedings of the ICLP '98, 24th International Conference on Lightning Protection, Birmingham, UK, 14–18 September 1998; pp. 826–831.

45. Li, H.; Wu, G. Fatigue evaluation of steel bridge details integrating multi-scale dynamic analysis of coupled train-track-bridge system and fracture mechanics. *Appl. Sci.* **2020**, *10*, 3261. [CrossRef]

46. Blandford, G.E. Progressive failure analysis of inelastic space truss structures. *Comput. Struct.* **1996**, *58*, 981–990. [CrossRef]

47. Blandford, G.E. Review of Progressive Failure Analyses for Truss Structures. *J. Struct. Eng.* **1997**, *123*, 122–129. [CrossRef]

48. Malla, R.B.; Agarwal, P.; Ahmad, R. Dynamic analysis methodology for progressive failure of truss structures considering inelastic postbuckling cyclic member behavior. *Eng. Struct.* **2011**, *33*, 1503–1513. [CrossRef]

49. Machaly, E.S.B. Buckling contribution to the analysis of steel trusses. *Comput. Struct.* **1986**, *22*, 445–458. [CrossRef]

50. Murtha-Smith, E. Alternate Path Analysis of Space Trusses for Progressive Collapse. *J. Struct. Eng.* **1999**, *114*, 1978–1999. [CrossRef]

51. Astaneh-Asl, A. Progressive collapse of steel truss bridges, the case of I-35W collapse. In Proceedings of the International Conference on Steel Bridges, Guimarães, Portugal, 4–6 June 2008; pp. 1–10.

52. Tanzer, A. High school gymnasium roof truss support collapse. *J. Fail. Anal. Prev.* **2011**, *11*, 208–214. [CrossRef]

53. Park, S.; Yun, C.B.; Roh, Y. Damage diagnostics on a welded zone of a steel truss member using an active sensing network system. *NDT E Int.* **2007**, *40*, 71–76. [CrossRef]

54. Riasat Azim, M.; Gül, M. Damage Detection of Steel-Truss Railway Bridges Using Operational Vibration Data. *J. Struct. Eng.* **2020**, *146*, 1–12. [CrossRef]

55. Wood, J.V.; Dawe, J.L. Full-scale test behavior of cold-formed steel roof trusses. *J. Struct. Eng.* **2006**, *132*, 616–623. [CrossRef]

56. Fülöp, A.; Iványi, M. Experimentally analyzed stability and ductility behaviour of a space-truss roof system. *Thin-Walled Struct.* **2004**, *42*, 309–320. [CrossRef]

57. Caglayan, O.; Yuksel, E. Experimental and finite element investigations on the collapse of a Mero space truss roof structure—A case study. *Eng. Fail. Anal.* **2008**, *15*, 458–470. [CrossRef]

58. Piroglu, F.; Ozakgul, K. Partial collapses experienced for a steel space truss roof structure induced by ice ponds. *Eng. Fail. Anal.* **2016**, *60*, 155–165. [CrossRef]
59. Pieraccini, L.; Palermo, M.; Trombetti, T.; Baroni, F. The role of ductility in the collapse of a long-span steel roof in North Italy. *Eng. Fail. Anal.* **2017**, *82*, 243–265. [CrossRef]

Publisher's Note: MDPI stays neutral with regard to jurisdictional claims in published maps and institutional affiliations.

Article

Study on Dynamic Behavior of Bridge Pier by Impact Load Test Considering Scour

Myungjae Lee [1], Mintaek Yoo [1], Hyun-Seok Jung [2], Ki Hyun Kim [1] and Il-Wha Lee [2,*

1   Railroad Structure Research Team, Korea Railroad Research Institute, 176 Cheoldobangmulgwan-ro, Uiwang-si, Gyeonggi-do 16105, Korea; myungjae@krri.re.kr (M.L.); thezes03@krri.re.kr (M.Y.); kimkh738@krri.re.kr (K.H.K.)
2   Advanced Infrastructure Research Team, Korea Railroad Research Institute, 176 Cheoldobangmulgwan-ro, Uiwang-si, Gyeonggi-do 16105, Korea; wjdgustjr601@krri.re.kr
*   Correspondence: iwlee@krri.re.kr

Received: 10 August 2020; Accepted: 24 September 2020; Published: 26 September 2020

Abstract: In this study, for the establishment of a safety evaluation method, non-destructive tests were performed by developing a full-scale model pier and simulating scour on the ground adjacent to a field pier. The surcharge load (0–250 kN) was applied to the full-scale model pier to analyze the load's effect on the stability. For analyzing the pier's behavior according to the impact direction, an impact was applied in the bridge axis direction, pier length direction, and pier's outside direction. The impact height corresponded to the top of the pier. A 1-m deep scour was simulated along one side of the ground, which was adjacent to the pier foundation. The acceleration was measured using accelerometers when an impact was applied. The natural frequency, according to the impact direction and surcharge load, was calculated using a fast Fourier transform (FFT). In addition, the first mode (vibratory), second mode (vibratory), and third modes (torsion) were analyzed according to the pier behavior using the phase difference, and the effect of the scour occurrence on the natural frequency was analyzed. The first mode was most affected by the surcharge load and scour. The stability of the pier can be determined using the second mode, and the direction of the scour can be determined using the third mode.

Keywords: prototype abutment; non-destructive test; surcharge load; mode number; scour

---

1. Introduction

The maintenance of railway bridges is required due to their special structures and socioeconomic roles. Therefore, technologies that can facilitate maintenance of the target performance based on reasonable inspections, measurements, evaluations, decision-making, repairs, and reinforcement procedures are required. However, the number of bridges that are damaged not only by structural problems but also by various environmental factors is gradually increasing. In particular, when flooding occurs, scouring occurs due to runoff in the ground adjacent to the bridge piers crossing the river, causing the bridge to collapse. This not only seriously affects the safety of people, but can also cause enormous losses to society and economy over a long period of time. Additionally, it is reported that the first cause of the bridge failure is not the structural defect of the bridge, but the destruction of the foundation due to scouring around the pier during flooding [1–4]. In general, railway bridges do not suddenly collapse; warnings are usually provided in advance. It is difficult to identify these abnormal signs via personnel-oriented irregular inspections or regular inspections at long-term intervals performed with inspection vehicles. Worldwide, studies have actively investigated the development of technologies for detecting the collapse of bridge piers in advance. In April 1987, a bridge collapsed at the Schoharie creek in New York, USA owing to the scour that occurred on the bottom surface of the pier footing. This accident resulted in more than ten human casualties as well as significant economic damage, and a research fund of 11 million USD was supported for the scour alone. Since then, research has been supported on a national level, led by the National Cooperation Highway Research Program (NCHRP). In addition, the Federal Highway Administration (FHWA) prepared the technical manuals of Hydraulic Engineering Circular (HEC)-18 [5], HEC-20 [6], and HEC-23 [7] for bridge scour, river stability, and countermeasures, respectively, due to active research and evaluation programs since 1987. These manuals are being used for the analysis and design of the bridge scour. Most of these studies, however, are focused on bridges that are built on sandy soils; thus, the characteristics of scour on soils other than sandy soils have not been considered. For scour analysis, the formula proposed by HEC-18 has been widely used. It is difficult to apply the formula to soils other than sand because the formula was obtained on the basis of experiments conducted on sand. Recently, a method that considered the scour rate and the influence of time [8] was proposed for clayey soils, and a new approach that used the erosion index [9] was attempted for rocky soils. In the Netherlands, systematic and comprehensive research on the scour pattern has been conducted as a national project by Dutch Delta Works since 1953. This research project was led by the Ministry of Transport, Public Works, and Water Management as well as Delft Hydraulics. Delft Hydraulics derived a semi-empirical scour formula as a function of time and location. This was achieved by performing numerous laboratory experiments while considering a variety of variables that are related to the hydraulic properties of the flow and the scour materials. They prepared a comprehensive technical manual that is referred to as the Breusers-equilibrium method on the scour phenomenon. This is based on the average flow velocity and the relative turbulence intensity of the flow and the dominant characteristics of time for the maximum scour depth [10].

As mentioned above, many projects have been conducted on the stability evaluation for piers and many studies have also been conducted. In previous research, the scour effect was simulated where the actual scour occurred by numerical analysis [11,12]. Cooley and Tukey [13] developed a fast Fourier transform (FFT) algorithm. Research on the bridge stability analysis and monitoring of the basis of this method has been conducted for many years [14–18]. In addition, many studies have been conducted to judge the state of the pier by its natural frequency. Sanayei and Maser [19] conducted research on the static measurement by using a vehicle load to estimate the ground stiffness of a bridge foundation. When a 200 kN truck passed each bridge that was built on a pile foundation and a footing, respectively, the stiffness ratio of the measured and theoretical values according to the foundation type were compared. Nishimura [20] introduced an impact vibration test method that measures the natural frequency of a pier by using the response waveform that was obtained by exciting its head with a weight of 300 N and it determines the stability from the changes in the natural frequency. Haya et al. [21] examined the possibility of estimating the natural frequency of a spread foundation pier with a microtremor measurement. They also compared the results of the impact excitation experiment and the microtremor measurement for a field bridge, and they reported that it was impossible to measure the natural frequency with a microtremor measurement. Samizo et al. [22], however, conducted research on a method for defining the natural frequency by measuring the microtremor of the existing bridge piers. Keyaki et al. [23] proposed a method that can identify the natural frequency by using the tremor measurement results alone without the impact vibration test results. Samizo et al. [24] developed a technique for evaluating the stability of the foundation. This was achieved by measuring the vibration of the pier using hydraulic power and analyzing the natural frequency change, and they also conducted research on soundness diagnosis indicators. Abe and Nozue [25] proposed a soundness diagnosis indicator that has a correlation with the natural frequency through a model test and the verification by a field measurement. Masahiro [26] proposed a statistical formula based on several measurement results by using the impact vibration test method. Japan's Ministry of Land, Infrastructure, and Transport [27] specified calculation formulas for each foundation type and the foundation soil type for railway maintenance standards, and they proposed a formula for the natural frequency of spread-foundation-type single-track piers.

As mentioned above, the stability of bridge substructures is closely related to the safety problem of bridges. The safety diagnosis and inspection, however, are focused on the materials for the bridge as well as structural problems, and there is no quantitative evaluation method for the substructures.

In this study, an impact load test was performed to analyze the effect of the surcharge load and scouring of the pier. This paper was focused on a bridge with shallow foundation and a plate girder deck because this is the most diffused typology in Korea. The full-scale model pier was built to analyze the effect of the surcharge load and confirm the mode shape and mode number of the bridge pier. Through the impact load experiment, it was possible to determine the three mode number of the pier according to the direction of the impact load. In addition, scouring was simulated using the pier of abandoned railway. The three mode number identified in the full-scale model experiment was derived and the effects of scouring were analyzed with natural frequencies.

2. Test Set-Up

2.1. Specifications of the Full-Scale Model Pier and the Cheongnyangcheon Bridge Pier

The full-scale model pier used the spread foundation type to evaluate the aged foundations. The pier foundation slab's dimensions were 5150 mm × 2420 mm × 50 mm (length × width × height), and the pier, which had a height of 4500 mm, was fabricated by repeating concrete pouring and curing with a height of 1500 mm for three times. The length and width of the pier were 4150 mm and 1420 mm, respectively.

The target pier of the field test was a shallow foundation type and an unreinforced concrete structure. The length and width of the top of the pier were 3900 and 1350 mm, respectively, and those of the bottom of the pier were 4680 and 2130 mm. The total length of the pier was 7800 mm; 3000 mm of the pier's length was embedded under the ground.

Figure 1 illustrates schematic view of the pier and Figure 2 shows the target pier for the impact load test.

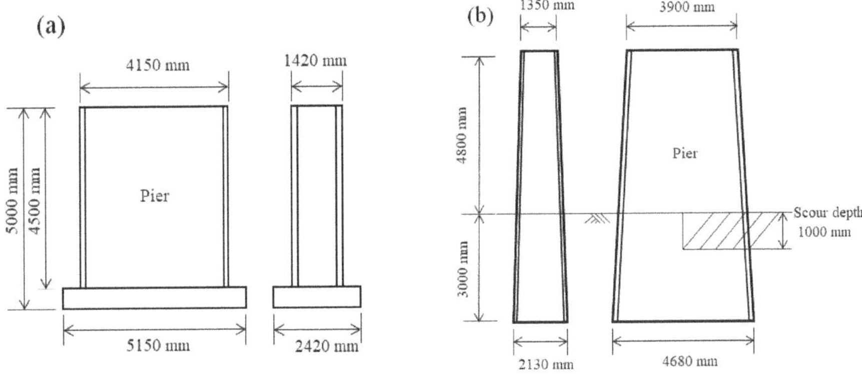

Figure 1. Schematic view of the pier: (**a**) full-scale model pier and (**b**) field bridge pier.

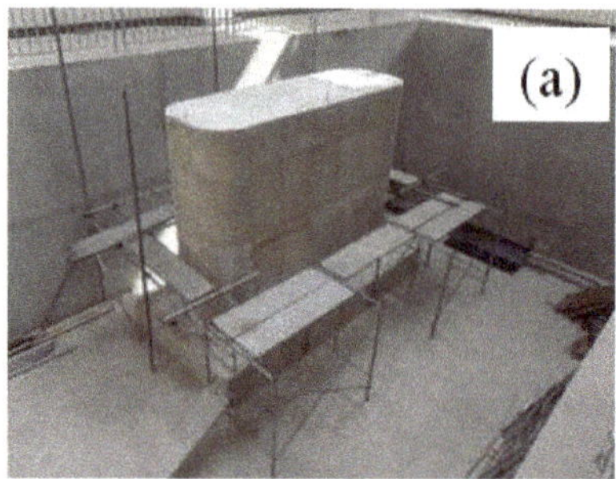

Figure 2. Target pier for the impact load test. (**a**) Full-scale model pier and (**b**) field bridge pier.

2.2. Non-Destructive Impact Vibration Test Method

The impact vibration test method can be used to evaluate the stability of the piers. The stability was evaluated based on the natural frequency that was derived when an impact was applied to the top of the pier in the pier length direction with a weight of approximately 0.3 kN. The impact vibration test method was proposed by Nishimura [20], who proposed a simple formula for the natural frequency of a sound pier by using the pier height, the girder weight, and the earth covering based on the natural frequency results of the piers that were derived through a series of tests.

Eight accelerometers were used to evaluate the stability of the pier through the impact load test, and a weight of 0.3 kN was used to apply an impact load. Figure 3 presents the measuring equipment and the weight that were used in the experiment. In the full-scale pier model test, the surcharge load slowly increased from 0 to 250 kN by 25 kN (a total of 11 loads) to analyze the influence of the surcharge load. Here, the surcharge load simulated the weight of the girder.

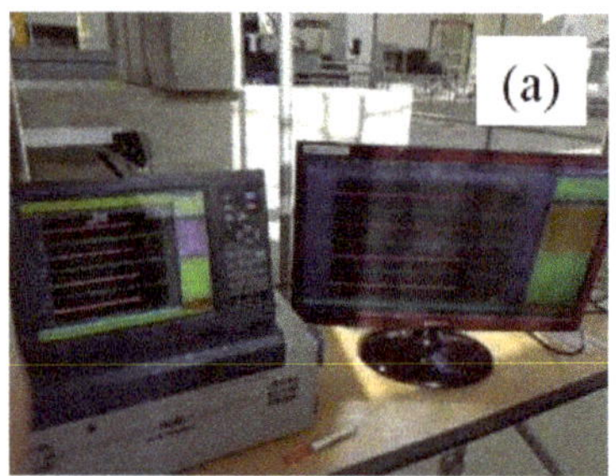

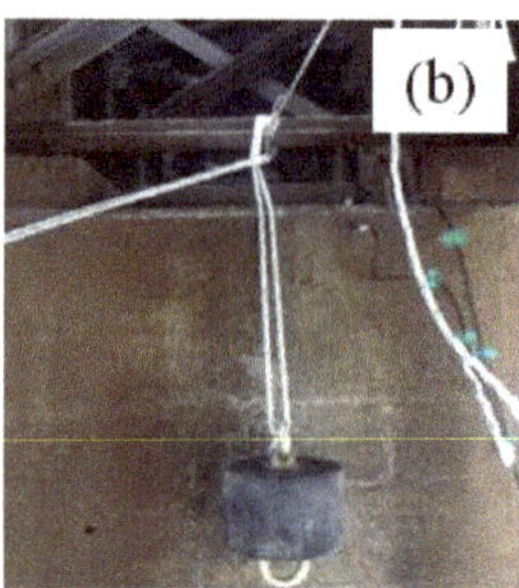

Figure 3. Equipment used in the experiment. (**a**) Data logger and (**b**) weight.

The field test was conducted for cases with or without a girder on the pier (Figure 4). In addition, the impact load test was conducted by simulating scour on one side of the ground that was adjacent to the pier to analyze the effect of the scour on the pier (Figure 5). The bridge pier in lab tests simulated the embedded in bedrock condition, and bridge pier in-situ condition simulated the embedded in weathered soil. The weather soil's SPT N value ranged 6–8. The full-scale model pier simulated a shallow foundation embedded weathered rock, in addition, therefore, the tests were performed without scour case.

Figure 4. Pier of field experiment (**a**) With girder on the pier and (**b**) without girder on the pier.

Figure 5. Simulated ground scour (scour depth = 1000 mm).

Figure 6 illustrates the test cases according to the impact direction. The accelerometers were attached to points 50 cm away from the top of the pier, a point 50 cm away from the bottom of the pier, and the center of the pier. Two accelerometers were attached to three points (a total of six accelerometers) to measure the acceleration in the bridge axis and pier length directions. Two accelerometers were attached to the outer surface of the pier to measure the acceleration in the pier length direction. An impact was applied to three points. The full-scale pier model test was repeated 33 times while considering the surcharge load, and the field test was repeated nine times. Table 1 summarizes the test cases.

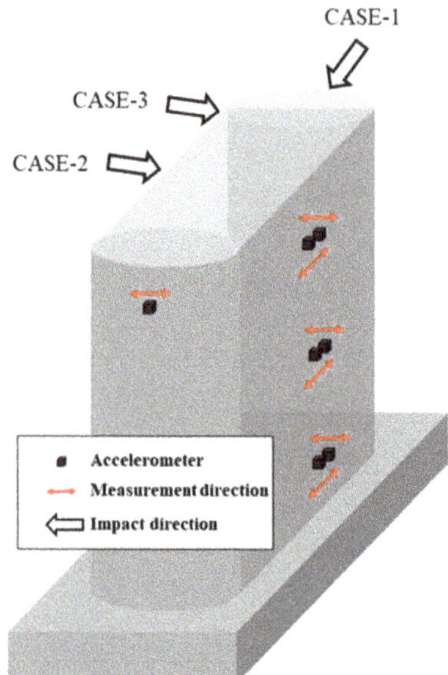

Figure 6. Conceptual diagram for the impact and measurement directions.

Table 1. Test cases.

Classification		Impact Direction	Surcharge Load	Scour	Impact Location
Full-scale pier model test	Case-1	Pier length direction	0–250 kN (increase by 25 kN)	N/A	
	Case-2	Bridge axis direction			
	Case-3	Bridge axis direction (outside)			Top of the pier
Field test	Case-1	Pier length direction	N/A	1-m depth	
	Case-2	Bridge axis direction			
	Case-3	Bridge axis direction (outside)			

3. Results

3.1. Mode Number Analysis for the Analysis of the Test Results

Prior to the full-scale pier model test and the field test, the eigenvalue of the pier was analyzed using Diana [28], which is a finite element software program, to analyze the behavior of the pier according to the impact direction. The finite element analysis only analyzed the mode number according to the direction of the impact load. Therefore, in order to reduce the variables in numerical analysis, the boundary condition between the pier bottom and the ground was set as a fixed condition. The size of the pier that was used for the analysis was the same as the full-scale pier. It was possible to analyze the behavior of the pier that corresponded to the first, second, and third modes according to the eigenvalue as shown in Figure 7. The pier exhibited displacement in the bridge axis direction in the first mode and in the pier length direction in the second mode. In the third mode, the torsional behavior of the pier was observed. According to the mode number analysis, the natural frequency corresponding to the second mode was caused by the impact in the pier length direction. In addition, the natural frequencies corresponding to the first and third modes were caused by the impact in the bridge axis direction. By analyzing the eigenvalue of the pier, the impact directions to derive the mode numbers of the full-scale model pier and the field pier could be selected. Based on this result, it was possible to determine the applicable impact load direction in the test, and full scale pier tests was establishment of the impact load test method to implement the pier behavior in the 1st–3rd modes.

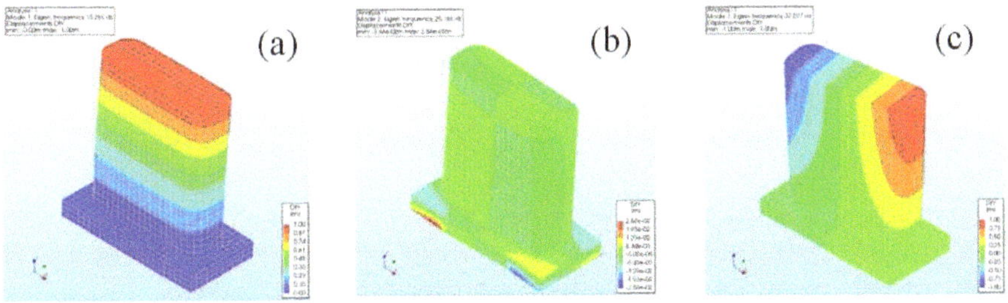

Figure 7. Mode number analysis. (**a**) First mode; (**b**) second mode, and (**c**) third mode.

3.2. Experimental Analysis Method

Figure 8a shows the acceleration values that were measured in the experiment. They are the measurement results of the impact in the bridge axis direction when the surcharge load was 0 kN. In the case of the impact vibration test, it is possible to determine the natural frequency and vibration mode by analyzing the spectrum of the repetitive waveform that was obtained from multiple recorded waveforms. For the spectrum analysis, it is desirable to use the FFT technique, which is capable of dividing the signals by the frequency. The natural frequency of the pier can be derived from the frequency domain using the FFT technique as shown in Figure 8b. In addition, the phase of the measured position can be represented as shown in Figure 8c. Based on this, the phase difference between the measuring instruments can be obtained, and the overall behavior of the pier can be analyzed. For example, in the case of the first mode and the second mode, the natural frequency occurs when the phase of the attached measuring instrument is in the same direction, so when the phase difference is 0°, it can be determined as the natural frequency of the first and second mode. In addition, in the case of the third mode, the natural frequency of the pier can be derived when the phase difference occurs 180° because it has torsional behavior.

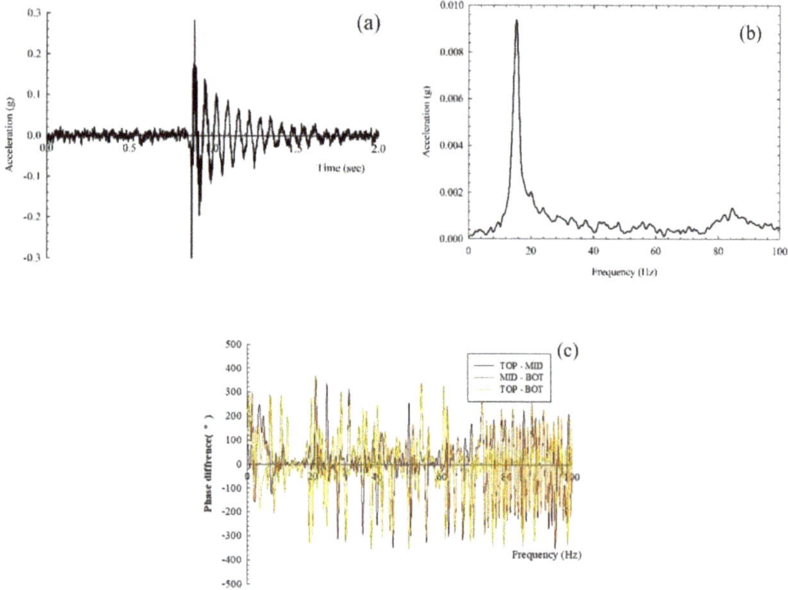

Figure 8. Natural frequency analysis method. (**a**) Representative signal; (**b**) representative natural frequency; and (**c**) representative phase.

3.3. Full-Scale Pier Test Results

Through a series of tests, the mode number of the pier was analyzed, and the natural frequency of the pier in each mode was derived. Figure 9a shows the natural frequency of the pier in case-2 where an impact was applied to the top of the pier in the bridge axis direction. The natural frequency was determined to be approximately 15.14 Hz. In addition, the phase was obtained for each height by using the measured values to analyze the behavior of the pier. Figure 9b shows the phase difference results that were obtained by using the phase for each position. It was determined that all of the phase differences where the natural frequency points occurred had a tendency to converge to 0°. This is because all the measuring instruments that were attached to each height were deformed in the same direction. In other words, the behavior of the pier occurred in the same direction. The behavior had a tendency to be similar to the first mode of the pier eigenvalue analysis.

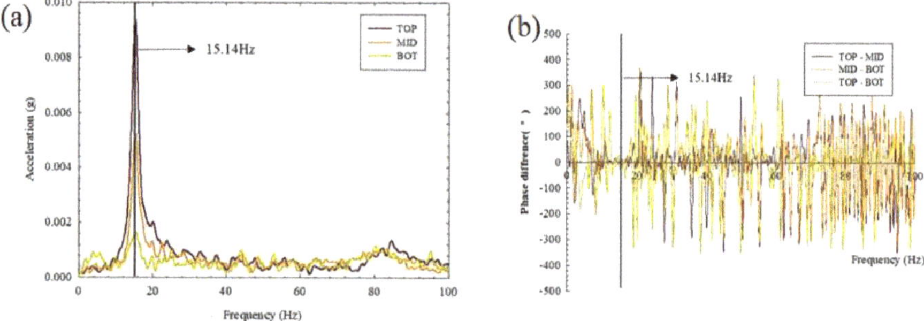

Figure 9. Results of the impact vibration load test in the bridge axis direction. (**a**) Natural frequency of the first mode and (**b**) phase difference of the first mode.

Figure 10a shows the results in case-1 where the natural frequency for each height was derived by applying an impact to the top of the pier in the pier length direction. The natural frequency of the pier was determined to be approximately 22.4 Hz, which is approximately 7 Hz higher in comparison to case-2 (bridge axis direction). This is because the stiffness of the pier was higher in the pier length direction than in the bridge axis direction. This confirms that the difference in the stiffness affected the natural frequency. To examine the behavior of the pier in the pier length direction, the phase difference was analyzed as shown in Figure 10b. In this instance, the phase difference converged to 0° when the natural frequency occurred along with the behavior in the first mode mentioned above. This appears to be similar to the behavior in the second mode of the pier eigenvalue analysis.

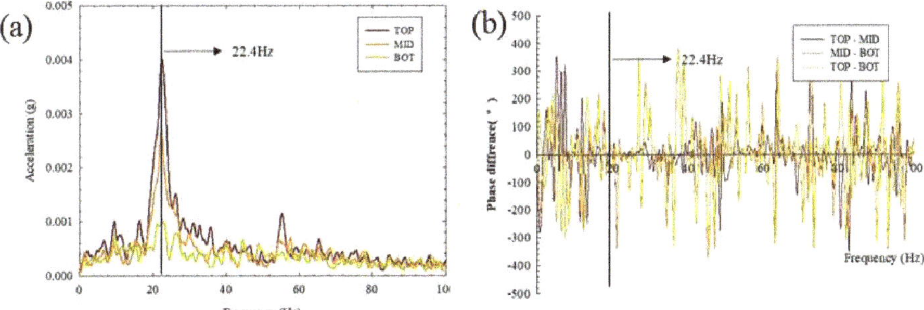

Figure 10. Results of the impact vibration load test in the pier length direction. (**a**) Natural frequency of the second mode and (**b**) phase difference of the second mode.

Figure 11a shows the natural frequency results in case-3 where the acceleration was measured by applying an impact to the upper outer point of the pier. Two clear natural frequencies were observed, and they were 15.14 and 54.19 Hz. The phase difference results in Figure 11b show that the phase difference converged to 0° at 15.14 Hz. This indicates that the first mode occurred, as the behavior and the natural frequency were the same as those of the first mode. Figure 11b, however, shows that the phase difference was 180° at 54.19 Hz. These results demonstrated that the outer part of the pier behaved in opposite directions; thus, indicating torsional behavior. This behavior was similar to the third mode, and it appeared that applying an impact to the outer point of the pier in the bridge axis direction led to the first mode and the third mode, which was the torsional behavior of the pier. Table 2 summarizes the natural frequency of the pier by the mode number according to the surcharge load.

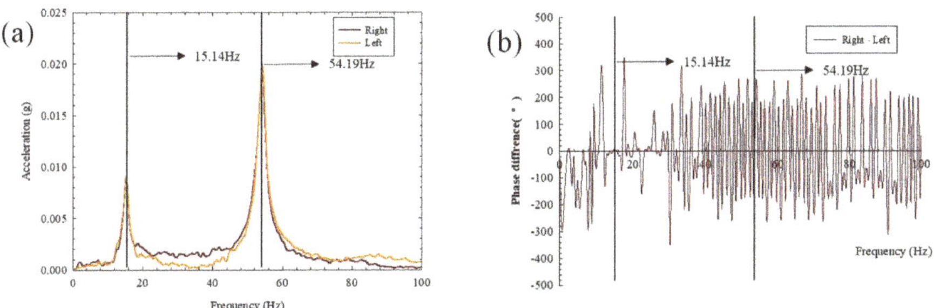

Figure 11. Results of the impact vibration load test in the bridge axis direction (outside). (**a**) Natural frequency of the third mode and (**b**) phase difference of the third mode.

Table 2. Natural frequency results of the full-scale model pier according to the surcharge load.

Surcharge Load (kN)	First Mode (Hz)	Second Mode (Hz)	Third Mode (Hz)
0	22.40	15.14	54.19
25	23.40	14.65	55.66
50	24.40	13.67	56.12
75	22.95	12.70	56.61
100	22.40	12.20	57.12
125	23.90	11.23	54.20
150	24.40	10.74	55.18
175	23.50	10.25	55.66
200	N/A	9.766	N/A
225	N/A	9.765	N/A
250	N/A	8.3	N/A

3.4. Field Pier Test Results

The conditions in the field pier test were mainly divided into three cases: (1) the pier with a girder, (2) the pier without a girder (removed), and (3) the 1 m deep scour on the ground that is adjacent to the pier. The natural frequency of the field pier was analyzed in the same way as the analysis method of the full-scale pier model test. Figure 10 shows the first mode natural frequency under the three field conditions. The natural frequencies under the field conditions were 21, 13, and 9.5 Hz, respectively, as demonstrated in Figure 12a–c. In the first mode, the natural frequency had the most sensitive change according to the field conditions (surcharge load and scour). This indicates that there are limitations in accurately evaluating the stability of the pier using the first mode natural frequency, which is affected by all of the variables.

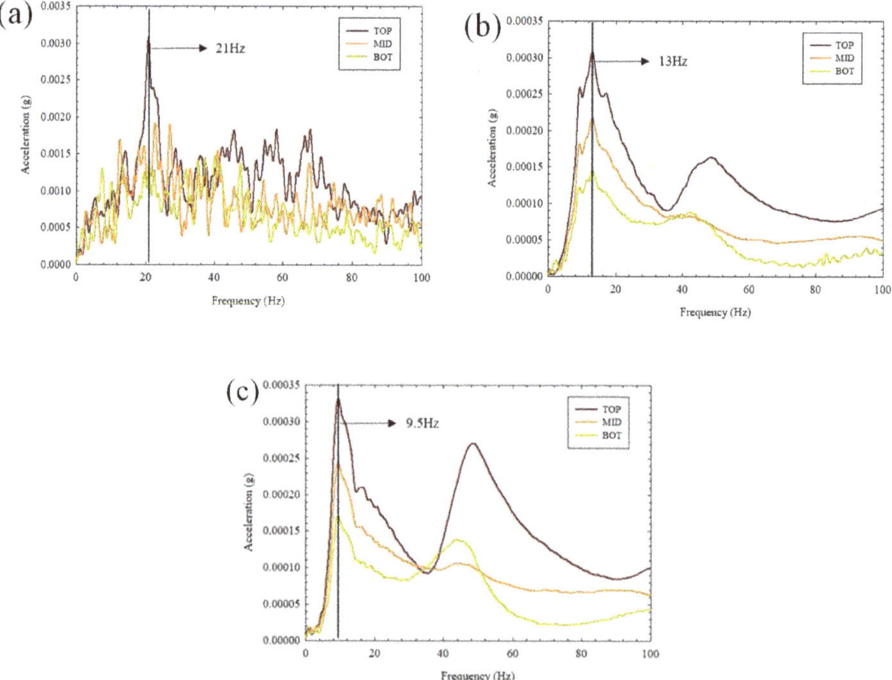

Figure 12. First mode natural frequency for each field condition. (**a**) With a girder; (**b**) Without a girder; and (**c**) 1 m deep scour on the ground adjacent to the pier.

Figure 13 shows the natural frequency of the second mode for each field condition. The natural frequencies under the field conditions were determined to be 20, 20, and 14 Hz, respectively, as displayed in Figure 13a–c. In the second mode, the natural frequency was identical regardless of the presence of a girder, unlike the first mode. This was similar to the result that the natural frequency of the second mode was not affected by the surcharge load in the full-scale pier model test. Due to the occurrence of scour, the natural frequency was reduced by approximately 6 Hz. Therefore, it is determined that the second mode natural frequency can be used as an indicator that can predict the condition of the ground that is adjacent to the pier without being affected by the surcharge load.

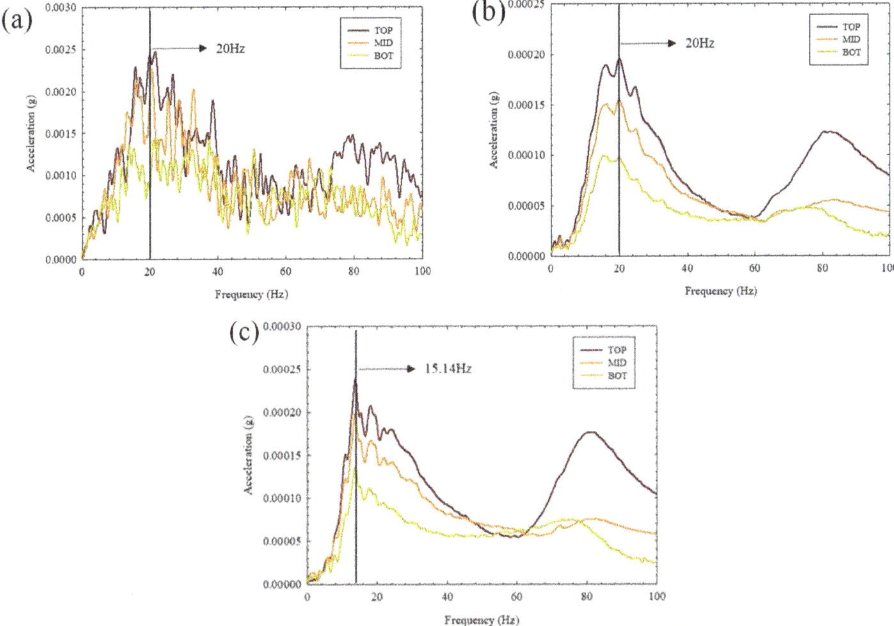

Figure 13. Second mode natural frequency for each field condition. (**a**) With a girder; (**b**) without a girder; and (**c**) 1 m deep scour on the ground adjacent to the pier.

Figure 14 shows the third mode natural frequency for each field condition when the torsional behavior of the pier occurred. The accelerometers were attached to the left and right of the top of the pier. The side that simulated scour was named left and the side that did not simulate scour was named right. There was no significant difference in the third mode natural frequency depending on the field conditions as in the first and second modes. Figure 14a shows the natural frequencies in the torsional behavior (third mode) before removing the girder. It was observed that very similar natural frequencies occur on both sides. In Figure 14b, however, a difference of approximately 2.5 Hz occurred between the natural frequencies that are measured on both sides after the girder removal. This is because the fixed end effect of the upper girder disappeared with the girder removal; thus, restraining the torsional behavior. Figure 14c shows the natural frequency results for the torsional behavior (third mode) when a 1 m deep scour was simulated on one side of the pier. The natural frequency values were determined to be similar to those in Figure 14b, but the acceleration of the side with a 1 m deep scour exhibited a sharp reduction in the amplitude. This indicates that it is possible to predict the location and degree of the scour. Table 3 summarizes the natural frequency results for each mode number of the pier according to the presence of scour. It was confirmed that the presence of scour decreases the natural [29–31].

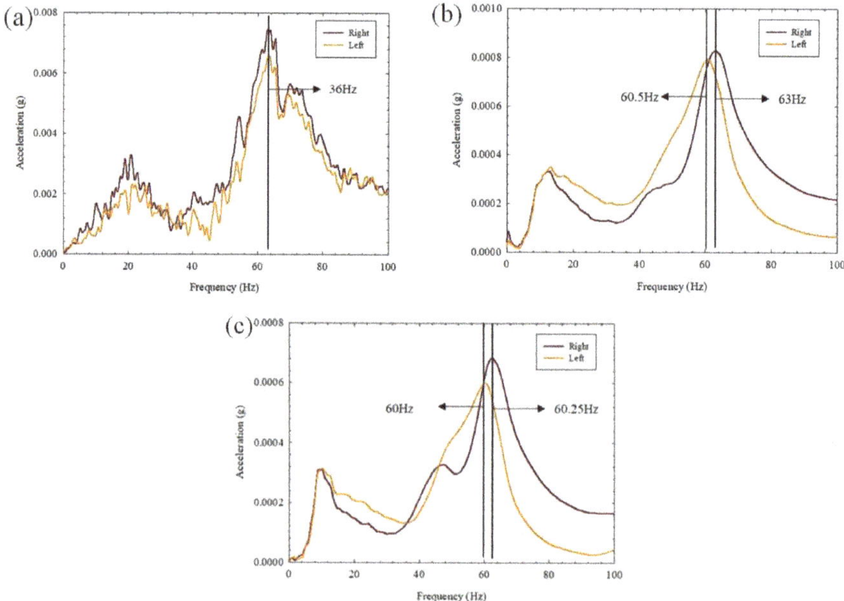

Figure 14. Third mode natural frequency for each field condition. (**a**) With a girder; (**b**) without a girder; and (**c**) 1 m deep scour on the ground adjacent to the pier.

Table 3. Natural frequency results of the field pier according to the presence of a scour.

Test Condition	Bridge Axis Direction (Hz) (First Mode)	Pier Length Direction (Hz) (Second Mode)	Torsion (Hz) (Third Mode)	
			Left (with Scour)	Right (without Scour)
With a girder	21	20	63	63
Without a girder	13	20	60.5	63
Without a girder and simulated 1 m deep scour	9.5	14	60	62.5

4. Discussion

4.1. Analysis of the Influence of the Surcharge Load through the Full-Scale Pier Test

Figure 15 presents the normalized natural frequency results of the full-scale pier for each mode number according to the surcharge load. As described above, the surcharge load was increased from 0 to 250 kN by 25 kN, and normalization was performed by assuming that the natural frequency that occurred for a surcharge load of 0 kN was 1. In the case of the first mode, the natural frequency showed a tendency to slowly decrease from 1 at 0 kN to 0.55 at 250 kN as the surcharge load increased. In the case of the second and third modes, however, the natural frequency was determined to be 1 or higher regardless of the surcharge load. Hence, the natural frequency of the structural system decreased with increasing mass, therefore the first mode of natural frequency increased eliminating the girder. The natural frequencies of the second and third mode are not significantly affected by the mass because the transverse direction is still stiffer than the longitudinal direction. For these modes, when the surcharge load was 200 kN or higher, it was not possible to calculate the natural frequency because clear signals could not be obtained. These results indicate that the mode number that was most affected by the surcharge load was the first mode.

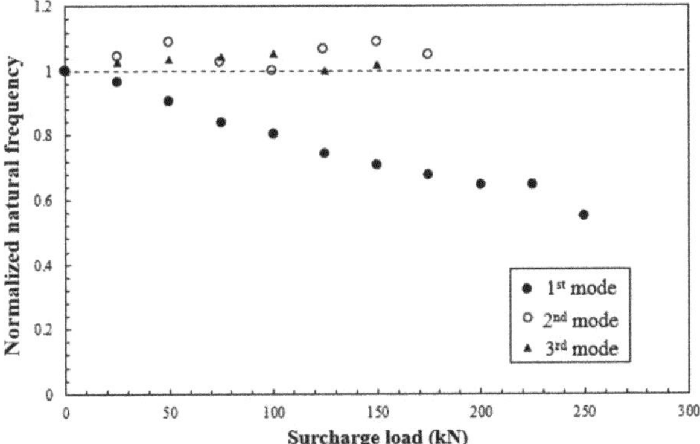

Figure 15. Normalized natural frequency according to the surcharge load (first, second, and third modes).

4.2. Analyzing the Influence of the Scour through the Field Pier Test

Figure 16 shows the normalized natural frequency results for each mode number according to the presence of a scour. Step 0 represents the test results with a girder and step 1 represents the test results after the girder removal. Step 2 represents the test results when the 1 m deep scour was simulated on one side of the ground that is adjacent to the pier. In the case of step 1, the natural frequency of the first mode decreased, but those of the second and third modes were similar. This is in agreement with the result of the full-scale model pier test that the second mode was not affected by the surcharge load. In the case of step 2, the natural frequencies of the first and second modes decreased, but the third mode remained similar. These findings indicate that the mode number that is most affected by the structural condition of the pier and the ground condition is the first mode. If the stability of a pier is evaluated using the first mode, it will be difficult to identify accurate problems. Therefore, it is reasonable to determine the boundary state of the ground that is adjacent to the pier using the second mode, which is affected by the scour in the ground, even though it is not affected by the surcharge load. In addition, the third mode is considered to be an effective method for determining the scour direction in the ground as described above in relation to Figure 12.

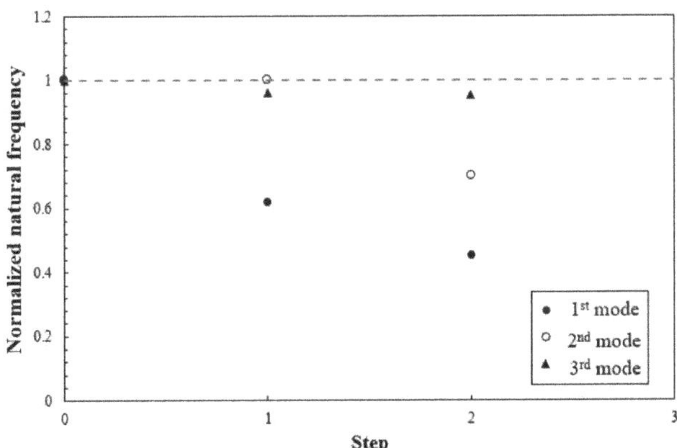

Figure 16. Normalized natural frequency according to the presence of the scour (first, second, and third modes).

5. Conclusions

In this study, the dynamic response analysis was performed on a shallow foundation used as a railroad pier through non-destructive tests. The full-scale model was conducted by constructing a full-scale model pier, and the effect of scouring was performed through the field pier test. Natural frequencies and phase differences were calculated by measuring the acceleration, and the modal number of piers was analyzed. The results of the study are summarized as follows.

1. Using the numerical analysis, the eigenvalue and mode number of the pier were derived according to the direction of impact. Based on this, the test method for the piers that can derive the first, second, and third modes was established through the full-scale model pier test and the field pier test.

2. Through the full-scale model pier, the natural frequencies of the first, second, and third modes were derived when the surcharge load on the pier increased. It was determined that the natural frequency of the first mode decreased as the surcharge load increased, and the second and third modes were not significantly affected by the surcharge load.

3. Through the field pier test, scour was simulated on the ground that was adjacent to the side of the pier to measure the natural frequency when the scour occurred. Due to the influence of the scour, the first mode exhibited the largest decrease in the natural frequency, followed by the second and third modes. In the case of the third mode, the amplitude of the acceleration was significantly small on the side that simulated the scour even though the natural frequency change was the smallest. This indicates that the direction of the scour can be determined through the third mode.

4. The results of the full-scale model pier test and the field pier test showed that the mode number that is most affected by the surcharge load and the scour is the first mode. If the stability of a pier is evaluated with the first mode, there are limitations in identifying accurate problems. Therefore, it is reasonable to determine the boundary state of the ground that is adjacent to the pier by using the second mode, which is not affected by the surcharge load.

5. These research results have a limitation for applying to other types of bridge piers and can be applicable to the deteriorated bridge with a shallow foundation and a plate girder. For further study, additional field tests and analysis will be performed and an indicator will be suggested for applying other type foundations, such as pile foundation.

Author Contributions: M.L. organized the paperwork, made a test plan, performed the impact load test; M.Y. and H.-S.J. helped the data analysis; K.H.K. performs numerical analysis; I.-W.L. supported making a test plan; all authors contributed to the writing of the paper. All authors have read and agreed to the published version of the manuscript.

Funding: This research was supported by a grant from R&D Program (PK2002A4) of the Korea Railroad Research Institute, Republic of Korea.

Conflicts of Interest: The authors declare no conflict of interest.

References

1. Shirole, A.M.; Holt, R.C. Planning for a comprehensive bridge safety assurance program. *Transp. Res. Rec.* **1991**, *1290*, 137–142.
2. Smith, D.W. Bridges failures. *Proc. Inst. Civ. Eng.* **1976**, *60*, 367–382. [CrossRef]
3. Ng, S.K.; Razak, R.A. Bridge hydraulic problems in Malaysia. In Proceedings of the 4th International Seminar on Bidges and Aqueducts 2000, Mumbai, India, 7–9 February 1998.
4. Lopardo, R.A. *Some Aspects of the Argentine Experience on Local Scour*; No. 19; Briaud, J.L., Ed.; ISSMGE Technical Committee-33 on Scour of Foundations: Melbourne, VIC, Australia, 2000; pp. 210–218.
5. Arneson, L.A.; Zevenbergen, L.W.; Lagasse, P.F.; Clopper, P.E. *Hydraulic Engineering Circular No. 18: Evaluating Scour at Bridges*; US Department of Transportation: Washington, DC, USA, 2012.

6. Lagasse, P.F.; Schall, J.D.; Richardson, E.V. *Hydraulic Engineering Circular No. 20: Stream Stability at Highway Structures*; Rep. No. FHWANHI-01; US Department of Transportation: Washington, DC, USA, 2001; Volume 2.

7. Lagasse, P.F.; Byars, M.S.; Zevenbergen, L.W.; Clopper, P.E. *Hydraulic Engineering Circular 23: Bridge Scour and Stream Instability Countermeasures: Experience. Selection and Design Guidance*; FHWA HI-97-030; US Department of Transportation: Washington, DC, USA, 1997.

8. Briaud, J.L.; Chen, H.C.; Kwak, K.W.; Han, S.W.; Ting, F.C.K. Multiflood and multilayer method for scour rate prediction at bridge piers. *J. Geotech. Geoenviron. Eng.* **2001**, *127*, 114–125. [CrossRef]

9. Annandale, G.W.; Wittler, R.J.; Ruff, J.F.; Lewis, T.M. Prototype validation of erodibility index for scour in fractured rock media. In *Water Resources Engineering '98*; ASCE: Reston, VA, USA, 1998; pp. 1096–1101.

10. Breusers, H.N.C. Conformity and time scale in two-dimensional local scour. In *Hydraulic Engineering Reports*; Hydraulic Research Laboratory: Poona, India, 1966.

11. Zampieri, P.; Zanini, M.A.; Faleschini, F.; Hofer, L.; Pellegrino, C. Failure analysis of masonry arch bridges subject to local pier scour. *Eng. Fail. Anal.* **2017**, *79*, 371–384. [CrossRef]

12. Scozzese, F.; Ragni, L.; Tubaldi, E.; Gara, F. Modal properties variation and collapse assessment of masonry arch bridges under scour action. *Eng. Struct.* **2019**, *199*, 109665. [CrossRef]

13. Cooley, J.W.; Tukey, J.W. An algorithm for the machine calculation of complex Fourier series. *Math. Comput.* **1965**, *19*, 297–301. [CrossRef]

14. Prendergast, L.J.; Gavin, K. A review of bridge scour monitoring techniques. *J. Rock Mech. Geotech. Eng.* **2014**, *6*, 138–149. [CrossRef]

15. Zhang, R.R.; King, R.; Olson, L.; Xu, Y.L. Dynamic response of the Trinity River Relief Bridge to controlled pile damage: Modeling and experimental data analysis comparing Fourier and Hilbert-Huang techniques. *J. Sound Vib.* **2005**, *285*, 1049–1070. [CrossRef]

16. Chen, C.C.; Wu, W.H.; Shih, F.; Wang, S.W. Scour evaluation for foundation of a cable-stayed bridge based on ambient vibration measurements of superstructure. *NDT E Int.* **2014**, *66*, 16–27. [CrossRef]

17. Dackermann, U.; Yu, Y.; Niederleithinger, E.; Li, J.; Wiggenhauser, H. Condition assessment of foundation piles and utility poles based on guided wave propagation using a network of tactile transducers and support vector machines. *Sensors* **2017**, *17*, 2938. [CrossRef]

18. Maroni, A.; Tubaldi, E.; Ferguson, N.; Tarantino, A.; McDonald, H.; Zonta, D. Electromagnetic sensors for underwater scour monitoring. *Sensors* **2020**, *20*, 4096. [CrossRef] [PubMed]

19. Sanayei, M.; Maser, K.R. Bridge foundation stiffness identification using static field measurements. In Proceedings of the 1999 Structures Congress, New Orleans, LA, USA, 18–21 April 1999; ASCE: Reston, VA, USA, 1999; pp. 320–323.

20. Nishimura, A. A study on the Integrity Assessment of Railway Bridge Foundations. *RTRI Rep.* **1989**, *3*, 41–49. (In Japanese)

21. Haya, H.; Sawada, R.; Nishimura, A.; Koda, M. Comparison of impact vibration test with microtremor measurement for spread foundation piers. *RTRI Q. Rep.* **1995**, *36*, 71–77.

22. Samizo, M.; Watanabe, S.; Fuchiwaki, A.; Sugiyama, T. Evaluation of the structural integrity of bridge pier foundations using microtremors in flood conditions. *Q. Rep. RTRI* **2007**, *48*, 153–157. [CrossRef]

23. Keyaki, T.; Yuasa, T.; Naito, N.; Watanabe, S. Soundness evaluation method of the ground around the Pier foundation against scouring by microtremor measurements at both sides of the bridge piers. *Geotech. J.* **2018**, *13*, 319–327. (In Japanese) [CrossRef]

24. Samizo, M.; Watanabe, S.; Fuchiwaki, A.; Sugiyama, T.; Okada, K. Proposal of an algorithm for estimating the natural frequency of railway bridge piers under flood conditions. *Proc. Jpn. Soc. Civ. Eng. F* **2010**, *66*, 524–535. (In Japanese) [CrossRef]

25. Abe, K.; Nozue, M. Monitoring method of soundness of railway bridge piers across a river. *Railw. Res. Rep. RTRI* **2016**, *30*, 29–34. (In Japanese)

26. Masahiro, S. Diagnostic technology of railway bridge substructure using vibration. *Acoust. Soc. Jpn.* **2013**, *69*, 133–138. (In Japanese)

27. Japan's Ministry of Land, Infrastructure and Transport. Railway technical research institute, standard for maintenance and management of railway structures, etc. In *Explanation (Structures), Foundations Anti-Earth Pressure Structures*; Maruzen: Tokyo, Japan, 2007. (In Japanese)

28. DIANA FEA. *Finite Element Analysis User's Manual—Release 10.2*; DIANA FEA: Delft, The Netherlands, 2017.

29. Prendergast, L.J.; Hester, D.; Gavin, K.; O'Sullivan, J.J. An investigation of the changes in the natural frequency of a pile affected by scour. *J. Sound Vib.* **2013**, *332*, 6685–6702. [CrossRef]

30. Prendergast, L.J.; Gavin, K.; Doherty, P. An investigation into the effect of scour on the natural frequency of an offshore wind turbine. *Ocean. Eng.* **2015**, *101*, 1–11. [CrossRef]

31. Ju, S.H. Determination of scoured bridge natural frequencies with soil-structure interaction. *Soil Dyn. Earth Eng.* **2013**, *55*, 247–254. [CrossRef]

Article

Mechanical Characterization of Timber-to-Timber Composite (TTC) Joints with Self-Tapping Screws in a Standard Push-Out Setup

Chiara Bedon [1],[*], Martina Sciomenta [2] and Massimo Fragiacomo [2]

[1] Department of Engineering and Architecture, University of Trieste, 34127 Trieste TS, Italy

[2] Department of Civil, Construction-Architectural & Environmental Engineering, University of L'Aquila, 67100 L'Aquila AQ, Italy; martina.sciomenta@univaq.it (M.S.); massimo.fragiacomo@univaq.it (M.F.)

[*] Correspondence: chiara.bedon@dia.units.it; Tel.: +39-040-558-3837

Received: 25 August 2020; Accepted: 16 September 2020; Published: 18 September 2020

Abstract: Self-tapping screws (STSs) can be efficiently used in various fastening solutions for timber constructions and are notoriously able to offer high stiffness and load-carrying capacity, compared to other timber-to-timber composite (TTC) joint typologies. The geometrical and mechanical characterization of TTC joints, however, is often hard and uncertain, due to a combination of various influencing parameters and mechanical aspects. Among others, the effects of friction phenomena between the system components and their reciprocal interaction under the imposed design loads can remarkably influence the final estimates on structural capacity, in the same way of possible variations in the boundary conditions. The use of Finite Element (FE) numerical models is well-known to represent a robust tool and a valid alternative to costly and time consuming experiments and allows one to further explore the selected load-bearing components at a more refined level. Based on previous research efforts, this paper presents an extended FE investigation based on full three-dimensional (3D) brick models and surface-based cohesive zone modelling (CZM) techniques. The attention is focused on the mechanical characterization of small-scale TTC specimens with inclined STSs having variable configurations, under a standard push-out (PO) setup. Based on experimental data and analytical models of literature, an extended parametric investigation is presented and correlation formulae are proposed for the analysis of maximum resistance and stiffness variations. The attention is then focused on the load-bearing role of the steel screws, as an active component of TTC joints, based on the analysis of sustained resultant force contributions. The sensitivity of PO numerical estimates to few key input parameters of technical interest, including boundaries, friction and basic damage parameters, is thus discussed in the paper.

Keywords: timber-to-timber composite (TCC) joints; push-out (PO) test setup; inclined self-tapping screws (STSs); finite-element (FE) method; cohesive zone modelling (CZM) method; boundaries; friction; sensitivity study

---

1. Introduction

Timber-to-timber composite joints are widely used in novel or existing buildings, with variable detailing (i.e., type of fasteners, detailing, spacing, arrangement, etc.). Among others, self-tapping screws (STSs) are particularly efficient due to their continuous thread, and their high withdrawal capacity allows one to realize connections with increased stiffness and load-carrying capacity. The benefit of STSs, compared to traditional TTC joints, can be clearly perceived, particularly when the screws are used with an inclined configuration with respect to the timber grain. On the other side, the arrangement of screws, requires the designer to account for several aspects that could directly affect the load transfer mechanism of a given TTC joint, including the bending capacity of screws, the embedment strength of wood, the withdrawal capacity of fasteners, the amount of friction phenomena between the involved components. Appropriate assessment methods and tools are thus required for their accurate mechanical characterization (Figure 1).

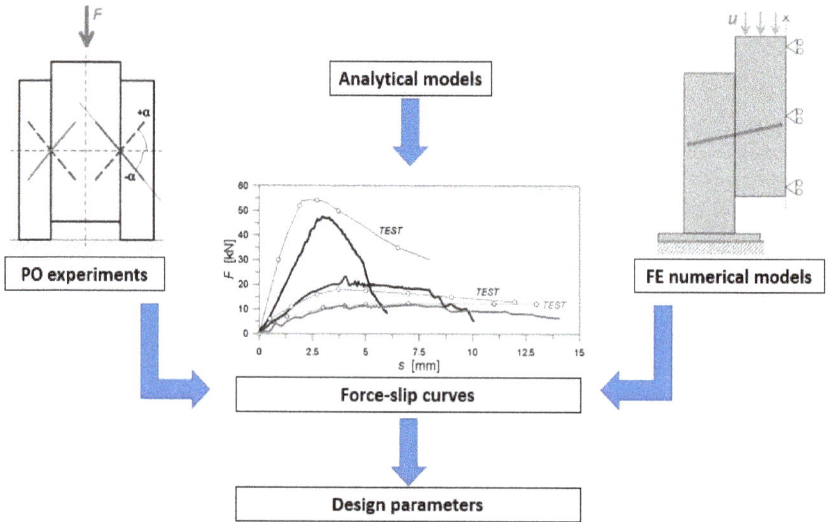

Figure 1. Available tools for the mechanical characterization and analysis of design parameters of timber-to-timber composite joints.

In the last years, analytical formulations have been proposed for the prediction of the expected stiffness and load-carrying capacity of TTC screwed connections [1–7]. However, it is generally recognized that both the joint features and the loading conditions can strongly affect the overall mechanical performance. This is in contrast with most of the design applications, that are commonly developed on simplified estimates of serviceability (or ultimate) stiffness and ultimate resistance values, and generally use a constant stiffness value for joints with variable spacing. In this regard, the more refined analytical methods of literature (i.e., [6,7]) are still partially capable to capture the actual performance of inclined STSs configurations. In the years, several research studies have been thus focused on more refined but cost/time consuming experimental investigations for the assessment of TTC joints, aiming at overcoming the actual gaps of design knowledge [8–12], and including also several timber-concrete composite (TCC) solutions [13,14], or novel hybrid techniques for TTC beams with inclined STSs [15–17].

This paper presents an extended Finite Element (FE) numerical study (ABAQUS/Explicit [18,19]) that takes into account a wide set of configurations for TTC joints with inclined STSs, under a standard push-out (PO) setup. The numerical investigation takes inspiration from past experimental results reported in [6] by Tomasi et al., where various joint prototypes of technical interest have been explored, as well as from [20,21], where an enhanced cohesive zone modelling (CZM) approach has been proposed and validated in support of an enhanced mechanical characterization of the fasteners. The reference modelling strategy is described in Section 3.

The numerically predicted resistance and stiffness parameters are thus briefly compared in Section 4, based on the available experimental and analytical data from [5,6]. As a reference, standard test procedures for timber joints are taken into account from [22,23]. Accordingly, the correlation of collected data is assessed with the derivation of empirical fitting curves. Successively, the load-bearing performance and sensitivity of TTC-PO specimens to some key input parameters is explored (Section 5).

The attention is focused on the effects of the base restraint, as well as friction phenomena and timber contact interfaces, and further on some basic CZM damage parameters for failure detection (Section 6). In doing so, major advantage is taken from the numerical derivation of the resultant forces that are sustained separately by the STSs or transferred by the timber components.

2. Background

2.1. Reference Experimental Approach

The short-term mechanical performance characterization of connections and joints of typical use for TTC (or TCC) systems according to Figure 2 generally depends on several uncertainties. Most of the parameters that are of primary need for analytical calculations and design are in fact sensitive to several geometrical features and loading conditions, thus affecting the corresponding serviceability stiffness and ultimate resistance values.

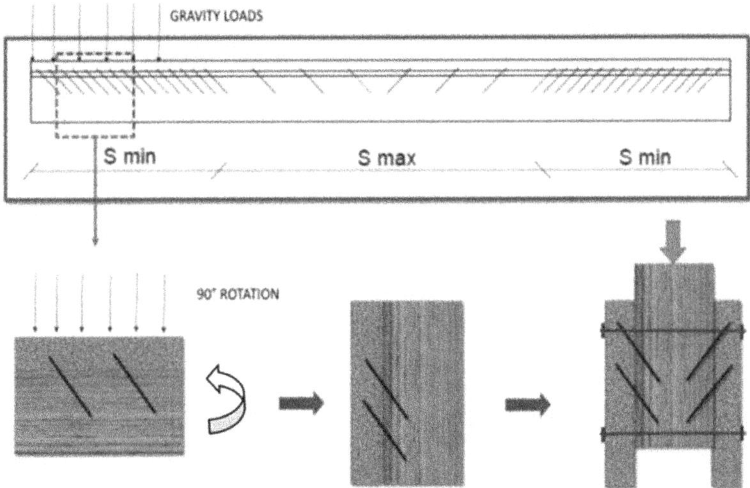

Figure 2. Conventional experimental procedure for the mechanical characterization of TTC joints with inclined STSs.

Moreover, most of the simplified methods of literature that can be used for TTC joints still lack consideration of several relevant aspects that especially in the case of timber, may have severe effects on the overall structural performance assessment (i.e., crushing or plasticity, time-dependent phenomena in the joint components, occurrence and evolution of local damage mechanisms, etc.). The actual result is thus represented by the need of extensive experimental testing in support of the required stiffness and resistance calculations [22,23]. The loading procedure recommended by the Eurocode 5 for timber structures [5], in this regard, requires for a standard PO test the repetition of 25 load cycles between 5% and 40% of the expected failure load. The specimen is then pushed further to collapse. Moreover, several test repetitions should be carried out for each joint configuration. Finally, a multitude of instruments is recommended to capture and control the specimen performance.

2.2. Selected Push-Out Specimens and Configurations

The numerical study discussed herein takes into account a series of TTC joints with inclined STSs characterized by geometrical variations in the inclination, the number and the position of fasteners, the loading direction (i.e., shear-compression ($\alpha < 0$) and shear-tension ($0 < \alpha \leq 45°$)). In accordance with Figure 3a–c, the typical TTC joint consists of three glued laminated timber elements classified as GL24h strength class (EN 1194 [24,25]). The mechanical connection of these spruce members is ensured by double-thread, carbon steel STSs (WT-T-8.2 type [26,27]), with a total length of L=190 mm or 220 mm (Figure 3d). Four joint typologies are thus numerically investigated in this paper (S#1-to-S#4 in Figure 3, with α = var), where [6]:

- S#1 = is a 2 + 2 screwed joint ($-45° \leq \alpha \leq 45°$)
- S#2 = represents a 4 + 4 screwed joint ($-45° \leq \alpha \leq 45°$), with a_1 = 70 mm $\approx$ 8d and d = D_3
- S#3 = is a 4 + 4 screwed joint like S#3 ($-45° \leq \alpha \leq 45°$), but a_1 = 160 mm $\approx$ 18d
- S#4 = denotes a 2 + 2, X-shaped screwed joint ($0° \leq \alpha \leq 45°$)

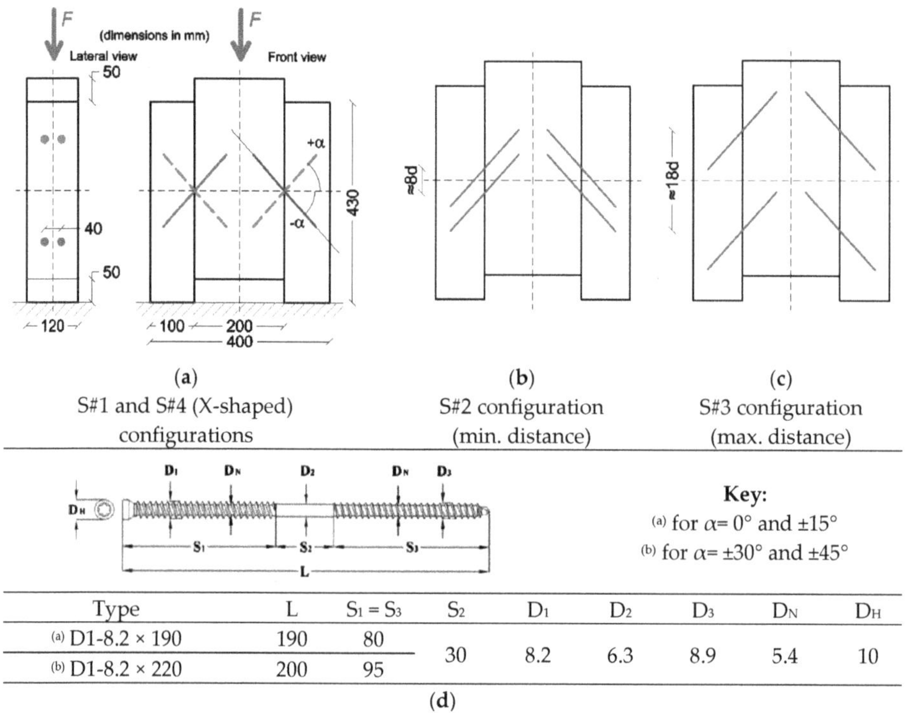

Type	L	$S_1 = S_3$	S_2	D_1	D_2	D_3	D_N	D_H
(a) D1-8.2 × 190	190	80	30	8.2	6.3	8.9	5.4	10
(b) D1-8.2 × 220	200	95						

(d)

Figure 3. TTC joints with inclined STSs under a standard PO setup. (**a**)–(**c**): selected configurations and (**d**) nominal dimensions (in mm).

3. Reference Numerical Modelling Approach

A key aspect of the full three dimensional (3D) solid models discussed in this paper is represented by the use of mechanical properties derived from [20,21]. Among others, a "soft layer" with CZM damage interactions is introduced at the interface between the steel STSs and the surrounding timber members. It was shown in [20,21] that the CZM damage modelling technique is particularly suitable for TTC joints with inclined STSs, where the region of fasteners can be sensitive to localized damage phenomena, with consequent relevant effects on the collected load-bearing responses. Compared to other modelling approaches, the CZM technique has well-known intrinsic advantages, since it does not need to pre-define potential cracks, or to introduce complex and computationally expensive adaptive mesh techniques, nor to define a very dense mesh pattern where cracks are expected.

3.1. Solving Strategy and Model Assembly

The numerical simulations are carried out with the ABAQUS/Explicit computer software [18,19], in the form of displacement-controlled, dynamic analyses with quasi-static deformations. All the FE assemblies are subjected to a linearly increasing vertical displacement, on the top face of the central member. The imposed vertical displacement is set in 20 mm, for all the examined configurations. Force-slip characteristic curves, as well as stress distributions and damage mechanisms in the TTC joint components are then monitored throughout the numerical investigation, a set of fixed FE assumptions is repeated for all the TTC joints under a standard PO setup, with major variations represented by trivial geometrical details.

The major simplification regards some basic symmetry considerations, thus 1/4th or 1/2nd the nominal geometry of each TTC specimen is taken into account. 8-node three-dimensional (3D) solid elements, (C3D8R-type) stress-strain bricks with reduced integration from the ABAQUS library are used for all the joint components. The reference FE model of TTC joint includes also a rigid base support made of steel, to allocate the lateral timber member in a standard PO setup. A swept (advancing font) meshing technique is then used to optimize the computational cost of simulations. The average edge size is minimized in the region of the STSs (0.3 mm-to-0.5 mm), and then maximized for the steel rigid base and the lateral portions of timber elements (5 mm-to-8 mm). Figure 4 shows an example of S#1 joint ($\alpha = -15°$). Based on [20,21], major efforts are then spent for the description of STSs and their mechanical interaction with the surrounding timber elements (Figure 5).

To this aim, each screw consists of an equivalent, circular cross-section with uniform diameter ($D_2 = 6.3$ mm from Figure 3d) and total length L.

A "soft layer" representative of STSs threads and timber fibers is then interposed between each screw and the surrounding timber ($D_1 = 8.1$ mm from Figure 3d).

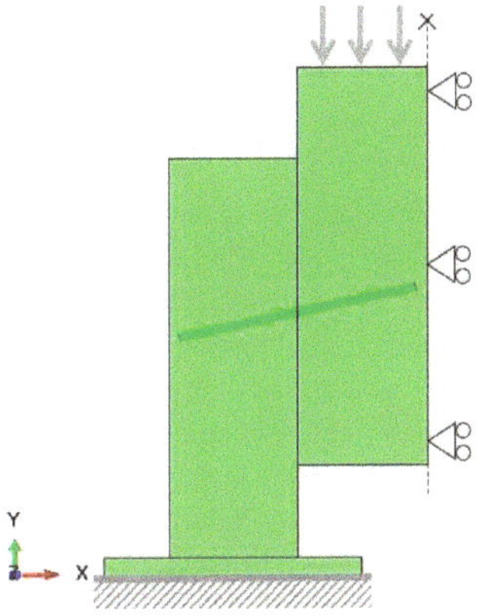

Figure 4. Reference FE numerical model for TTC joints with inclined STSs under a standard PO setup (S#1, $\alpha = -15°$).

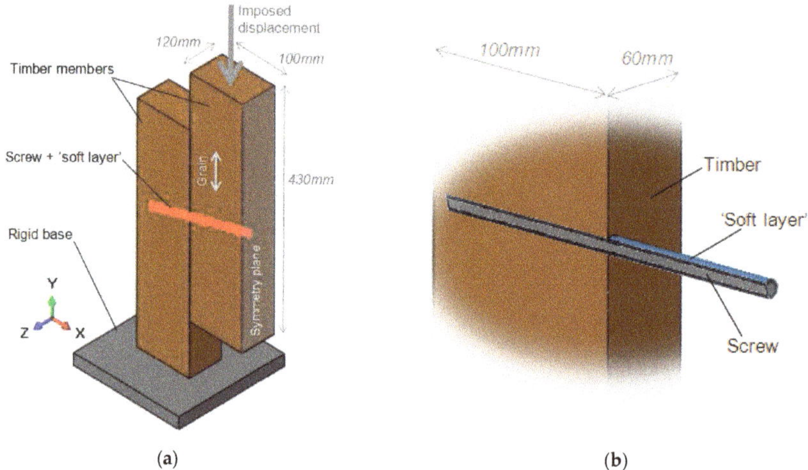

Figure 5. Example of (**a**) typical FE assembly (1/4th the S#1 geometry, with = −15°) and (**b**) detail of the STS region (ABAQUS/Explicit, hidden mesh pattern). Reproduced from [21] with permission from Elsevier®, Copyright license number 4895820420991, August 2020.

3.2. Material Properties

The mechanical response of both timber and steel components is described with the support of constitutive laws and material models available in the ABAQUS library. An isotropic, elastic-plastic Von Mises constitutive law is used for carbon steel screws (10.9 the nominal resistance class), with $E = 210$ GPa and $v = 0.3$ as nominal modulus of elasticity (MoE) and Poisson' ratio. Based on [6], the yielding/ultimate stress values are set in $f_y = f_u = 940.3$ MPa. An ultimate strain $\delta_u = 0.5\%$ is considered. At the same time, an orthotropic constitutive law is used for spruce [21]. To ensure more realistic behaviours for the timber members, this constitutive law is integrated with a Hill plastic criterion and a brittle failure parameter. The Hill criterion allows to account for different resistance values in the principal directions of timber, and thus for different potential critical mechanisms for the examined PO setup. The additional brittle failure law, moreover, is used to include possible crushing phenomena in the timber close to the fasteners. Once attained the ultimate resistance $f_{c,90}$, a linear propagation of compressive damage is taken into account for timber. Due to lack of more detailed experimental feedback, this material degradation is set to maximize at a failure deformation of $\delta_u = 4$mm. This value is adapted from [20], where C24 timber members have been investigated, based on the similarity in the resistance parameters for GL24h spruce. The final input parameters for timber are listed in Table 1, based on the nominal mechanical properties for GL24h strength class [24,25].

Table 1. Input mechanical properties for GL24h strength class timber (ABAQUS/Explicit).

Elastic Moduli (**Mean Values, in MPa**)	Parallel to the grain $E_\perp$	11600
	Perpendicular to the grain $E_\parallel$	390
	Radial E	390
	Longitudinal shear modulus G	690
Resistance (**Mean Values, in MPa**)	Compression parallel to the grain $f_{c,0}$	37.5
	Compression perpendicular to the grain $f_{c,90}$	3.57
	Shear f_v	3.85
Failure (**Mean Values**)	Maximum stress $f_{c,90}$ (MPa)	5
	Damage evolution	Linear
	Failure displacement δ_u (mm)	4

Finally, the equivalent soft layer in Figure 5 is mechanically characterized in accordance with [20,21], in the form of an indefinitely linear elastic material with GL24h input properties (Table 2).

Table 2. Input mechanical properties for the soft layer and for the CZM contact interaction (ABAQUS/Explicit).

Soft Layer	Elastic moduli (mean values)	Longitudinal (screw axis)	MPa	370
		Tangential	MPa	370
		Shear	MPa	720
		Radial	MPa	50
	Failure	Maximum shear (MPa)	MPa	5
		Damage evolution	-	Linear
		Failure displacement δ_u (mm)	Mm	4
CZM Contact Interaction	Resistance (mean values)	Longitudinal	MPa	37.55
		Transverse	MPa	3.85
		Shear	MPa	3.85
		Rolling shear	MPa	3.5

3.3. Mechanical Interactions and CZM Properties

The reliable mechanical performance of TTC numerical models is offered by an accurate calibration of material properties, but especially by the combination of multiple mechanical interactions between the involved load-bearing components (Figure 6).

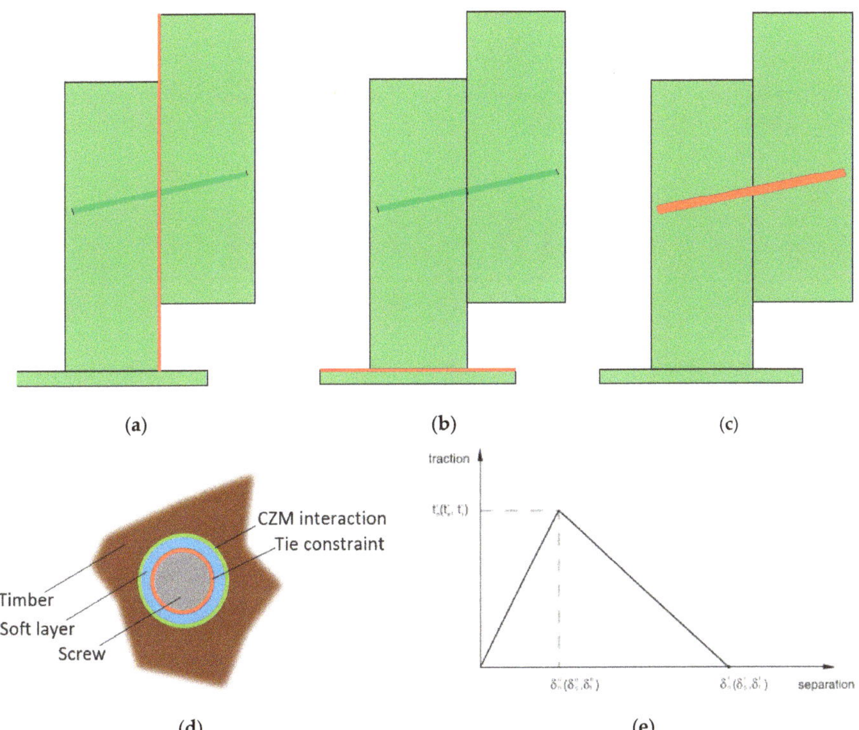

Figure 6. FE modelling TTC joints with inclined STSs under a standard PO setup. In evidence, the mechanical contact interactions for: (**a**) timber-to-timber, (**b**) base support and (**c,d**) STS details (hidden mesh pattern), with (**e**) traction-separation law.

For the typical FE model in Figure 6a, tangential "penalty" and normal "hard" surface-to-surface behaviors are first defined for the timber surfaces (μ_{timber} = 0.5 [28] the static friction coefficient). A second surface-to-surface contact interaction, see Figure 6b, is introduced between the bottom face of timber (lateral member) and the base steel support (μ_{base} = 0.2 [28]). Finally, see Figure 6c,d, a double restraint is used in the region of the steel fasteners. Each screw is first rigidly connected with the surrounding soft layer via a distributed "tie" constraint, so that relative rotations and displacements among the interested surfaces could be avoided. The external surface of the soft layer and the timber elements are then interconnected by a surface-based CZM behaviour, that is conventionally defined in its basic features (linear elastic traction-separation model (Figure 6e), damage initiation criterion, damage evolution law). In this study, the "default contact enforcement method" of ABAQUS library is used for the definition of the interface stiffness parameters prior to damage onset. The "Damage initiation", in this regard, is set to coincide with timber failure, based on Tables 1 and 2. This limit condition is implemented in the form of a maximum nominal stress (MAXS) criterion:

$$max\left\{ \frac{t_n}{t_n^0}, \frac{t_s}{t_s^0}, \frac{t_t}{t_t^0} \right\} = 1 \tag{1}$$

with t_n^0, t_s^0 and t_t^0 representing the allowable nominal stress peaks corresponding to normal deformations (n) to the bonding interface or in the first (s) or second (t) shear directions (GL24h resistance values). For the examined PO setup, any kind of rate-dependent behaviour for the traction-separation elasticity law is disregarded in this study. The damage evolution is finally set as "linear", that is:

$$t = (1 - D)\bar{t} \tag{2}$$

where D is a scalar damage variable ($0 \leq D \leq 1$) that interrelates the contact stress value t (in any direction), compared to its value predicted by the elastic traction-separation behaviour for the separation without damage. A null residual CZM contact stiffness is thus achieved at the first attainment of an ultimate displacement equal to δ_u = 4 mm (Table 2).

3.4. Analysis of Force Contributions

The derivation of relevant FE results is carried out on the basis of the collected numerical force-slip curves, for each one of the examined TTC configurations. More in detail, the attention of the post-processing stage is first focused on the shear force contributions that are sustained by the timber members and in the STSs. According to Figure 7, for an imposed vertical displacement, the total vertical reaction force F and the horizontal reaction force H at the base of each TTC joint are separately monitored. In the case of the vertical reaction F, moreover, the shear force terms sustained by the steel screws or at the timber-to-timber interface (by contact) are separately calculated, given that the overall load-bearing capacity of the TCC joint in the PO setup can be expressed as:

$$F = F_{screw} + F_{timber} \tag{3}$$

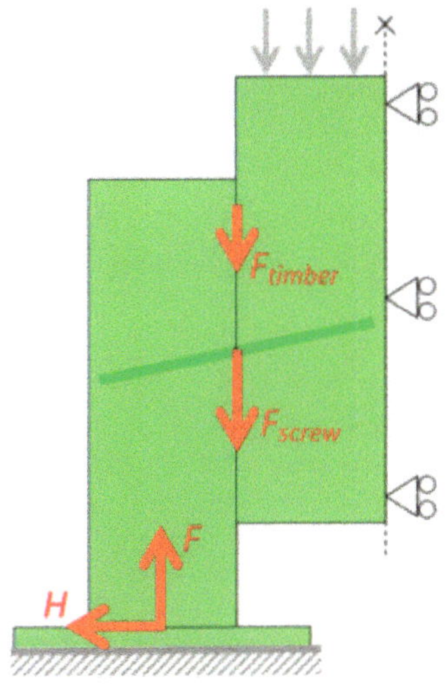

Figure 7. Monitored force contributions for the mechanical analysis and characterization of TTC joints with inclined STSs under a standard PO setup.

4. Discussion of FE Results

4.1. Force-Slip Curves

Generally, the FE modelling strategy herein presented proved to offer relatively good correlation with the selected literature data, both in quantitative and qualitative terms. For few configurations (especially for the TCC joints characterized by high α values), a major scatter was observed and justified by local damage phenomena that compromised the overall load-bearing performance of the FE assembled components.

Selected examples are shown in Figure 8 for S#1 specimens and different STS inclinations, while the corresponding test results are derived from [6].

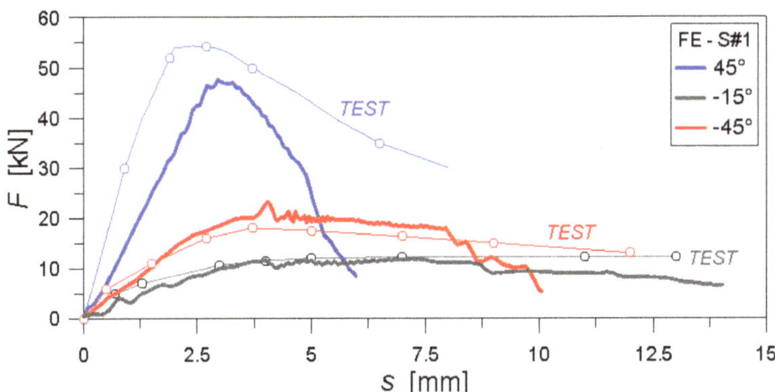

Figure 8. Force-slip curves of selected TTC joints with inclined STSs under a standard PO setup (ABAQUS/Explicit, S#1, α = var) and corresponding experimental results (data from [6]).

The typical PO analysis was thus stopped due to convergence issues, in the very late damaged stage. In most of the cases, see Figure 8, this ultimate collapse configuration was achieved for relatively small slip amplitudes ($s < 10$ mm), compared to the imposed displacement of 20 mm. For the FE predictions agreeing with Figure 8, the maximum resistance F_{max} can be conventionally detected as the first condition between the attainment of the (a) actual maximum force or (b) a total force corresponding to a joint slip s = 15 mm (if any). The corresponding serviceability stiffness K_{ser} is then given by [22,23]:

$$K_{ser} = \frac{0.4\,F_{max}}{\frac{4}{3}\left(s_{04} - s_{01}\right)} \tag{4}$$

with s_{04} and s_{01} the measured sliding amplitudes at the 40% and 10% part of the maximum resistance F_{max}.

4.2. Damage Mechanism

For the global and local analysis of parametric FE results, a key role is assigned to the detailing of stress peaks distributions and damage initiation/evolution in all the joint components. For the examined TTC configurations, the collapse detection of PO specimens was typically associated to a combination of:

- crushing phenomena in timber (in the region of screws);
- progressive yielding of screws and
- damage of the CZM contact (screw-to-timber interface).

A relevant example is proposed in Figure 9 (S#1 specimen with $\alpha = -15°$, at an imposed slip s = 12 mm). In the direction of the grain, the wooden fibers were generally subjected to high stresses peaks in a limited region only, when moving far away from the fasteners. This can be noticed in the crushed (red) regions of Figure 9a. The STSs in use, moreover, commonly failed due to the occurrence of two plastic hinges (Figure 9b), and this is in line with the experimental observations reported in [6]. For all the examined TTC joints, finally, a primary role was recognized for the CZM contact, being responsible of the final slope for the collected force-slip curves. In Figure 9c, in this regard, the non-dimensional CSMAXCRT parameter is shown (1 = fully damaged or 0 = undamaged interface). This parameter, for most of the examined joints, was observed to reach its maximum unitary value of failure with an extension up to $\approx$ 1/3rd the nominal length L for the STSs in use.

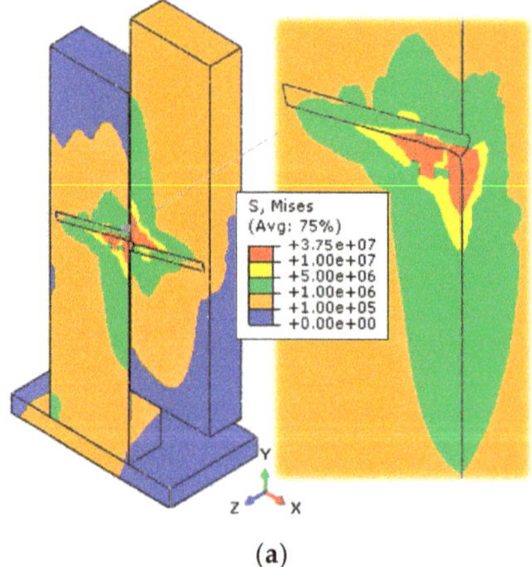

(**a**)

Figure 9. *Cont.*

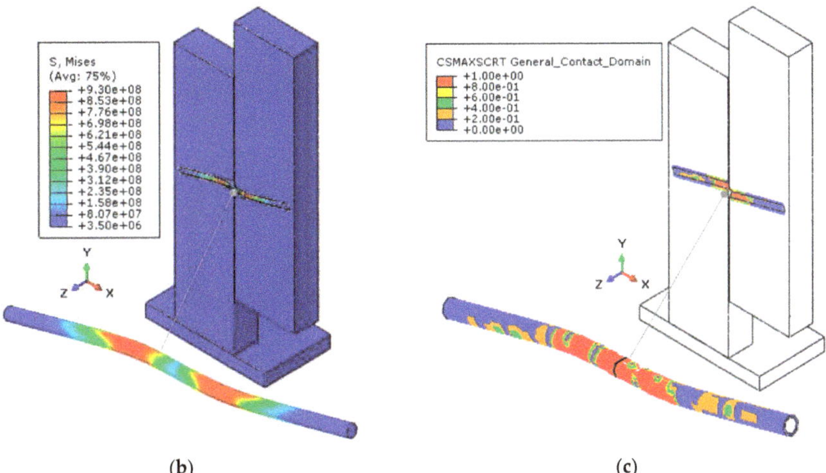

(b)   **(c)**

Figure 9. Typical damage propagation in the selected TTC joints with inclined STSs, under a standard PO setup (ABAQUS/Explicit). Example for the S#1 specimen with $\alpha = -15°$, with evidence of: (**a**) local damage of timber (stress values in Pa), (**b**) yielding of screws (stress values in Pa) and (**c**) CZM damage parameter. Reproduced from [21] with permission from Elsevier®, Copyright license number 4895820420991, August 2020.

5. Mechanical Characterization of TTC Joints

5.1. Experimental Assessment of Maximum Force Predictions (F_{max})

In Figure 10, some comparisons are proposed in terms of maximum resistance values for the S#1-to-S#4 TTC joints, grouped by series of specimens, with the support of experimental data from [6]. As a general outcome of the overall parametric numerical simulations, the FE models generally gave evidence of a mostly stable variation of the estimated F_{max} values with. Despite such a stable numerical dependency of F_{max} estimations on α, however, in some cases the scatter between numerical and past experimental predictions was found to be in the order of ±30%. The numerical results were in fact found to either underestimate or overestimate the corresponding experiments, depending on the number and inclination of STSs. For the majority of the examined TTC joints, the FE results proved to be non-conservative especially for the specimens under shear-tensile loads ($0 < \alpha \leq 45°$).

A possible motivation of such a kind of comparative outcomes could lie in localized numerical issues (i.e., numerical singularities, local damage phenomena), and this is especially the case of TTC joints with STSs characterized by high inclination values. In any case, given also the lack of a detailed experimental characterization for the mechanical properties of the material in use, the FE modelling approach herein discussed proves to offer reasonable estimations for the expected maximum force of TTC joints with inclined STSs, and thus to represent a valid support for design.

Worth of interest in Figure 11, in this regard, is the general trend of the calculated percentage scatter for the so-derived maximum force values F_{max}. In the figure, the scatter F_{max} is calculated as:

$$\Delta = 100 \cdot \frac{x_{FE} - x}{x} \tag{5}$$

where x_{FE} denotes the numerical force peak for each *FE* analysis and x the corresponding experimental average value (for each test series), as derived from [6].

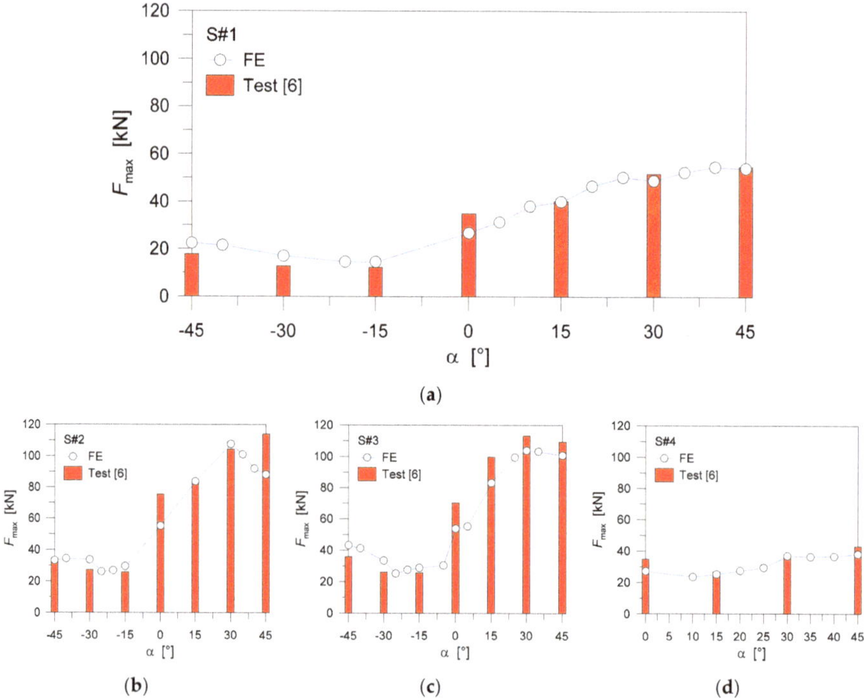

Figure 10. Comparison of numerical (ABAQUS/Explicit) and experimental [6] maximum force estimates for TTC joints with inclined STSs: (**a**) S#1, (**b**) S#2, (**c**) S#3 and (**d**) S#4 joints.

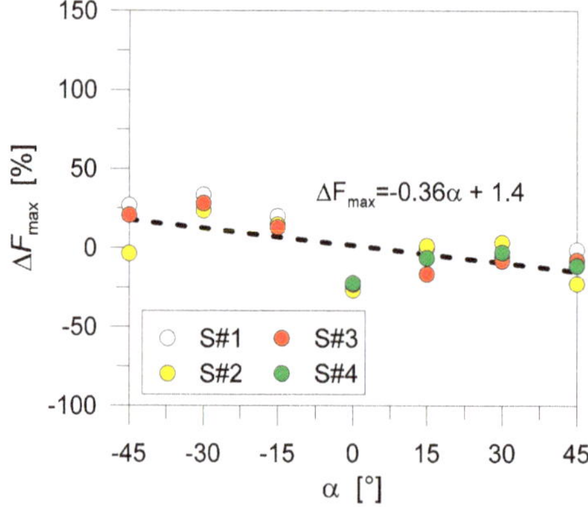

Figure 11. Percentage scatter of maximum force values for TTC joints with inclined STSs (Equation (5)), as obtained from the FE numerical analyses (ABAQUS/Explicit) and by the experiments in [6].

It is thus possible to notice that as far as the number and arrangement of the STSs in use modifies, the calculated F_{max} is mostly regular, for all the examined series of TTC specimens. This can be also perceived by the linear fitting curve that is proposed in Figure 11, as a function of the screw inclination α.

5.2. Analytical Assessment of Maximum Force Predictions (F_{max})

A further assessment of the collected FE numerical results can be carried out with the support of two suitable analytical models of literature, namely the Eurocode 5 provisions [5] or the enhanced analytical model proposed by in [6] by Tomasi et al. Figure 12 presents the so-collected comparative data, grouped by series of TTC joints. Disregarding the joint configuration and the number/inclination of STSs, the Eurocode 5 generally manifests a weak reliability of maximum force estimates. On the other side, Figure 12 shows an improved correlation between the analytical model from [6] and the FE numerical predictions discussed herein.

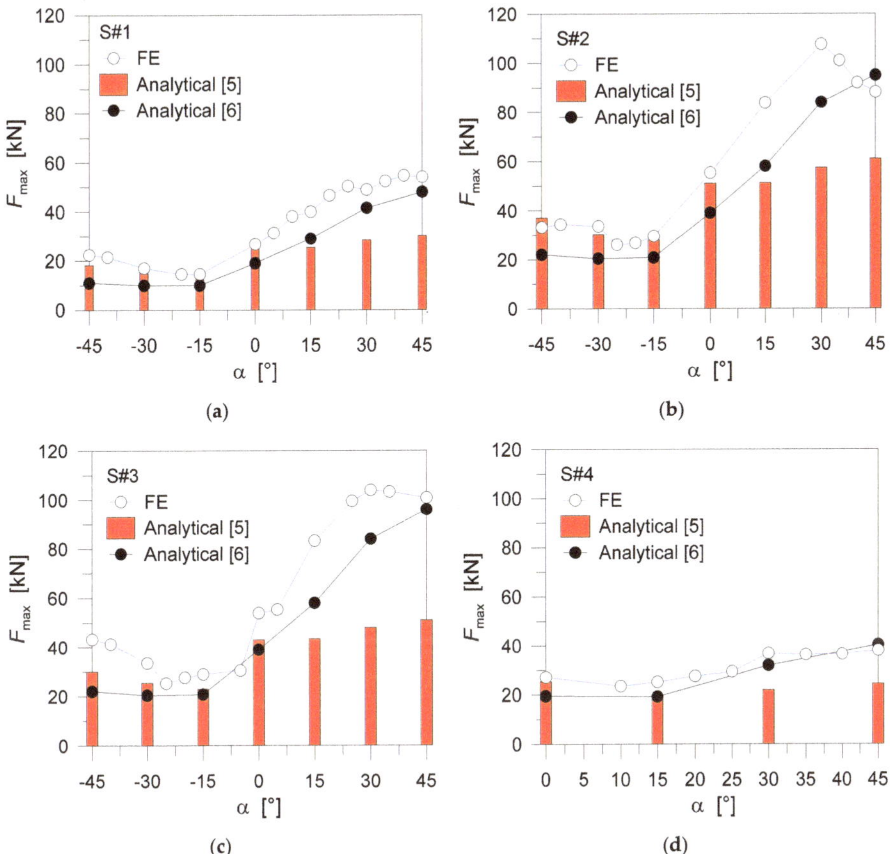

Figure 12. Comparison of numerical (ABAQUS/Explicit) and analytical (Eurocode 5 [5] or Tomasi et al. [6]) maximum force estimates for TTC joints with inclined STSs: (**a**) S#1, (**b**) S#2, (**c**) S#3 and (**d**) S#4 joints.

As far as the percentage scatter in Equation (5) is taken into account for the analytical assessment of numerical output data, the typical result takes the form of Figure 13. For most of the examined TTC configurations, it is important to notice that the FE models in use typically manifested a marked overestimation of the analytical predictions. Following the experimental validation in Section 5.1, this suggests that both the Eurocode 5 analytical approach and the enhanced analytical model in [6] are able to only roughly capture the complex mechanical behaviour of TTC joints with inclined STSs. In any case, Figure 13 shows that the calculated percentage scatter is generally less pronounced and regular for the FE numerical results towards the analytical formulation in [6], thus confirming the weakness of the Eurocode 5 approach.

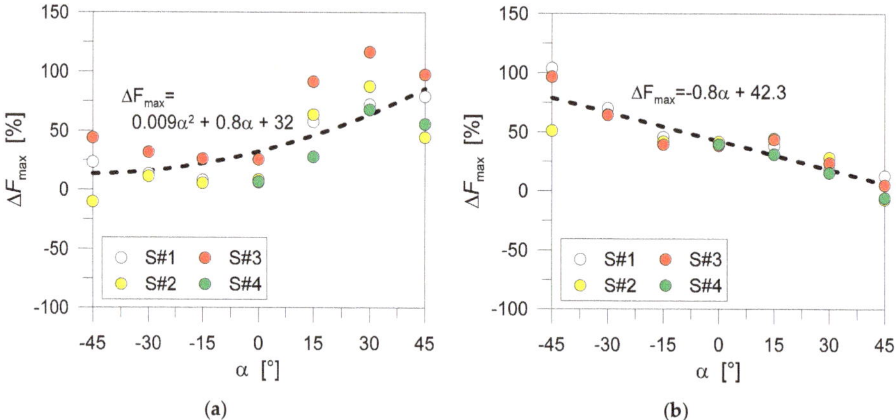

Figure 13. Percentage scatter of maximum force values for TTC joints with inclined STSs (Equation (5)), as obtained from the FE numerical analyses (ABAQUS/Explicit) and by literature analytical models: (**a**) Eurocode 5 [5] and (**b**) Tomasi et al. [6].

5.3. Experimental and Analytical Assessment of Serviceability Stiffness Predictions (K_{ser})

The elastic stiffness K_{ser} is then estimated for the examined TTC joints, based on Equation (3) and the collected numerical force-slip curves. In Figure 14, the FE stiffness values are reported for the S#1-to-S#3 type (average) or S#4 type of specimens, as a function of α. Comparisons are proposed towards the past experimental data from [6], as well as the enhanced analytical formulation proposed in [6].

In general, a rather close correlation can be observed for the stiffness trend of most of the S#1, S#2 and S#3 configurations in Figure 14. However, major scatter of the FE predictions to the experiments can be still observed especially for high α values, both for shear-tension and shear-compression loading conditions. In the case of X-shaped joints, even a more pronounced sensitivity can be observed in terms of stiffness estimations as a function of α, compared to the S#1-to-S#3 joints and the respective experimental data from [6]. In terms of analytical assessment, finally, the same numerical predictions in Figure 14 are comprised, for the majority of joint configurations, between the single/double stiffness predictions derived from [6].

As far as the percentage scatter is calculated from Equation (5), it is interesting to notice in Figure 15 that such a variation is less regular than in the case of maximum force predictions, when the inclination α modifies. For the experimental data in Figure 15a, the scatter trend is mostly regular for all the S#n joints, thus suggesting a certain stability of material properties and mechanical assumptions for the FE models in use. Major sensitivity can be perceived in Figure 15b,c, as far as the single stiffness or double stiffness analytical model from [6] is taken into account.

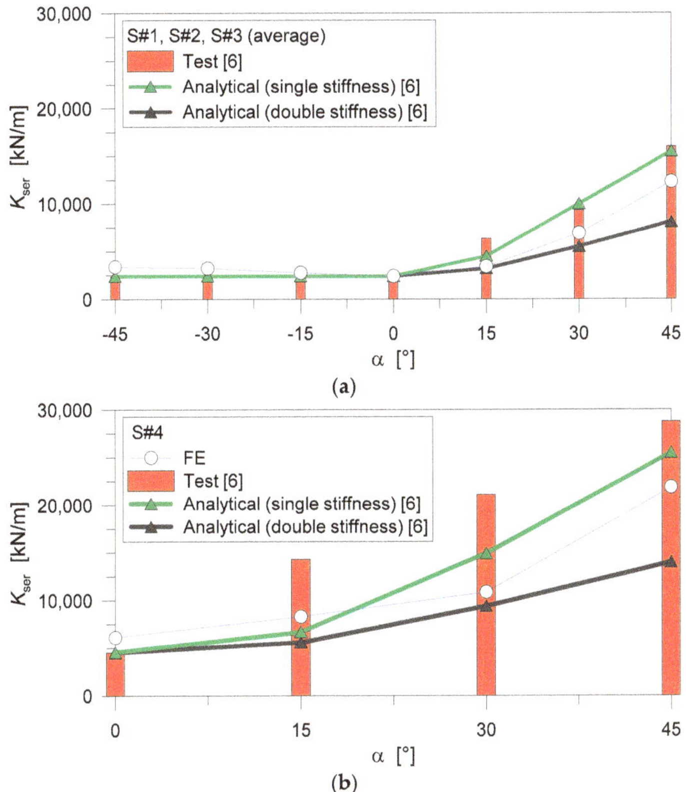

Figure 14. Comparison of numerical (ABAQUS/Explicit) and analytical [6] serviceability stiffness estimates for TTC joints with inclined STSs: (**a**) S#1-to-S#3 (average) or (**b**) S#4 joints.

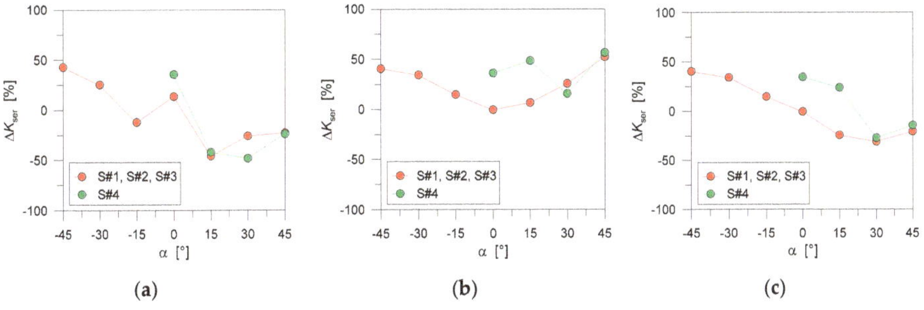

Figure 15. Percentage scatter of stiffness values for TTC joints with inclined STSs (Equation (5)), as obtained from the FE numerical analyses (ABAQUS/Explicit) and by literature [6]: (**a**) experimental data, (**b**) single stiffness analytical model, (**c**) double stiffness analytical model.

6. Parametric FE Investigation

Besides the rather close correlation in Section 5 for the FE predictions and the experimental and analytical results of literature, the sensitivity of the modelling technique to some influencing properties was further assessed. As a reference configuration, the typical FE model herein considered is thus characterized by input properties according to Section 3.

6.1. Mechanical Interactions and CZM Damage Parameters

A first insight is dedicated to the effects of mechanical interactions, with a special care of the STSs in use. It was shown in Section 3 the key role of the soft layer with the CZM contact interaction, as well as of reliable material property definitions. In this sense, the "upper limit condition" for the parametric study is assumed as a "tie" rigid constraint that is used to replace the CZM interaction for the soft layer (Figure 6). In other words, any kind of possible damage propagation in the region of screws (with the exception of possible material degradation in the timber and steel components) is fully disregarded. The "lower limit condition", at the same time, is set to coincide with the CZM formulation in Section 3 (with δ_u = 4 mm, Table 2). Among these two conditions, further FE analyses are carried out with the CZM input parameters of Section 3 (Table 2), but progressively increasing the reference failure displacement δ_u in the range from 4mm to a maximum of $10 \times 4 = 40$ mm. From a practical point of view, such a variation in δ_u represents a residual capacity of the soft-layer to provide a certain mechanical interaction between each STS and the surrounding timber. Such an input value was in fact magnified so as to reproduce an ideal bonding condition with a weak mechanical degradation for the soft-layer interface, even under large slip amplitudes (Figure 6e). Selected numerical results are proposed in Figure 16 for two different screw arrangements (S#1 joints), in terms of measured vertical (F) and horizontal (H) base reaction forces as a function of the measured slip s.

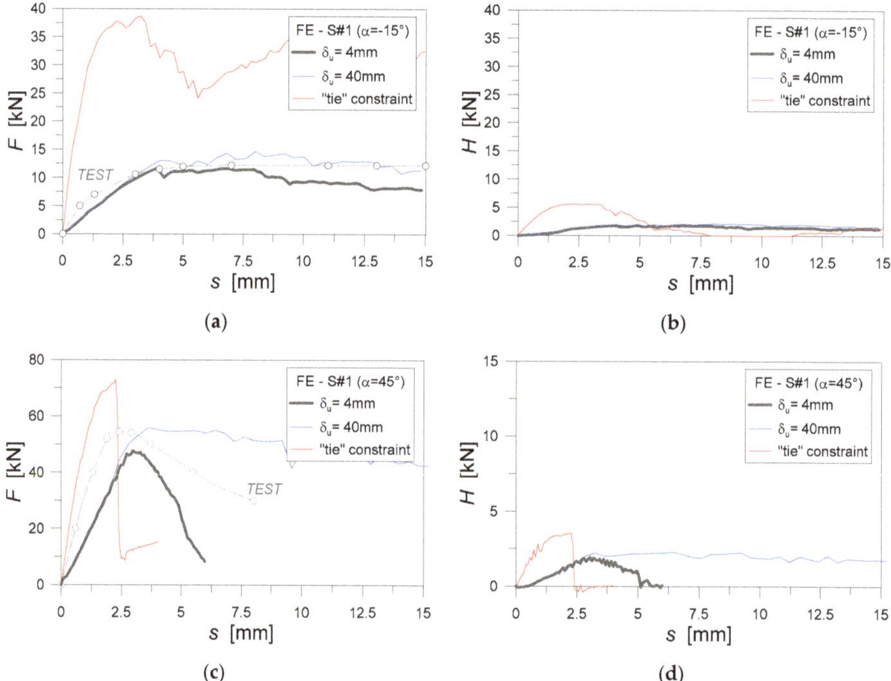

Figure 16. Analysis of mechanical interaction and CZM effects on the PO numerical response of TTC joints with inclined STSs (S#1). Vertical (F) and horizontal (H) reaction forces as a function of slip, for (**a**,**b**) $\alpha = -15°$ and (**c**,**d**) $\alpha = 45°$ (ABAQUS/Explicit).

Major variations between the comparative plots are represented by the screw inclination, with $\alpha = -15°$ and $+45°$.

The use of CZM interfaces, as also expected, proved to have a key effect on the collected mechanical responses for the selected STS configurations. This was observed especially towards the "upper limit condition" characterized by the use of rigid "tie" constraint. In the latter case, the FE outcomes were in fact typically associated to unreliable local and global effects for the examined TTC joints, with a consequent marked increase of both the calculated serviceability stiffness K_{ser} and ultimate resistance F_{max}. Such a combination of phenomena, finally, was also found associated to a remarkable modification of the measured reaction forces (see for example Figure 16b–d).

Regarding the CZM interaction and failure, on the other side, major issues were represented by the accurate calibration of input parameters for damage initiation and evolution. While the nominal resistance values of timber (Table 1) can be reasonably taken into account for the CZM damage initiation, the characterization of its damage law would in fact necessarily require dedicated studies at the component level, and possibly the support of small-scale experiments.

Under the assumption of Table 2, the modification in δ_u was commonly associated to a rather constant elastic response for the examined TTC joints, but to a marked decrease of residual resistance and stiffness for most of the tested configurations. Such an effect can be notice in Figure 16. As far as the critical displacement δ_u for the CZM interaction increases, a reduced slope can be observed for the descending arm of the collected force-slip curves. Compared to the available experimental data from [6], a reliable fitting of degreasing arms for the comparative force-slip curves was observed in the range of δ_u = 6–7mm. This fitting value δ_u, however, results from a numerical calibration in which the nominal mechanical properties of timber are taken into account (Tables 1 and 2). Accordingly, further refined, multi-objective and multi-parameter calculations should be carried out in this direction. Moreover, given that the CZM failure data were found to do not affect the initial stage of the collected force-slip curves (and thus the calculated serviceability stiffness and ultimate resistance for the examined joints), the reference value δ_u = 4mm could be taken into account for preliminary conservative calculations on timber members with similar mechanical properties/class.

6.2. Base Restraints

The actual boundary condition of timber members (and in particular the base restraint of the lateral members for the PO setup in Figure 3) represents, in the same way of mechanical interactions, a relevant influencing parameter for the examined joints. In this paper, three different boundary conditions are thus taken into account for the typical TTC specimen, including:

- BC#1: a base contact interaction, as described in Section 3 (Figure 17a);
- BC#2: a distributed, rigid restraint at the base of the lateral timber members (Figure 17b); and
- BC#3: a mixed restraint, as obtained with a surface contact interaction between the timber member and the rigid base (to avoid possible compenetration) and an additional linear simply support (external edge of the timber member, see Figure 17c).

The BC#1 model, in this sense, coincides with the reference modelling strategy described in Section 3 and validated in Section 4. Variations for the BC#2 and BC#3 models are then represented by the restraint detailing only, with identical material properties and contact formulations.

In Figure 18, selected FE results are proposed for a given joint under different boundary conditions. According to Section 6.2, no relevant variations were observed for the elastic stage of the collected force-slip curves of Figure 18a, as well as for most of the examined joint configurations. However, depending on the arrangement of STSs in use, variations in the vertical reaction force were numerically predicted with up to a +30% increase of the BC#1 value, for the BC#2 and BC#3 conditions.

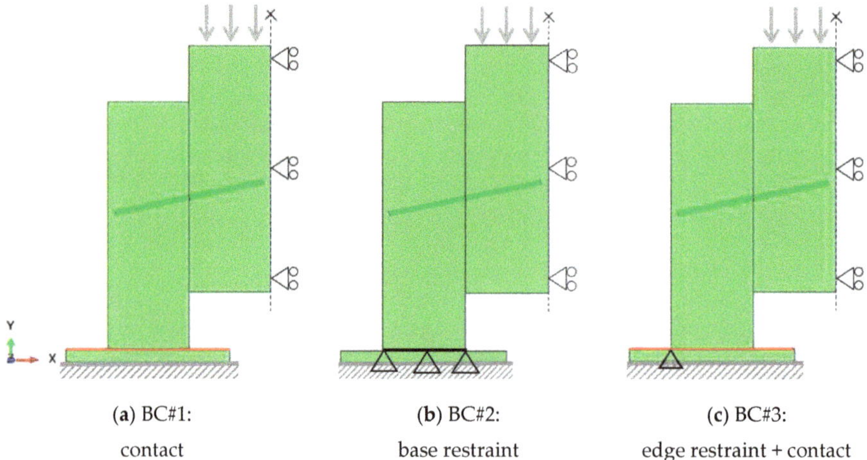

(a) BC#1:

contact

(b) BC#2:

base restraint

(c) BC#3:

edge restraint + contact

Figure 17. Analysis of boundary condition effects on the PO performance of TTC joints.

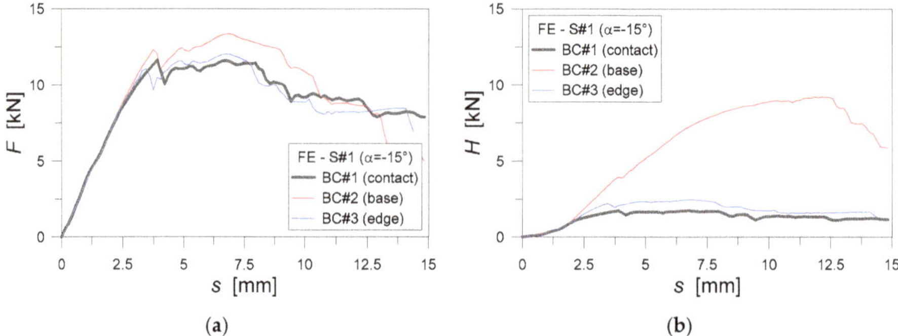

(a)

(b)

Figure 18. Analysis of boundary effects on the PO numerical response of TTC joints with inclined STSs. In evidence, the **(a)** vertical and **(b)** horizontal base reaction force, as a function of the measured slip for a selected TTC joint (ABAQUS/Explicit).

In addition, major variations were observed especially in terms of reaction forces in the horizontal direction for the BC#2 restraint, due to the use of unreliable boundary conditions for the standard PO setup (see Figure 18b).

6.3. Friction Coefficient

At a final stage of this concise sensitivity study, the effects of friction phenomena are explored, with a particular attention for the timber-to-timber interface (Figure 6a). To this aim, the reference FE models in Section 3 are still taken into account, while the static friction coefficient for the mechanical contacts in use are progressively modified. According to [5], it is in fact known that friction at the timber interfaces should be considered for screws subjected to shear-tension stresses only (i.e., TTC joints with positive inclination α for the STSs, based on the convention of this paper). On the other side, any friction mechanism should be disregarded for STSs under shear-compression stresses (STSs with negative α). In this case, the central and lateral timber members of the PO setup are in fact expected to separate from each other, and thus enabling the development of possible attritive interactions.

Following Figure 7 and Equation (2), the force contributions are thus separately analyzed in this paper, for the examined TTC joints. For comparative purposes, more in detail, the input value for μ_{timber} is progressively modified in the range from 0 and 0.8.

As expected, major variations of μ_{timber} for TTC joints with an imposed shear-compression stress regime ($\alpha < 0$) were found to have negligible effects on the collected force-slip contributions, given that:

$$F = F_{screw} \tag{6}$$

with:

$$F_{timber} \approx 0 \tag{7}$$

Figure 19 presents an example of the so-measured force-slip curves, with a focus on the S#1 joint with $\alpha = -15°$. As shown in Figure 19a, minimum variations can be observed in the collected curves, even in presence of marked modifications for μ_{timber}. Regarding the force contribution F_{timber} sustained by the timber members, see Figure 19b, this is estimated as a limited part of the total F, thus agreeing with Equations (6) and (7).

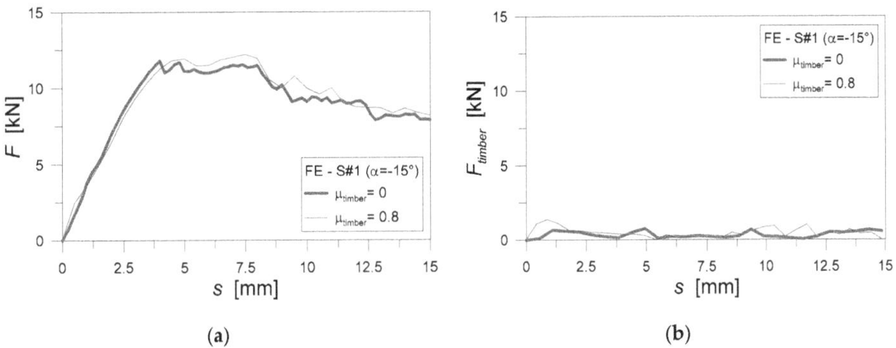

Figure 19. Analysis of static friction effects on the PO numerical response of TTC joints with inclined STSs. In evidence, the (**a**) vertical and (**b**) horizontal base reaction force, as a function of the measured slip (ABAQUS/Explicit).

In order to further investigate such an effect for different screw arrangements, finally, the FE parametric study (with $\mu_{timber} = 0$) was extended to several TTC joints under shar-compressive stresses. In Figure 20, comparative numerical results are proposed for the S#1 specimens as a function of μ_{timber}. The parametric numerical results were post-processed from the collected force-slip curves according to Figure 8. As far as the relevant mechanical parameters are taken into account for them, their trend with μ_{timber} can be investigated.

Figure 20, more in detail, shows the percentage variation Δ given by Equation (5), in terms of:

- ultimate total force F_{max} for a TTC given joint, as a function of μ_{timber};
- shear force contribution F_{screw} taken up by the STSs only, as a function of μ_{timber}.
- and serviceability stiffness K_{ser} (calculated in accordance with Equation (4)), as a function of μ_{timber}.

For clarity of presentation, the FE models with $\mu_{timber} = 0$ are set as a reference condition for Δ calculations.

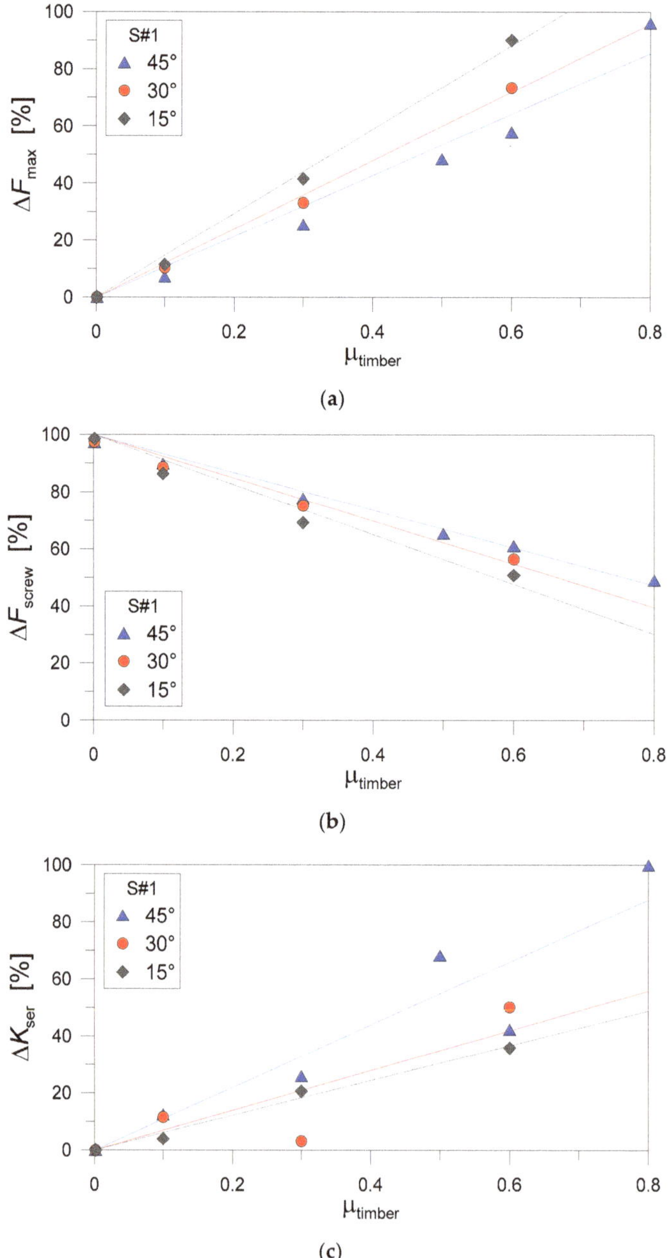

Figure 20. Percentage variation (Equation (4)) of performance indicators for TTC joints with inclined STSs, as a function of the timber-to-timber friction coefficient (S#1 joints under shear-compression): (**a**) maximum force; (**b**) load-bearing contribution of the STSs and (**c**) serviceability stiffness (ABAQUS/Explicit).

Regarding the total ultimate resistance F_{max} of S#1 joints in Figure 20a) for example, it is possible to see that F_{max} progressively increases as far as μ_{timber} increases, for a given α. A relatively regular trend can be observed for the collected FE dots, as also suggested by the linear fitting curves. At the same time, however, it is possible to see that the increase of is indirectly proportional to α, thus maximum benefits deriving from additional frictional phenomena can be expected for STSs with limited inclination α only ($\alpha = 15°$, in this study).

Such an outcome is strictly related to the occurrence, at failure, of local damage mechanisms in timber that can be further magnified especially for high α values (see also Section 4). As far as the typical static friction coefficients of interest for TTC systems are taken into account (i.e., $\mu_{timber} \approx$ 0.25–0.5), moreover, it is interesting to notice that the predicted F_{max} values show a mean + 20–30% variation for STSs joints under shear-compression loads. This result from Figure 20a is thus a further confirmation of the relatively high sensitivity of ultimate resistance predictions for the examined TTC joints, under a standard PO setup.

When the shear force contribution that is sustained by the STSs only is taken into account, see Figure 20c, the same order of percentage variation is observed for various α values. In the figure, in particular, $\Delta F_{screw} = 100\%$ coincides with F_{max} for the whole TTC specimen when $\mu_{timber} = 0$. Otherwise, the progressive increase of frictional effects with μ_{timber} lead to a mostly linear increase of the total resistance F_{max} in Figure 20a. As a result of such a kind of phenomenon, the load-bearing contribution of the STSs (in percentage terms) progressively decreases with μ_{timber}, with variations that can be expected up to −20% compared to frictionless TTC joints (Figure 20b).

Finally, when the serviceability stiffness K_{ser} is taken into account in Figure 20c, an opposite trend can be noticed for the collected FE results, as a function of and μ_{timber}. This is in line with the general expectations and past literature efforts on the topic, where the serviceability stiffness of a given TTC joint reasonably increases when increasing the inclination α of the STSs. As a further remark for the FE results in Figure 20c, it can be noticed a relatively scattered variation of K_{ser} estimates with μ_{timber}, as far as α increases (i.e., Figure 14).

7. Conclusions

In this paper, the structural performance of timber-to-timber composite (TTC) joints with inclined self-tapping screws (STSs) was numerically investigated. The finite element (FE) numerical modelling assumptions were validated towards past experimental results of literature, by taking into account different arrangement and features for the STSs joints, including serviceability stiffness and ultimate resistance comparisons with analytical methods of literature. Through the FE parametric investigation, as shown, a key role was assigned to timber material properties but especially interface damage contacts in the region of fasteners. Major advantage was taken from the use of a surface-based cohesive zone modelling (CZM) damage interaction, so as to capture possible local effects and damage mechanisms in the examined TTC joint components.

For the examined small-scale TTC specimens under a standard push-out (PO) setup, in particular, an average scatter of −25% or +10% was generally observed for the load-bearing estimations in shear-compression and shear-tension respectively. Major deviations of FE models from the literature tests were mainly observed for the TTC specimens characterized by the presence of STSs with high inclination α ($\pm40°$ or $\pm45°$, in the current study), hence suggesting possible numerical issues due to mostly local effects, as well as possible uncertainties on the material properties and on the idealized description of the reference PO test setup.

On the other side, the collected FE estimations were always found to offer enhanced predictions for various STSs arrangements, compared to analytical models of literature. The ultimate resistance values, in particular, were generally strongly underestimated by analytical calculations, for various inclinations of STSs.

Based on extended parametric FE calculations, the actual sensitivity of PO numerical predictions to a series of relevant input parameters (and in particular the CZM damage parameters, the actual boundary condition of timber members and the effect of friction phenomena) was further emphasized. In doing so, major advantage was taken from the analysis of resultant forces that are expected to be sustained (through the whole PO monotonic loading stage) by the steel screws or by timber components. The reciprocal mechanical interaction of the involved load-bearing members was thus explored.

Author Contributions: This paper results from a joint collaboration of all the involved authors. C.B.: conceptualization, analysis, validation, draft preparation, review; M.S. and M.F.: investigation, draft preparation, review. All authors have read and agreed to the published version of the manuscript.

Funding: This research received no external funding.

Conflicts of Interest: The authors declare no conflict of interest.

References

1. Johansen, K.W. Theory of timber connections. International association of bridge and structural engineering. *Bern* **1949**, *9*, 249–262.
2. Blaß, H.J.; Bejtka, I.; Uibel, Y. *Tragfähigkeit von Verbindungen mit Selbstbohrenden Holzschrauben Mit Vollgewinde*; Karlsruher Berichte zum Ingenieurholzbau; KIT Scientific Publishing: Karlsruhe, Germany, 2006; Volume 4.
3. Hansen, K. Mechanical properties of self-tapping screws and nails in wood. *Can. J. Civ. Eng.* **2002**, *29*, 725–733. [CrossRef]
4. Ringhofer, A.; Brandner, R.; Schickhofer, G. A universal approach for withdrawal properties of self-tapping screws in solid timber and laminated timber products. In Proceedings of the 2nd International Network on Timber Engineering Research (INTER 2015), Sibenik, Croatia, 24–27 August 2015. paper INTER/48-7-1 (USB Drive).
5. EN 1995-1-1. *Design of Timber Structures-Part 1-1: General-Common Rules and Rules for Buildings*; European Committee for Standardization (CEN): Brussels, Belgium, 1995.
6. Tomasi, R.; Crosatti, A.; Piazza, M. Theoretical and experimental analysis of timber-to-timber joints connected with inclined screws. *Constr. Build. Mater.* **2010**, *24*, 1560–1571. [CrossRef]
7. Girhammar, U.A.; Jacquier, N.; Källsner, B. Stiffness model for inclined screws in shear-tension mode in timber-to-timber joints. *Eng. Struct.* **2017**, *136*, 580–595. [CrossRef]
8. Giongo, I.; Piazza, M.; Tomasi, R. Out of plane refurbishment techniques of existing timber floors by means of timber to timber composite structures. In Proceedings of the WCTE 2012-World Conference on Timber Engineering, Auckland, New Zealand, 16–19 July 2012.
9. Giongo, I.; Piazza, M.; Tomasi, R. Cambering of timber composite beams by means of screw fasteners. *J. Herit. Conserv.* **2012**, *32*, 133–136.
10. Opazo, A.; Bustos, C. Study of the lateral strength of timber joints with inclined self-tapping screws. In Proceedings of the 51st International Convention of Society of Wood Science and Technology, Concepcion, Chile, 10–12 November 2008. paper WS-39 (USB Drive).
11. Ringhofer, A.; Schickhofer, G. Investigations concerning the force distribution along axially loaded Self-Tapping Screws. In *Materials and Joints in Timber Structures*; Aicher, S., Reinhardt, H.W., Garrecht, H., Eds.; RILEM Bookseries; Springer: Dordrecht, The Netherlands, 2015; Volume 9.
12. Hossain, A.; Popovski, M.; Tannert, T. Cross-laminated timber connections assembled with a combination of screws in withdrawal and screws in shear. *Eng. Struct.* **2018**, *168*, 1–11. [CrossRef]
13. Berardinucci, B.; di Nino, S.; Gregori, A.; Fragiacomo, M. Mechanical behavior of timber-concrete connections with inclined screws. *Int. J. Comput. Methods Exp. Meas. Spec. Issue Timber Struct.* **2017**, *5*, 807–820. [CrossRef]
14. Symons, D.; Persaud, R.; Stanislaus, H. Slip modulus of inclined screws in timber-concrete floors. *Struct. Build.* **2010**, *163*, 245–255. [CrossRef]
15. Jacquier, N.; Girhammar, U.A. Evaluation of bending tests on composite glulam-CLT beams connected with double-sided punched metal plates and inclined screws. *Constr. Build. Mater.* **2015**, *95*, 762–773. [CrossRef]
16. Schiro, G.; Giongo, I.; Sebastian, W.; Riccadonna, D.; Piazza, M. Testing of timber-to-timber screw-connections in hybrid configurations. *Constr. Build. Mater.* **2018**, *171*, 170–186. [CrossRef]
17. Silva, C.; Branco, J.M.; Ringhofer, A.; Lourenco, P.B.; Schickhofer, G. The influence of moisture content variation, number and width of gaps on the withdrawal resistance of Self Tapping Screws inserted in Cross Laminated Timber. *Constr. Build. Mater.* **2016**, *125*, 1205–1215. [CrossRef]
18. *ABAQUS Computer Software*; v.6.14, Simulia; Dassault Systemes: Providence, RI, USA, 2020.
19. *ABAQUS Documentation and Uses' Guide*; v.6.14, Simulia; Dassault Systemes: Providence, RI, USA, 2020.

20. Avez, C.; Descamps, T.; Serrano, E.; Léoskool, L. Finite Element modelling of inclined screwed timber to timber connections with a large gap between the elements. *Eur. J. Wood Wood Prod.* **2016**, *74*, 467–471. [CrossRef]

21. Bedon, C.; Fragiacomo, M. Numerical analysis of timber-to-timber joints and composite beams with inclined self-tapping screws. *Compos. Struct.* **2019**, *207*, 13–28. [CrossRef]

22. EN 12512:2001. *Timber Structures-Test Methods-Cyclic Testing of Joints Made with Mechanical Fasteners*; European Committee for Standardization (CEN): Brussels, Belgium, 2001.

23. EN 26891:1991. *Joints Made with Mechanical Fasteners-General Principles for the Determination of Strength and Deformation Characteristics*; European Committee for Standardization (CEN): Brussels, Belgium, 1991.

24. EN 338:2003. *Structural Timber-Strength Classes*; European Committee for Standardization (CEN): Brussels, Belgium, 2003.

25. EN 1194:1999. *Timber Structures-Glued Laminated Timber-Strength Classes and Determination of Characteristic Values*; European Committee for Standardization (CEN): Brussels, Belgium, 1999.

26. ETA-11/0190. *Würth Self-Tapping Screws-Self-Tapping Screws for Use in Timber Constructions*; European Organization for Technical Approvals (EOTA): Brussels, Belgium, 2016.

27. Rothoblaas-WT Double thread connector-Technical Data Sheet. Available online: https://www.rothoblaas.com/products/fastening/screws (accessed on 22 February 2020).

28. Murase, Y. Friction of wood sliding on various materials. *J. Fac. Agric. Kyushu Univ.* **1984**, *28*, 147–160.

Review

An Overview of Progressive Collapse Behavior of Steel Beam-to-Column Connections

Iman Faridmehr [1] and Mohammad Hajmohammadian Baghban [2,*

[1] South Ural State University, Lenin Prospect 76, Russian Federation, 454080 Chelyabinsk, Russia; imanfaridmehr@gau.edu.tr

[2] Department of Manufacturing and Civil Engineering, Norwegian University of Science and Technology (NTNU), 2815 Gjovik, Norway

* Correspondence: mohammad.baghban@ntnu.no; Tel.: +47-48-351-726

Received: 19 July 2020; Accepted: 27 August 2020; Published: 29 August 2020

Abstract: Local failure of one or more components due to abnormal loading can induce the progressive collapse of a building structure. In this study, by the aid of available full-scale test results on double-span systems subjected to the middle column loss scenario, an extensive parametric study was performed to investigate the effects of different design parameters on progressive collapse performance of beam-to-column connections, i.e., beam span-to-depth ratio, catenary mechanism, and connection robustness. The selected full-scale double-span assemblies consisted of fully rigid (welded flange-welded web, SidePlate), semi-rigid (flush end-plate, extended end-plate), and flexible connections (top and seat angle, web cleat). The test results, including load-deformation responses, development of the catenary mechanism, and connection robustness, are presented in detail. The finding of this research further enables a comprehensive comparison between different types of steel beam-to-column connections since the effects of span-to-depth ratio and beam sections were filtered out.

Keywords: progressive collapse; steel beam-to-column connections; catenary mechanism; double-span assemblies; stiffness degradation

---

1. Introduction

Progressive collapse refers to a devastating phenomenon in which failure of one key structural component, due to abnormal events, leads to chain reaction and spreads to other structural members, causing disproportionate or even entire collapse of the structure [1]. Vehicle impact, terrorist attack, and gas explosions are among incidents that can produce progressive collapse in structures. By the growth of tall and complex structures throughout the world and the increasing number of terrorist- or accident-induced catastrophic events, progressive collapse has received extensive attention from scientists and structural engineers in recent decades [2–7]. The vast majority of structural design codes provide general mitigation strategies to deal with the effects of progressive collapse on structural components that may experience a relatively high demand capacity ratio (DRC).

The design of structures to resist against progressive collapse was first introduced by the UK construction regulations [8] in the aftermath of the Ronan Point building collapse in 1968. Many research and scientific efforts were concerned with this phenomenon, especially after the 9/11 terrorist attack on the World Trade Center in the US. Several design codes including general service administration (GSA) [9] and department of defense (DoD) [10] proposed the alternate load path method as an important design measure to mitigate progressive collapse. In this technique, the structural component is allowed to experience local damage when subjected to extreme loading events. However, it seeks to afford alternate paths so the damage will be localized without spreading to the surrounding areas.

During sudden column removal from the frame, high attention should be paid to the load-carrying capacity of the double-span assembly above the removed column since it has an essential function in the progressive collapse prevention. Generally, as the middle column becomes ineffectual, the catenary action develops in connected beams and slabs, leading to large deformation in beam-to-column connections. Since the robustness of connections preserves the integrity in a double-span column removal scenario, there is a necessity to investigate the beam-to-column connection performance under the simultaneous presence of moment, shear, and tension in conjunction with high ductility demand. Such a complex loading protocol negatively affects beam-to-column connection performance and poses the risk of unexpected brittle failure.

In steel structures with fully rigid and semi-rigid beam-to-column connections, the catenary mechanism plays a major role to resist progressive collapse through the axial tension in the connected beams. In fact, the catenary mechanism is the fundamental resistance source of structures to vertical loads in the large deformation stage. Beam-to-column connections with appropriate robustness and reliable axial resistance are compulsory for developing the catenary actions where this stage is the final line of defense against progressive collapse. F Wang et al. investigated the behavior of bolted and welded flange plate connections subjected to progressive collapse [11]. They concluded that the connection with welded flange plates can lead to greater flexural strength than that with bolted angles and the application of welded haunch plates can arrest fracture failure on the welds within the beam-column joint. Hao Wang et al. performed experimental tests of steel frames with different beam-column connections under falling debris impact [12]. They concluded that the majority of the external work applied to the system was absorbed by bending deformation, especially by the plastic rotation at mid-span of the beam. Besides, they concluded that the catenary action was shown to significantly improve the load-carrying capacity and energy absorption in specimens with high levels of rotational ductility. Alrubaidi et al. investigated the behavior of different steel intermediate moment frame connections under a column-loss scenario [13]. Performance of different connections was compared based on their modes of failure and load-displacement response in both flexural and catenary action stages. They concluded that significant axial tensile forces were generated in the beams and the catenary action stage was then fully mobilized, providing an increase in the progressive collapse resistance.

It has been well documented that beam-to-column connections play a vital function in mitigating progressive collapse potentials in steel structures. There is a large volume of published studies investigating the connection performances under sudden column removal, where flexible, semi-rigid [14–18], and rigid [19–22] connections have been studied in detail.

The connected beam's axial forces significantly contribute to the development of the catenary mechanism. In the case that a double-span assembly experiences large deformation, the beam-span-to-depth ratio, R_i, is also a major parameter that has been studied by several researchers [23–25]. The effect of the R_i ratio on the mitigation of progressive collapse in steel moment frames was investigated by Rezvani et al. [26]. They concluded that the vertical resistance of frames increases as the R_i ratio decreases.

A large and growing body of literature has also investigated the robustness effects of steel beam-to-column connections to mitigate the progressive collapse [27–30]. More recently, several attempts have been made to investigate the influence of the seismic design of beam-to-column connections on an anti-progressive collapse mechanism [23,31,32]. The behavior of welded unreinforced flange-bolted web and reduced beam section connections subjected to column removal were investigated by Chen et al. [33]. Using an experimental test, Yang and Tan investigated the performance of flexible and semi-rigid connections including different types of bolted beam-to-column connections [34]. They concluded that maximum tensile resistance of the connection significantly contributes to the development of catenary action after large rotations. Driver et al. [35] reported experimental results of several shear connections including 15 bolted single-angle and 6 double-angle specimens subjected to double-span assembly. They came to the conclusion that rupture or tearing of the cross-section in the vicinity of the angle heel leads to sudden failure. Qin et al. investigated the progressive collapse behavior of conventional and reinforced welded flange-bolted web connections using numerical simulations validated by experimental tests [36]. Their study confirmed that the reinforced flange-bolted connection possesses higher ductility and robustness compared to the conventional connection, leading to more reliable collapse performance. Many researchers, such as Oosterhof and Drive [37] and Shen and Astaneh [38], have established or implemented several mechanical spring techniques for bolted-angle beam-to-column connections. Stylianidis and Nethercot [39] investigated the progressive collapse performance by using component-based connection models.

Overall, the previous literature is mainly concerned with the anti-progressive collapse behavior of typical beam-to-column connections using experimental tests or numerical simulations. However, the comprehensive comparisons of common practice beam-to-column connection performance subjected to column removal addressing the load transfer mechanisms requires robustness and ductility, and R_i effects are still very limited. Therefore, this research comprehensively investigates the anti-collapse behavior of double-span assemblies with flexible, semi-rigid, and fully rigid beam-to-column connections. This is done with the aid of available test results on steel beam-to-column connections including top-seat angle and welded unreinforced flange-bolted web. Meanwhile, for a reliable comparison of different types of double-span assemblies and to evaluate the anti-collapse performance of steel beam-to-column connections, the vertical pushdown load and equivalent rotation were normalized against connected beams' plastic hinge and plastic rotation, respectively.

2. R_i Effects on Progressive Collapse

To evaluate the effects of the R_i ratio, one set of pushdown analyses was performed in this research using structural analysis program (SAP) 2000 academic version 21. In this case study, three steel double-span assemblies with R_i ratios of 5, 10, and 15 were considered. All specimens possess fully rigid connections which were designed according to the strong column–weak beam theory in compliance with The American. Institute of Steel Construction (AISC) seismic design [40]. Figure 1 shows the topology of one studied frame with double-span assemblies.

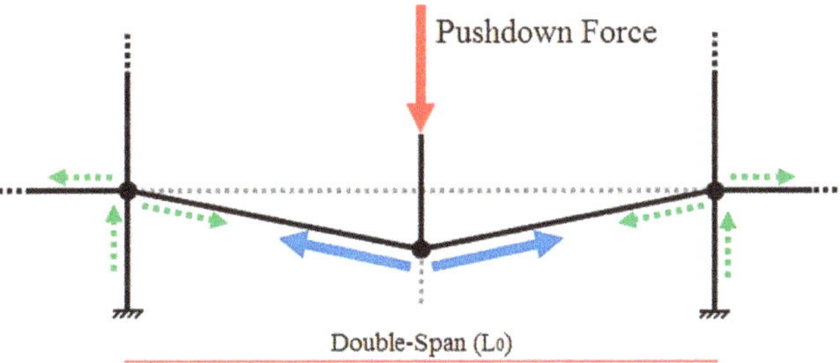

Figure 1. Double-span assembly.

In Figure 1, the pushdown force, *F*, represents the concentrated load in the pushdown analysis applied at the top of the removed column. This concentrated load is equivalent to the progressive collapse resistance of the double-span assembly. Table 1 shows the beam and column section properties used in progressive collapse analysis.

The plastic hinge capacity of connected beams, F_p, of double-span assembly can be calculated from the following equation:

$$F_p = \frac{2\,M_p}{\frac{L_0}{2}} = \frac{4\,W_p\,f_y}{L_0} \tag{1}$$

where M_p is the plastic moment of the connected beam, L_0 is the double-span assembly length, W_p is the plastic modulus of the connected beam, and f_y is the yield stress.

Table 1. Beam and Column Sections of the Steel Double-Span Assemblies (AISC Database).

Span-to-Depth Ratio, R_i	Beam Section	Column Section
5	W 40×149	W 40×531
10	W 21×50	W 21×248
15	W 12×35	W 12×50

To facilitate the comparison of different types of double-span assemblies against each other independently and regardless of connected beam plastic capacity, the pushdown force, F, should be normalized against the plastic hinge capacity of connected beams, F_p, as in the following equation:

$$\frac{F}{F_p} = \frac{F\,L_0}{4\,W_p\,f_y}$$ (2)

Besides, to highlight the differences of different beam sections during the development of the catenary action, it is necessary to normalize the chord rotation, θ, over the plastic rotation, θ_p, based on the following equations:

$$\theta = \frac{\delta}{L_0}$$ (3)

$$\theta_p = \frac{\delta_p}{\frac{L_0}{2}} = \frac{F_p\,2}{K_e\,L_0} = \frac{\frac{4\,W_p\,f_y}{L_0}}{\frac{48\,E\,I_b}{L_0^3}}\,\frac{2}{L_0} = \frac{W_p\,f_y\,L_0}{6\,E\,I_b}$$ (4)

where I_b is the moment of inertia of a connected beam, and K_e is the elastic stiffness of a simply supported beam subjected to pushdown force. Figure 2 shows the plots of the pushdown analysis results of double-span assemblies with different R_i ratio, in which the vertical and horizontal axes are the normalized force, $\frac{F}{F_p}$, and the normalized rotation, θ/θ_p, respectively. Figure 2 shows that the frame with the larger R_i has a higher capacity to develop progressive collapse resistance. Also, the double-span assembly with the larger R_i possesses relatively high initial stiffness.

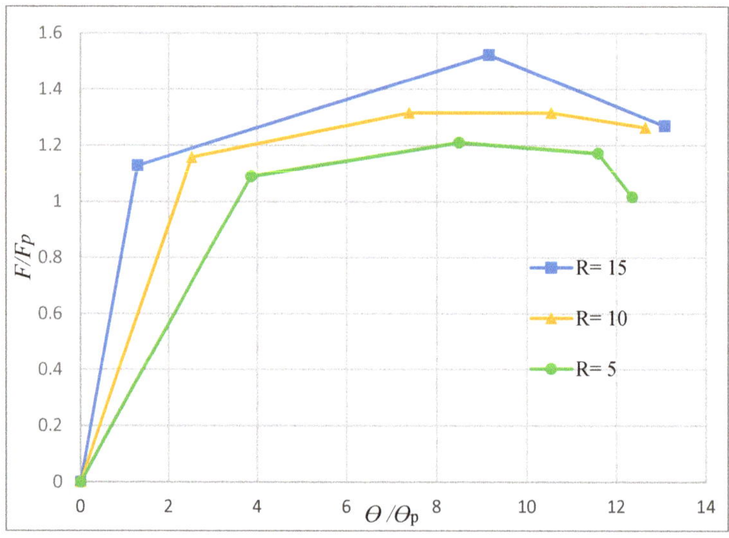

Figure 2. Normalized resistance against normalized rotation.

3. Development of Data Bank

Having recognized that different beam-to-column connections have different behavior and pushdown resistance capacity, especially the catenary mechanism, it is necessary to develop a database of test results. In this section, with the aid of available full-scale test results on double-span assemblies, this is comprehensively investigated in the performance of different steel beam-to-column connections against progressive collapse. The specimens consist of fully rigid (welded flange-welded web, SidePlate), semi-rigid (flush end-plate, extended end-plate), and flexible connections (top and seat angle, web cleat). Figure 3 depicts the configuration and geometric details of the studied connections.

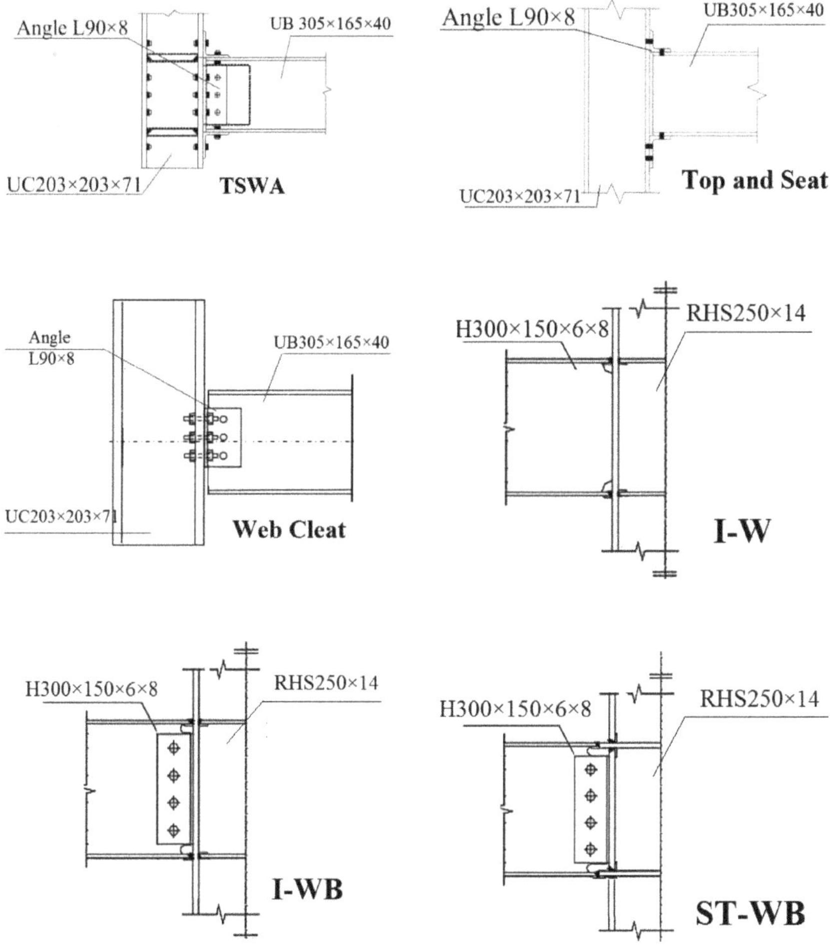

Figure 3. *Cont.*

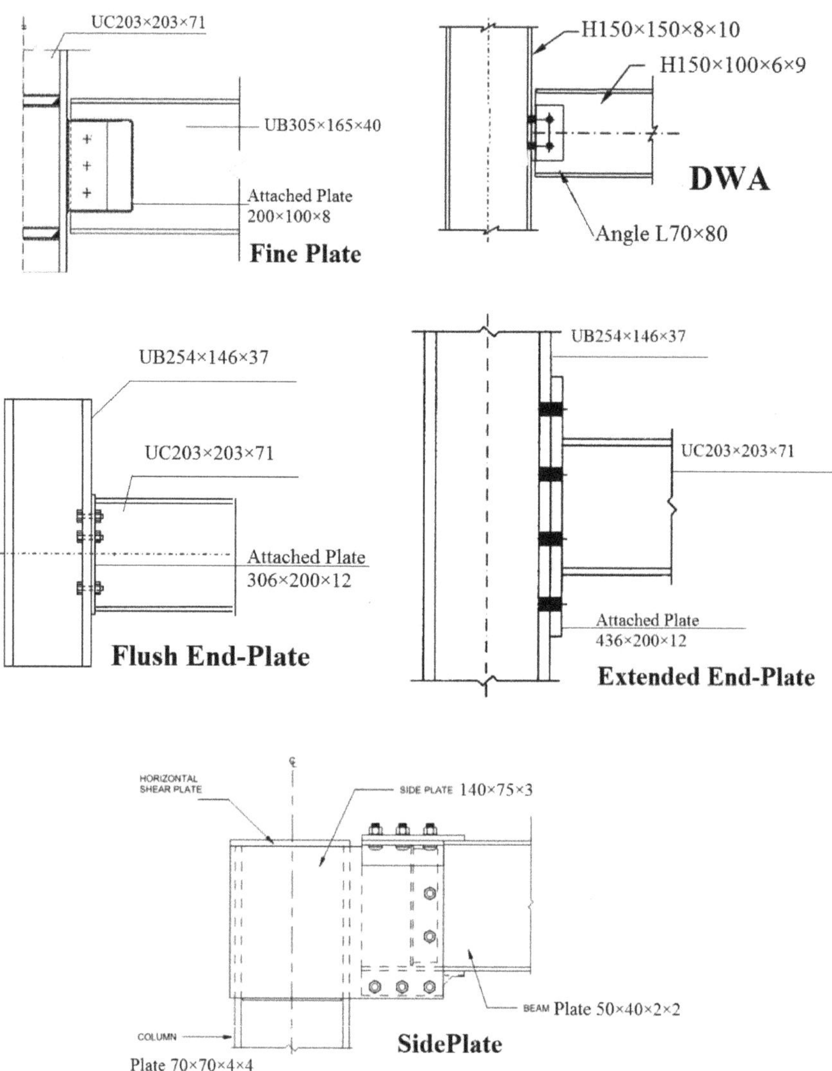

Figure 3. Configuration and geometric details of studied connections.

Table 2 shows the information of the tested specimens, including the connection type, connection category, and relevant references. In this research, there was an emphasis to select double-span assemblies having similar span-to-depth ratios, R_i. One important implication of this procedure is that it enables an adequate comparison between different types of beam-to-column connections since the effects of different spans and beam sections were filtered out.

Table 2. Summary of Tested Specimens.

Specimen	Connection Type	Connection Category	Span-to-Depth Ratio R_i	Reference
WUF	Welded Unreinforced Flange-Welded Web	Fully rigid	10	W Zhong et al. [41]
SidePlate	SidePlate		20	Faridmehr et al. [42]
I-W	Welded Flange-Weld Web Connection		7.5	W Wang et al. [43]
I-WB	Welded Flange-Bolted Web Connection		7.5	W Wang et al. [43]
ST-WB	Welded Flange-Bolted Web Connection with Shear Diaphragm		7.5	W Wang et al. [43]
DWA	Double Web Angle	Flexible	10	W Zhong et al. [41]
TSWA (L = 8 mm)	Top and Seat-Web Angle		7.9	B Yang et al. [44]
Web Cleat	Web Cleat		7.9	B Yang et al. [44]
Top and Seat angle	Top and Seat Angle		7.9	B Yang et al. [44]
Fine Plate	Fine Plate		7.9	B Yang et al. [44]
TSWA (L = 10 mm)	Top and Seat-Web Angle	Semi-rigid	10	W Zhong et al. [41]
TSWA (L = 12 mm)	Top and Seat-Web Angle		9.5	B Yang et al. [44]
Extended End-Plate	Extended End-Plate		9.5	B Yang et al. [44]
Flush End-Plate	Flush End-Plate		9.5	B Yang et al. [44]

4. Test Results

Table 3 shows the summary of test results, including maximum vertical loads, displacement, connection rotation, and failure mode.

Table 3. Summary of Test Results.

Specimen	$max\,\frac{F}{F_p}$	$max\,\frac{\theta}{\theta_p}$	Failure Mode
		Fully rigid	
WUF	1.45	11.77	Fracture at the flange
SidePlate	6.38	8.3	Beam failure
I-W	1.25	7.27	Bottom flange and beam web fractured, weld fracture
I-WB	1.25	16	Fracture of shear plate and bottom flange
ST-WB	1.25	15.4	Fracture of shear plate and bottom flange
		Semi-rigid	
TSWA (L = 10 mm)	2.24	21.3	Fracture at the angle and failure of the web bolt
TSWA (L = 12 mm)	1.72	17.3	Fracture bolt
Extended end-plate	0.91	14.22	Fracture of weld
Flush end-plate	1.17	12.6	Bolt thread stripping
		Flexible	
DWA	1.2	20.4	Fracture at the shear angles
TSWA (L = 8 mm)	0.62	10.9	Fractured web angle
Web Cleat	0.68	10.54	Fractured angle
Top and Seat angle	0.25	16.4	Fractured bottom angle
Fine Plate	0.46	7.6	Bolt fractured in shear

Since the pushdown force is applied to the middle column, the vertical displacement gradually increases. In simple connections such as top and seat angle where the connection possesses a limited capacity to develop the full plastic moment of connected beams, the specimen rotates at both ends following a major deflection below the removed column. The normalized force versus normalized rotation for all tested simple connections is shown in Figure 4. Figure 4 indicates that at the initial pushdown stage, there is no substantial loading resistance. The average F/F_p is around 0.2 at the normalized rotation of 5, until developing axial tensile force in the connected beam as a result of large deflection, indicating the beginning of catenary action. Besides, the shear fracture of bolts causes several jumps in the curve in which the connections experience major localized bearing deformations in the vicinity of bolt holes. Overall, the previous literature indicated that axial tensile force has a major contribution toward progressive collapse resistance in flexible double-span assemblies in which the catenary mechanism takes place before connection component rupture or shear fracture of bolts. Also, the important feature of progressive collapse performance of flexible connections is that although double-span assemblies experienced large deformation, the connected beam remains in the elastic region and connection components experience large plastic strain.

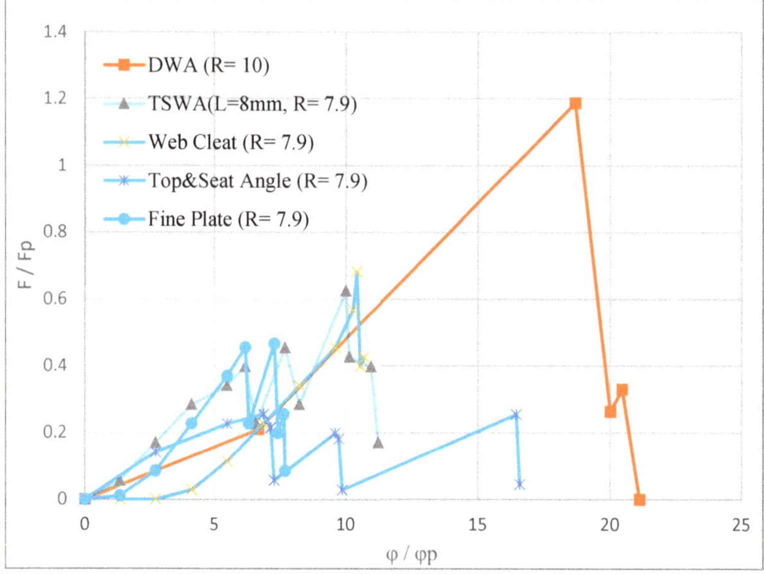

Figure 4. Normalized pushdown resistance against normalized rotation of flexible connections.

The normalized force versus normalized rotation for all tested semi-rigid connections is shown in Figure 5. Figure 5 indicates that semi-rigid connections have higher initial stiffness as a result of high flexural capacity. Nevertheless, after normalized rotation of 5, the stiffness experiences a decrease in most of the specimens as a result of limited capacity connections' components, i.e., top and seat angles. Overall, the previous literature indicates that connections in this category mainly fail due to bolt thread stripping and fracture or fracture at the web angles. Besides, connection failure is categorized in two phases, in which, at the first phase, the connection resists vertical pushdown force through flexural action, and after large plastic rotation, the connections go into the catenary mechanism, indicating phase 2.

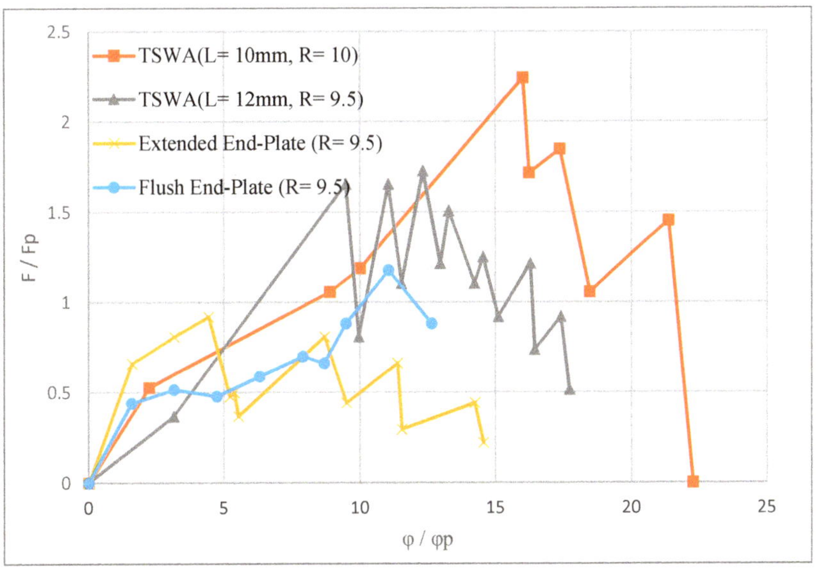

Figure 5. Normalized pushdown resistance against normalized rotation of semi-rigid connections.

The normalized force versus normalized rotation for all tested fully rigid connections is shown in Figure 6. This category possesses the highest stiffness, where in the normalized rotation of around 2, almost all specimens develop the full plastic moment of connected beams. In addition, Figure 6 shows that by considering unique configuration and large R_i, SidePlate has the highest stiffness and ultimate strength. Generally, the previous literature indicated that ductility demand for traditional rigid connections in the case of column removal is mainly controlled by the column shear panel zone, while in the SidePlate connection, it is controlled by a connected beam. Also, in fully rigid connections, the fracture at beam flange or shear plate are the main reasons for failure mode without premature weld or connection's component failure.

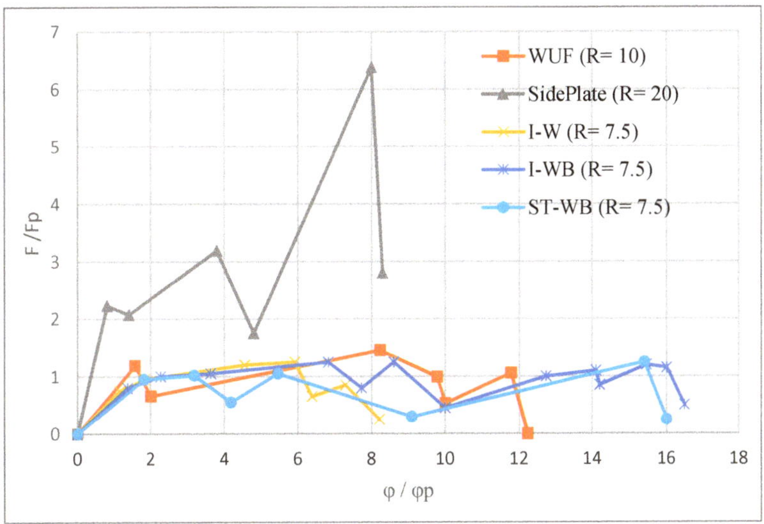

Figure 6. Normalized pushdown resistance against normalized rotation of fully rigid connections.

Yield mechanisms and failure modes are the factors that control both the resistance and ductility or rotational capacity of the connection. Failure modes and yield mechanisms are related but are inherently different. Failure modes cause fracture, loss of deformation capacity, or significant loss of resistance. Yield mechanisms induce inelastic deformation and result in dissipation of energy and changes in stiffness without inducing fracture or excessive loss of resistance. In fully rigid connections, the ratio of rotational capacity of connection to rotational capacity of beam ($\frac{\varphi}{\varphi_p}$) is around 12, where the failure mode is controlled by fracture at shear plate and beam flange. For some specimens in this category, such as I-W, weld fracture causes significantly less ductility, energy dissipation, and plastic rotational capacity. The ratio of $\frac{\varphi}{\varphi_p}$ is around 16 and 13 for semi-rigid and flexible connections, respectively. The failure modes in semi-rigid connections are governed by local buckling and deterioration caused by the large inelastic deformation web cleats and bolts. So, it can be concluded that semi-rigid connections provide large ductility and a ductile failure mode compared to fully rigid connections.

5. Discussions of Test Results

5.1. Catenary Mechanism

A steel beam is mainly subjected to the combined effect of shear force and bending moment under the gravity loading condition, whereas there is no considerable axial force. In the case of sudden column removal, the bending moment at both ends and mid-span of the beam will increase accordingly, resulting in large deflection, developing plastic hinges in these particular locations. In this situation, the beam will experience significant axial force. As the axial force increases gradually, the whole cross-section will experience yielding, resulting in reducing bending moment at plastic hinge locations. At the final stage, the resistance to the external loading only depends on the axial capacity of the beam. This process is generally recognized in the engineering community as tensile catenary action.

The pushdown force versus vertical deflection for a beam with axially restrained support is plotted in Figure 7 [44]. Figure 7a is representative of the behavior of all tested specimens in which failure takes place at a different stage of their response depending on rotation and ductility capacity. In the elastic range, the behavior is generally controlled by connection stiffness, while at the post-elastic stage (beyond point B), the behavior largely depends on geometric and material nonlinearity. Depending on connection categories, different behavior is observed during the different stages, as shown in the typical graph of Figure 7b. Generally, the connected beams and connections are subjected to relatively high axial compression and bending moment at the compressive arching stage, leading to premature instability in the vicinity of compressive components. After point C, the compressive arching impact steadily declines, and the beam axial compression force is converted to tensile (after point D). By increasing the axial tension, the connection's bending moment becomes less significant and they experience severe tensile deformation. Finally, following the point E, the catenary tensile action becomes the governing force-carrying mechanism.

Considering a tensile catenary stage in the beam in the case of sudden column removal (point D onward in Figure 7b), a simplified model for plastic interaction between axial force and bending moment is developed according to the following assumptions:

i.   The beam has an I-shaped section with elastic/perfectly plastic material.
ii. Plastic hinges only appear at the mid-span and both ends, and hinges follow the rigid-plastic model.
iii. The beam has symmetric restraint with the elastic response.
iv. The restraint of the steel beam is symmetric and in elastic response.
v.   Once the whole I-section experiences yielding, the correlation between axial force and bending moment controls the behavior of the catenary mechanism.

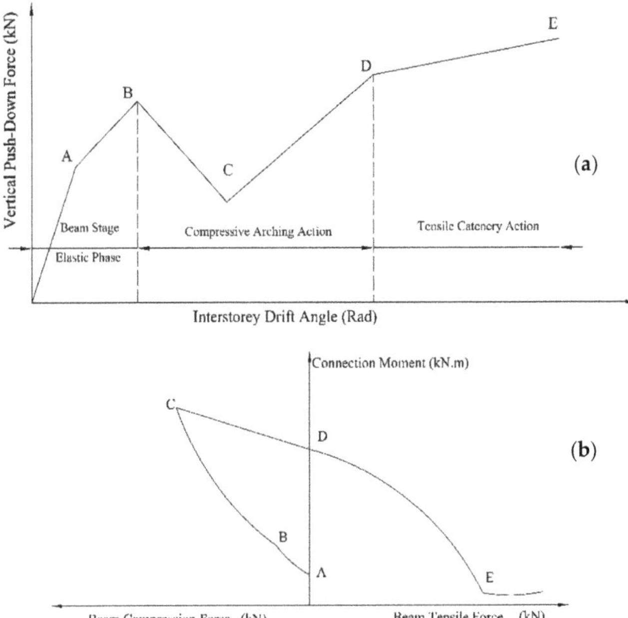

Figure 7. Beam and connection performance following column removal. (**a**) beam nonlinear load-deflection response, (**b**) beam axial load - connection moment interaction

Assuming the entire cross-section yield, the bending moment and tensile force correlation is given by the following equation in the case where the neutral axis is in the web:

$$\frac{M}{M_y} + \frac{(1+\alpha)^2}{\alpha[2(1+\beta)+\alpha]}\left(\frac{p}{p_y}\right)^2 = 1 \tag{5}$$

where M_y and p_y are the plastic moment capacity and ultimate plastic axial force of the beam respectively, calculated by:

$$M_y = Zf_{ye} \quad p_y = \left(2A_f + A_w\right)f_{ye}$$

where f_{ye} is the expected material yield strength, Z is the plastic section modulus, A_f is the beam's flange area, and A_w is the beam's web plate area. β is the ratio of flange thickness to web height, and α is also defined by the following equation:

$$\alpha = A_w/\left(2A_f\right)$$

In the case that the plastic neutral axis appears in the beam's flange, the moment and tensile correlation is given by the following equation:

$$\frac{1-\gamma}{1-\frac{(1+\alpha)^2\gamma^2}{\alpha[2(1+\beta)+\alpha]}}\frac{M}{M_y} + \frac{P}{P_y} = 1 \tag{6}$$

where

$$\gamma = A_w/\left(2A_f + A_w\right)$$

Since the web height is much bigger than flange thickness, it is acceptable to assume $\beta = 0$. Accordingly, Equations (5) and (6) can be rewritten as:

$$\frac{M}{M_y} + \zeta \left(\frac{P}{P_y}\right)^2 = 1 \tag{7}$$

$$\lambda \frac{M}{M_y} + \frac{P}{P_y} = 1 \tag{8}$$

where

$$\zeta = \frac{(1+\alpha)^2}{[\alpha(2+\alpha)]}; \quad \lambda = \frac{1-\gamma}{1-\zeta\gamma^2}$$

Figure 8a presents the interaction of Equations (7) and (8), indicating that the curves of Equations (7) and (8) intersect at point $(1 - \zeta\gamma^2.\gamma)$. For simplicity, the nonlinear domain of Equation (7) can be replaced by a straight line, using the following equation:

$$\frac{M}{M_y} + \zeta\gamma \frac{P}{P_y} = 1 \tag{9}$$

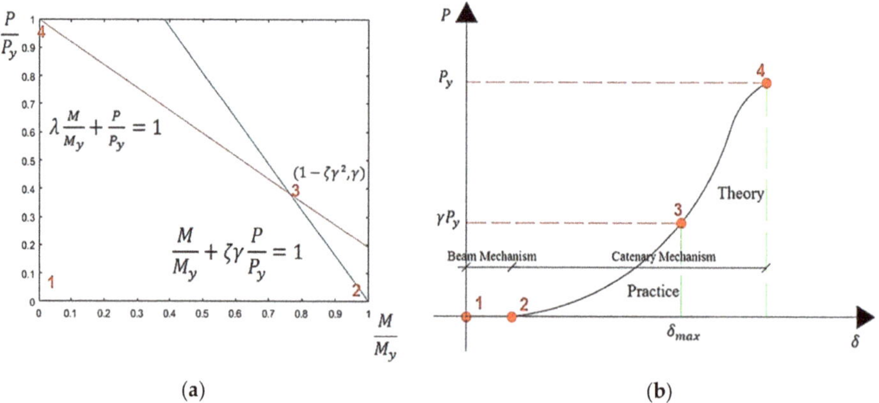

Figure 8. (**a**) Plastic interaction between bending moment and tensile force, and (**b**) beam axial force versus mid-span deflection.

Figure 8b shows the correlation between the beam's axial force and vertical displacement in the case of middle column removal. According to the experimental results, two main different failure mechanisms in the beam are recognized, as presented in Figure 8b. These two failure mechanisms are introduced as the beam mechanism (phase 1 to phase 2) and the catenary mechanism (phase 2 to phase 4), respectively. Experimental results indicate that the tensile force in the beam is smaller than the ultimate plastic axial force before the failure. Accordingly, it is reasonable to assume that the moment of the beam is equal to $1 - \zeta\gamma^2 M_y$ and the axial force is equal to γP_y before the beam experiences failure (refer to practice failure point 3 in Figure 8b).

Figure 9 shows the vertical reaction, V_R, of connection, which can be calculated by Equation (10).

$$V_R = V_i \cos\varphi_i + P_i \sin\varphi_i = F_f + F_c \tag{10}$$

where φ_i, V_i, and P_i are rotation of deflected beam, transverse shear, and axial force in the case of middle column removal, respectively. The internal transverse shear and axial force are measured through installed strain gauges distributed on a connected beam. F_f in Equation (10) represents flexural mechanism resistance while F_c is the resistance component due to the catenary mechanism.

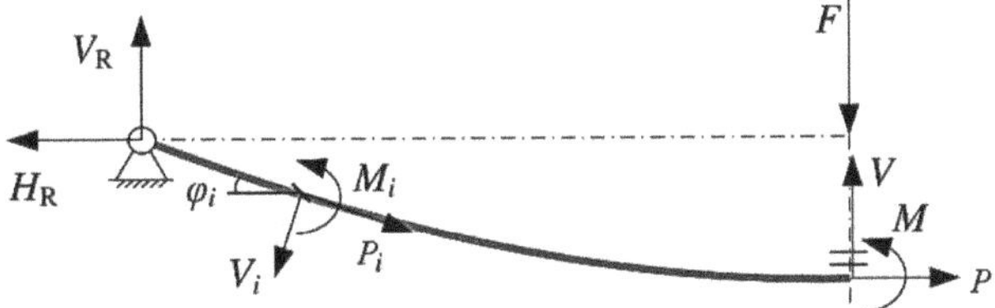

Figure 9. Development of internal forces in the beam-to-column assembly in the case of middle column removal.

Figure 10 shows the development of axial force and bending moment in the case of middle column removal for three tested specimens in reference [43]. According to Figure 10a, at the preliminary phases, the behavior of the beam is controlled by flexural resistance, and the tensile force is almost zero. With increased downward displacement, the axial tension also increases in the beams, developing catenary mechanism until the beam or connection can no longer bear the combined flexural stresses and tensile force and fails (see Figure 10b).

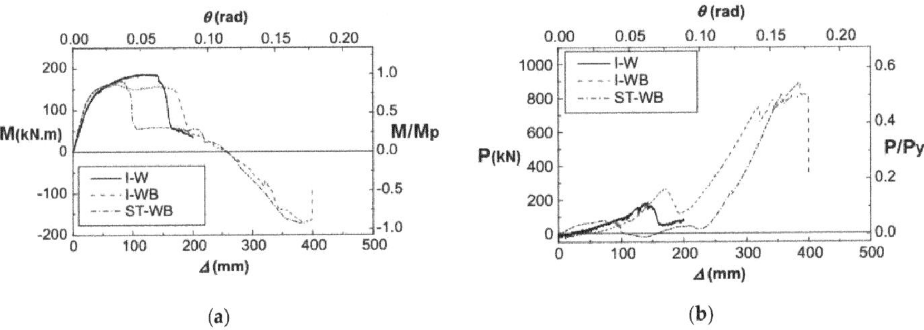

Figure 10. Load resistance mechanism: (**a**) development of a flexural mechanism F_f, and (**b**) development of a catenary mechanism F_c.

Figure 10 also indicates that fully rigid connections, I-WB and ST-WB, can develop relatively high axial force and a bending moment of connected beam compared to the I-W specimen. In other words, the type of beam-to-column connection plays an important role in developing the catenary mechanism. This issue is also acknowledged in Figure 11, where the flexible connections, i.e., top and seat angle and TSWA = 8 mm, fail to develop the catenary mechanism of the connected beam. This issue confirms the fact that the failure is mainly concentrated at the connection components in a flexible and semi-rigid connection, and therefore, these types of connections have limited capacity to develop the catenary mechanism.

5.2. Maximum Rotation and Ductility Capacity

In steel structures subjected to sudden column removal, the maximum deformation and rotation capacities of connections within the affected areas have a major contribution to the ultimate load-carrying capacities of the system. The main progressive collapse guidelines, i.e., the Unified Facilities Criteria (UFC) 4 023 03 [45], specify a series of plastic rotation angles for several beam-to-column connection categories based on a nonlinear modeling simulation, as shown in Table 4.

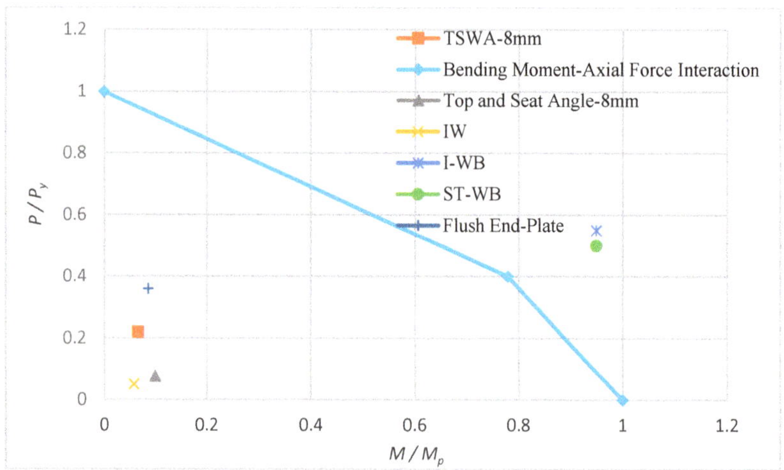

Figure 11. Correlation between axial force and bending moment in studied specimens at the final stage of the pushdown test.

Table 4. Modeling Parameters and Acceptance Criteria for Nonlinear Modeling of Steel Frame Connections.

Connection Type	Nonlinear Modeling Parameters			Nonlinear Acceptance Criteria
	Plastic Rotation Angle (Rad)		Residual Strength Ratio	Plastic Rotation Angle (Rad)
	a	b	c	Primary Component
Fully Restrained Moment Connections				
WUF	$0.0284 - 0.0004d$	$0.043 - 0.0006d$	0.2	$0.0284 - 0.0004d$
SidePlate®	$0.089 - 0.0005d$	$0.169 - 0.0001d$	0.6	$0.089 - 0.0005d$
Partially Restrained Moment Connections (Relatively Stiff)				
Shear in Bolt	0.036	0.048	0.2	0.03
Tension in Bolt	0.016	0.024	0.8	0.013
Tension in Tee	0.012	0.018	0.8	0.010
Flexure in Tee	0.042	0.084	0.2	0.035
Partially Restrained Simple Connections (Flexible)				
Flexure in Angles	$0.1125 - 0.0027d_{bg}$	$0.150 - 0.0036d_{bg}$	0.4	$0.112 - 0.0027d_{bg}$
Simple Shear Tab	$0.0502 - 0.0015d_{bg}$	$0.072 - 0.0022d_{bg}$	0.2	$0.0502 - 0.0015d_{bg}$

d_{bg} = depth of bolt group, inch; d = depth of beam, inch.

Figure 12 shows the nonlinear force–vertical displacement curve of several fully rigid tested specimens along with idealized multi-linear curves. For a welded unreinforced flange WUF connection, represented by I-WB and ST-WB in this research, the results indicate large ductility when subjected to mid-column removal. The plastic rotation angle b for I-WB and ST-WB is around 0.16 rad, well beyond the recommended acceptance criteria by the UFC presented in Table 4. This indicates that tested specimens can resist large plastic rotation without a significant decline in resisting vertical force.

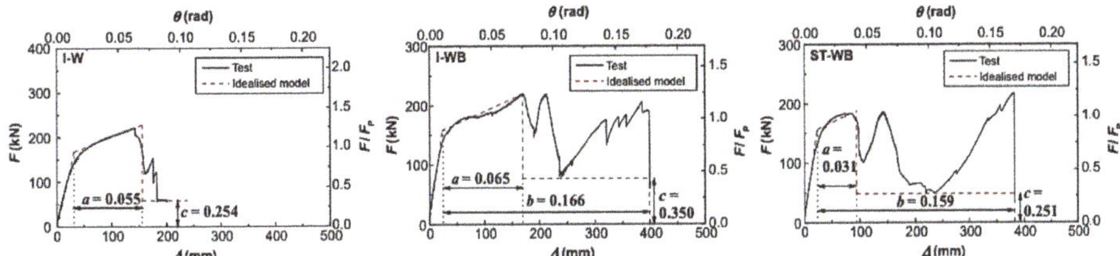

Figure 12. Ductility assessment of fully rigid tested specimens.

In this section, the plastic rotation angle of all tested specimens is compared to the acceptance criteria, estimated based on the depth of the connection, recommended by the UFC. Figures 13–15 show the maximum rotational capacity of studied connections versus connection depth along with the acceptance criteria line recommended by the UFC, as presented in Table 4.

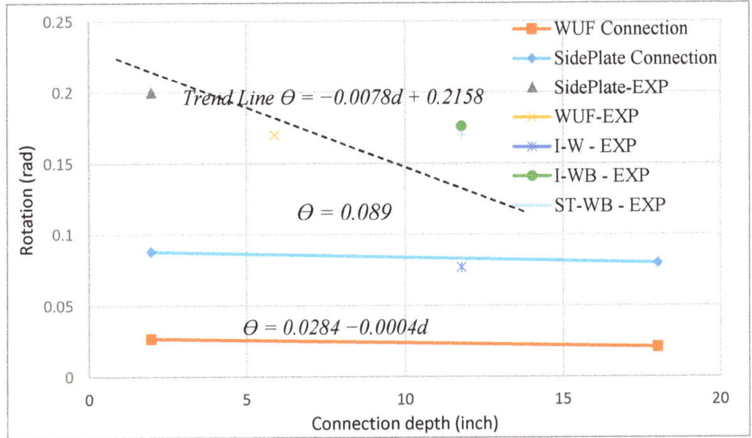

Figure 13. Maximum rotation capacities versus connection depth for fully rigid connections.

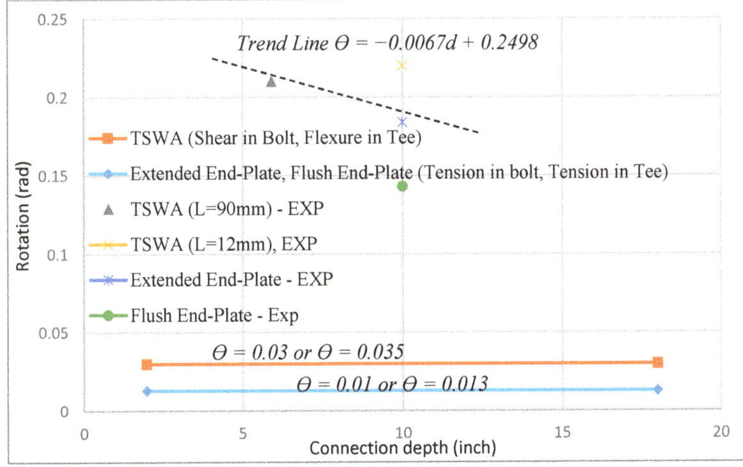

Figure 14. Maximum rotation capacities versus connection depth for semi-rigid connections.

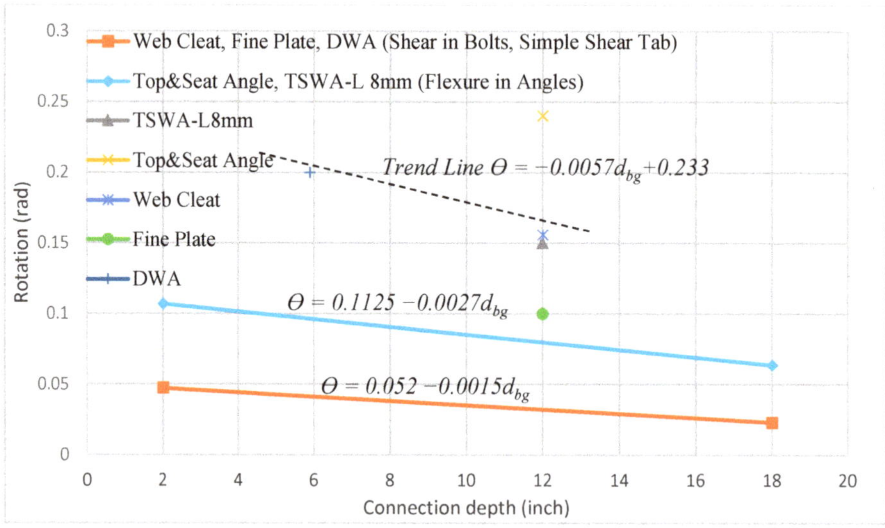

Figure 15. Maximum rotation capacities versus connection depth for flexible connections.

Overall, Figures 13–15 indicate that the average plastic rotation angle varies from 0.1 to 0.2 rad, which significantly surpasses the UFC recommended acceptance criterion. This indicates that all three beam-to-column connection categories can address adequate plastic rotation during sudden column removal. In addition, the results reveal that the suggested ductility acceptance criteria are on the conservative side for all three beam-to-column connection categories. The current in-practice acceptance criteria employed in progressive collapse analysis are based on cyclic simulations proposed by the American Society of Civil Engineers (ASCE/SEI) 41 13 [46] that do not consider large axial demands imposed over sudden column removal. Actually, the results show that the connection depth alone is not a reliable indicator to predict the rotational capacity of a connection, where different connection types with the same R_i result in totally different maximum rotation capacity. Moreover, the results show that although the stiffness of fully rigid connections is higher than flexible connections, both categories result in almost the same rotation capacity, by around 0.15 rad.

The previous literature indicates that the rotational ductility of connections is significantly affected by bolt rupture and brittle failure of welding [47]. In other words, a variation in failure mode will affect the maximum rotational ductility of connections. To meet the criterion for rotation capacity, in this research, an acceptance criterion is proposed by considering initial stiffness and the design flexural resistance. To this end, the maximum rotation capacity of the connection, φ_u, is defined by the rotation at the point where either (i) the connection resistance has dropped to 0.8 M_n, or (ii) the deformation is more than 0.03 rad. The first yielding rotation φ_y is also defined as follows:

$$\varphi_y = \frac{\frac{2}{3} M_y}{S_{j.ini}} \tag{11}$$

The above-mentioned parameters have been used to determine the ductility of studied connections. The ability to sustain large inelastic deformations without significant loss in strength is referred to as ductility, μ, and is defined as follows:

$$\mu = \frac{\varphi_u}{\varphi_y} \tag{12}$$

Figure 16 illustrates the nonlinear idealization of the moment-rotation curve of a typical steel beam-to-column connection.

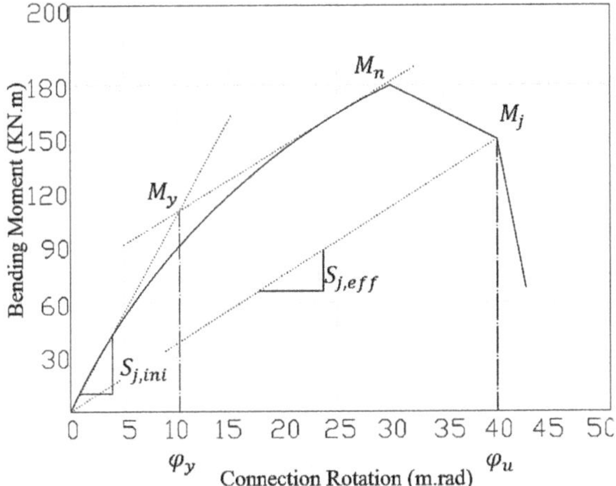

Figure 16. Typical moment-rotation response.

Table 5 illustrates the yielding rotation, maximum rotation capacity, and ductility of studied specimens. Table 5 indicates that fully rigid and semi-rigid connections result in bigger ductility compared to flexible connections. Generally, in a situation where the connection strength exceeds the beam strength, the ductility of the whole system is controlled by the beam, and the connection remains elastic. For instance, the behavior and overall ductility of the SidePlate moment connection system are defined by the plastic rotational capacity of the beam, where the limit state is ultimately the failure of the beam flange, away from the connection. On the other hand, if the beam capacity exceeds the connection capacity, then deformations only take place in the connection itself.

Table 5. Yielding Rotation, Maximum Rotation Capacity, and Ductility of Studied Specimens.

Specimen	φ_y (rad)	φ_u (rad)	μ
WUF	0.023	0.176	7.6
SidePlate	0.02	0.19	9.5
I-W	0.018	0.067	3.7
I-WB	0.02	0.168	8.4
ST-WB	0.019	0.17	8.9
TSWA (L = 10 mm)	0.033	0.32	9.6
TSWA (L = 12 mm)	0.025	0.23	9.2
Extended end-plate	0.02	0.184	8.9
Flush end-plate	0.02	0.143	7.1
DWA	0.065	0.36	5.53
TSWA (L = 8 mm)	0.061	0.149	2.44
Web Cleat	0.041	0.156	3.8
Top and Seat angle	0.041	0.246	6
Fine Plate	0.03	0.108	3.6
Average			6.73

5.3. Stiffness Degradation

Steel beam-to-column connections will experience stiffness degradation when subjected to cyclic loading. It has conclusively been shown that the stiffness will decrease by increasing the inter-storey drift angle in steel beam-to-column connections [48]. Meanwhile, there is reliable evidence that the metal material behaves differently when subjected to monotonic and impact loading [49]. However, far too little attention has been paid to stiffness degradation of steel material under strong impact loading. Since the available data solely consider the monotonic loading for simulating sudden column removal, in this study, initial stiffness, $S_{j.ini}$, and secant stiffness, $S_{j.eff}$, were taken into account to evaluate the stiffness degradation. Secant or effective stiffness, $S_{j.eff}$, is defined as the ratio of bending moment before substantial loss of strength to the maximum rotation φ_u, as shown in Figure 16.

$$S_{j.eff} = \frac{M_j}{\varphi_u} \tag{13}$$

Table 6 shows the normalized initial and secant stiffness as well as the connected beam stiffness for all studied specimens before substantial strength degradation. The initial and secant stiffness are normalized with respect to the connected beam stiffness. Table 6 clearly indicates that fully rigid specimens, i.e., SidePlate and I-WB, possess much higher stiffness compared to semi-rigid and flexible specimens. It is also evident that although the initial stiffness in flexible connections is negligible, they can maintain the stiffness where the secant stiffness is almost three times higher than the initial stiffness (consider the web cleat and fine plate specimens).

Table 6. Stiffness Properties of the Studied Specimens.

Specimen	$\frac{EI_b}{L_b}$ $(\frac{KN\cdot m}{rad})$	Normalized $S_{j.ini}$	Normalized $S_{j.eff}$
WUF	1295.3	3.02	0.35
SidePlate	20.2	17.3	4.95
I-W	5457.6	1.37	0.36
I-WB	5457.6	2.02	0.19
ST-WB	5457.6	2.55	0.24
TSWA (L = 10 mm)	1295.3	0.93	0.26
TSWA (L = 12 mm)	4420	0.35	0.16
Extended end-plate	4420	1.08	0.09
Flush end-plate	4420	0.79	0.27
DWA	1295.3	0.12	0.06
TSWA (L = 8 mm)	6787.7	0.07	0.06
Web Cleat	6787.7	0.019	0.06
Top and Seat angle	6787.7	0.09	0.02
Fine Plate	6787.7	0.014	0.06

The normalized stiffness degradation versus inter-storey drift angle plots for all specimens are shown in Figures 17–19. Figure 17 generally indicates that by increasing inter-storey drift angles, the normalized stiffness decreases, although the SidePlate connection has considerably higher initial stiffness as well as lower degradation slope. This stiffer behavior is attributed to the distribution of the hinge formation mechanism due to the presence of two side plates wrapping around the shear panel joint region, resulting in increased stiffness of the subassembly. Generally, in this category, flexural action controls the stiffness of the specimens in the early stage of the response.

Figure 18 shows that semi-rigid connections have an irregular pattern for stiffness degradation, where elementary step stiffness experiences degradation, whereas at higher inter-storey drift angle, the stiffness remains constant or even increases before the failure of the specimen. This irregular pattern can be explained by desirable features of extremely high ductility (rotational capacity) and developing catenary action.

Figure 19 indicates that although the flexible connections have considerably lower stiffness compared to fully rigid and flexible connections, they can develop the initial stiffness as the inter-storey drift angle increases. On average, the initial stiffness increases up to 100 percent when the inter-storey drift angles reach around 0.1 rad. This behavior can be explained by the connection geometry that allows rotation at preliminary steps, whereas the tensile capacity of connections' components, i.e., web cleat, bolts, etc., along with stiffness hardening at higher drift angles, resists against rotation and subsequently develops the stiffness.

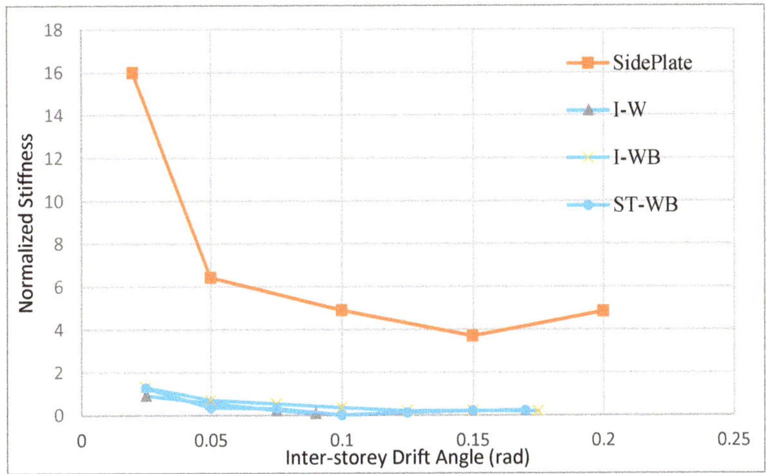

Figure 17. Stiffness degradation versus inter-storey drift angle for fully rigid connections.

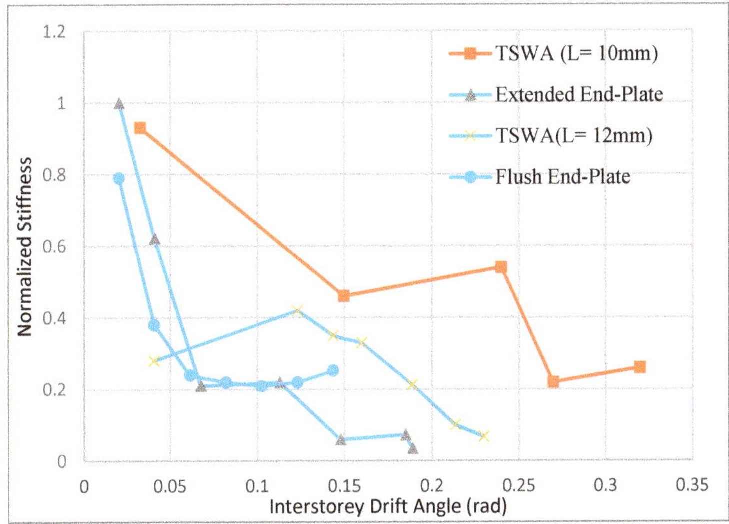

Figure 18. Stiffness degradation versus inter-storey drift angle for semi-rigid connections.

Stiffness degradation is an important aspect of seismic design since a large deformation leads to P-Delta effects, eventually destabilizing the structure. Several codes and regulations prescribe recommendations for stiffness and strength degradation as a measure in seismic design. For instance, the AISC seismic provisions recommend that the connection must sustain an inter-storey drift angle of at least 0.04 rad, while the flexural resistance at the column face must be equal to at least 0.80 M_p of the connected beam. Overall, according to the results of this study, it is recommended that the stiffness at an inter-storey drift angle of 0.05 rad should be larger than 75% of the initial stiffness to develop a reliable catenary mechanism.

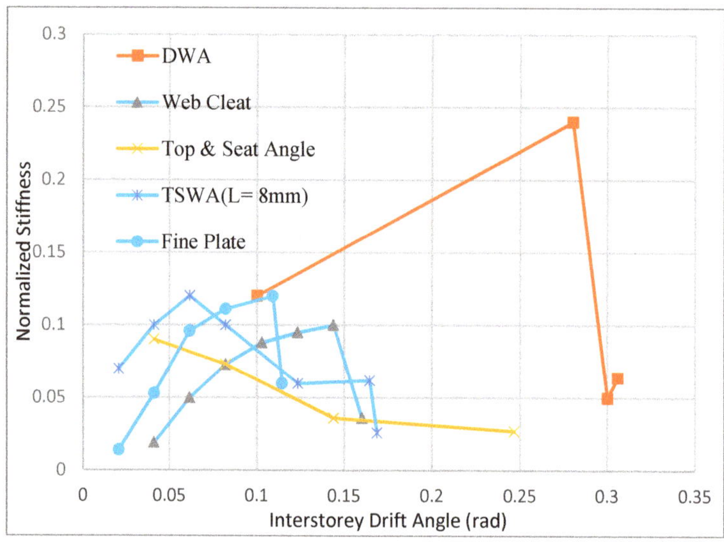

Figure 19. Stiffness degradation versus inter-storey drift angle for flexible connections.

5.4. Evaluation of Catenary and Flexural Mechanisms Under Different R_i

This section investigates the influence of different R_i on flexural and catenary mechanisms. As discussed in Section 5.1, under the middle column removal condition, the vertical resistance of the beam-to-column connection is controlled by the contribution of flexural and catenary mechanisms. Figure 19 illustrates the gravity resistance development for the beam-column assembly of two specimens, having welded unreinforced flange-bolted web (WUF-BW) connections with the same beam section but different R_i, as investigated by Li et al. [25]. The flexural (f_f) and the catenary (f_c) mechanisms, recognized as two components of the gravity resistance, are separately plotted in Figure 20a respectively, and subsequently, their resultant is plotted in Figure 20b. Three distinctive phases are recognized, as shown in the line graphs, introduced by the flexure action-dominated phase I, the combination of flexure-catenary mechanism phase II, and finally, the catenary-dominated mechanism phase III. These three phases are normally separated from each other by the formation of a plastic hinge followed by an initial fracture in the connection's components.

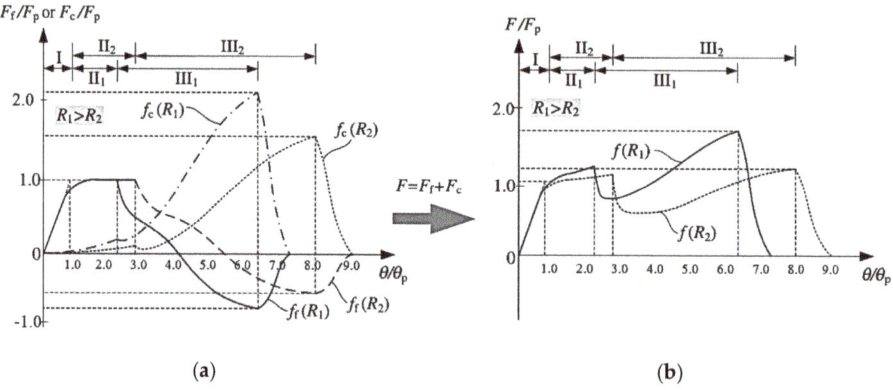

(a)

(b)

Figure 20. Gravity resistance development for beam-to-column assembly: (**a**) flexural and catenary mechanism, and (**b**) overall gravity resistance.

It is evident that a specimen with larger R_i (R_1) can provide a higher vertical resistance as a result of higher catenary mechanism response. However, the specimen with lower R_i (R_2) results in a larger chord rotation ratio, leading to a more ductile response.

6. Summary and Conclusions

This paper presented the descriptions and experimental results of available full-scale double-span systems subjected to the middle column loss scenario. Several parameters and features including beam span-to-depth ratio, catenary mechanism, stiffness, and ductility have been investigated for fully rigid, semi-rigid, and flexible connections. The following conclusions can be drawn:

I. After middle column removal at the preliminary phases, the behavior of the beam is controlled by flexural resistance, and the tensile force is almost zero, recognized as a flexure action-dominated phase. With increased downward displacement, the axial tension also increases in the beams, developing a catenary mechanism recognized as a catenary-dominated mechanism phase. The results of this research show that the magnitude of axial force in the flexible connections, i.e., top and seat angle, is significantly small compared to fully rigid connections. This phenomenon can be justified by the failure mechanism that develops in the connection's components rather than the connected beam, preventing the catenary mechanism development.

II. The maximum rotation capacity versus connection depth for almost all beam-to-column connection categories significantly surpasses the DoD's recommended acceptance criterion. The suggested acceptance criteria are on the conservative side as it only considers pure flexural resistance. Therefore, connection depth alone is not a reliable indicator to predict the rotational capacity of beam-to-column connections.

III. The stiffness in fully rigid and semi-rigid connections generally experiences a decrease by increasing inter-storey drift angle. On the other hand, the flexible connections have a potential to develop the initial stiffness as the inter-storey drift angle increases. Such behavior can be explained by the geometry of these types of connections that allows rotation at preliminary steps, while the stiffness can be developed at higher drift angles depending on tensile capacity of connections' components and stiffness hardening.

Author Contributions: Original draft preparation, numerical models, validation, and methodology: I.F.; supervision, review, and editing: M.H.B. All authors have read and agreed to the published version of the manuscript.

Funding: This work was supported by funding from the Federal State Autonomous Educational Institution of Higher Education, South Ural State University (National Research University).

Conflicts of Interest: The authors declare no conflict of interest.

References

1. Pantidis, P.; Cao, L.; Gerasimidis, S. Partial Damage Distribution and Progressive Collapse of Buildings. In *Structures Congress 2020*; American Society of Civil Engineers: Reston, VA, USA, 2020.
2. Brunesi, E.; Nascimbene, R.; Parisi, F.; Augenti, N. Progressive collapse fragility of reinforced concrete framed structures through incremental dynamic analysis. *Eng. Struct.* **2015**, *104*, 65–79. [CrossRef]
3. Masoero, E.; Wittel, F.K.; Herrmann, H.J.; Chiaia, B.M. Progressive collapse mechanisms of brittle and ductile framed structures. *J. Eng. Mech.* **2010**, *136*, 987–995. [CrossRef]
4. Agarwal, A.; Varma, A.H. Fire induced progressive collapse of steel building structures: The role of interior gravity columns. *Eng. Struct.* **2014**, *58*, 129–140. [CrossRef]
5. Kim, J.; Park, J. Design of steel moment frames considering progressive collapse. *Steel Compos. Struct.* **2008**, *8*, 85–98. [CrossRef]
6. Cassianola, D.; D'Aniello, M.; Rebelo, C.; Landolfo, R.; da Silva, L.S. Influence of seismic design rules on the robustness of steel moment resisting frames. *Steel Compos. Struct.* **2016**, *21*, 479–500. [CrossRef]
7. Zhu, F.Y.; Chen, C.H.; Yao, Y.; Keer, L.M.; Huang, Y. Dynamic increase factor for progressive collapse analysis of semi-rigid steel frames. *Steel Compos. Struct.* **2018**, *28*, 209–221.

8. Byfield, M.; Mudalige, W.; Morison, C.; Stoddart, E. A review of progressive collapse research and regulations. *Proc. ICE Struct. Build.* **2014**, *167*, 447–456. [CrossRef]

9. Gsa, U. *Progressive Collapse Analysis and Design Guidelines for New Federal office Buildings and Major Modernization Projects*; The US General Services Administration: Washington, DC, USA, 2003.

10. Department of Defense. *Design of Buildings to Resist Progressive Collapse*; Unified Facilities Criteria (UFC)-DoD. 2005. Available online: https://www.wbdg.org/FFC/DOD/UFC/ARCHIVES/ufc_4_023_03_2009_c2.pdf (accessed on 18 July 2020).

11. Wang, F.; Yang, J.; Pan, Z. Progressive collapse behaviour of steel framed substructures with various beam-column connections. *Eng. Fail. Anal.* **2020**, *109*, 104399. [CrossRef]

12. Wang, H.; Tan, K.H.; Yang, B. Experimental Tests of Steel Frames with Different Beam–Column Connections under Falling Debris Impact. *J. Struct. Eng.* **2020**, *146*, 04019183. [CrossRef]

13. Alrubaidi, M.; Elsanadedy, H.; Abbas, H.; Almusallam, T.; Al-Salloum, Y. Investigation of different steel intermediate moment frame connections under column-loss scenario. *Thin Walled Struct.* **2020**, *154*, 106875. [CrossRef]

14. Guo, L.; Gao, S.; Fu, F. Structural performance of semi-rigid composite frame under column loss. *Eng. Struct.* **2015**, *95*, 112–126. [CrossRef]

15. Zhang, J.; Jiang, J.; Xu, S.; Wang, Z. An investigation of the effect of semi-rigid connections on sudden column removal in steel frames. In *Structures*; Elsevier: Amsterdam, The Netherlands, 2018.

16. Xu, G.; Ellingwood, B.R. Disproportionate collapse performance of partially restrained steel frames with bolted T-stub connections. *Eng. Struct.* **2011**, *33*, 32–43. [CrossRef]

17. Dinu, F.; Marginean, I.; Dubina, D. Experimental testing and numerical modelling of steel moment-frame connections under column loss. *Eng. Struct.* **2017**, *151*, 861–878. [CrossRef]

18. Zahrai, S.M.; Ezoddin, A. Cap truss and steel strut to resist progressive collapse in RC frame structures. *Steel Compos. Struct.* **2018**, *26*, 635–647.

19. Liu, C.; Tan, K.H.; Fung, T.C. Component-based steel beam–column connections modelling for dynamic progressive collapse analysis. *J. Constr. Steel Res.* **2015**, *107*, 24–36. [CrossRef]

20. Tay, C.G.; Koh, C.G.; Liew, J. Efficient progressive collapse analysis for robustness evaluation of buildings experiencing column removal. *J. Constr. Steel Res.* **2016**, *122*, 395–408. [CrossRef]

21. Qin, X.; Wang, W.; Chen, Y.; Bao, Y. Experimental study of through diaphragm connection types under a column removal scenario. *J. Constr. Steel Res.* **2015**, *112*, 293–304. [CrossRef]

22. Faridmehr, I.; Osman, M.H.; Nejad, A.F.; Tahir, M.M.; Azimi, M. Seismic and Progressive Collapse Assessment of New Proposed Steel Connection. *Adv. Struct. Eng.* **2015**, *18*, 439–452. [CrossRef]

23. Li, L.; Wang, W.; Chen, Y.; Teh, L.H. A basis for comparing progressive collapse resistance of moment frames and connections. *J. Constr. Steel Res.* **2017**, *139*, 1–5. [CrossRef]

24. Stylianidis, P.; Nethercot, D.; Izzuddin, B.; Elghazouli, A.Y. Study of the mechanics of progressive collapse with simplified beam models. *Eng. Struct.* **2016**, *117*, 287–304. [CrossRef]

25. Li, L.; Wang, W.; Teh, L.H.; Chen, Y. Effects of span-to-depth ratios on moment connection damage evolution under catenary action. *J. Constr. Steel Res.* **2017**, *139*, 18–29. [CrossRef]

26. Rezvani, F.H.; Yousefi, A.M.; Ronagh, H.R. Effect of span length on progressive collapse behaviour of steel moment resisting frames. In *Structures*; Elsevier: Amsterdam, The Netherlands, 2015.

27. Sun, R.R.; Burgess, I.W.; Huang, Z.; Dong, G. Progressive failure modelling and ductility demand of steel beam-to-column connections in fire. *Eng. Struct.* **2015**, *89*, 66–78. [CrossRef]

28. Kuhlmann, U.; Rölle, L.; Izzuddin, B.A.; Pereira, M.F. Resistance and Response of Steel and Steel–Concrete Composite Structures in Progressive Collapse Assessment. *Struct. Eng. Int.* **2012**, *22*, 86–92. [CrossRef]

29. Liu, Y.; Málaga-Chuquitaype, C.; Elghazouli, A.Y. Behaviour of open beam-to-tubular column angle connections under combined loading conditions. *Steel Compos. Struct.* **2014**, *16*, 157–185. [CrossRef]

30. Goldaran, R.; Turer, A. Application of acoustic emission for damage classification and assessment of corrosion in pre-stressed concrete pipes. *Measurement* **2020**, *160*, 107855. [CrossRef]

31. Hadidi, A.; Jasour, R.; Rafiee, A. On the progressive collapse resistant optimal seismic design of steel frames. *Struct. Eng. Mech.* **2016**, *60*, 761–779. [CrossRef]

32. Kordbagh, B.; Mohammadi, M. Influence of seismicity level and height of the building on progressive collapse resistance of steel frames. *Struct. Des. Tall Spec. Build.* **2017**, *26*, e1305. [CrossRef]

33. Chen, J.; Shu, W.; Huang, H. Rate-Dependent Progressive Collapse Resistance of Beam-to-Column Connections with Different Seismic Details. *J. Perform. Constr. Facil.* **2017**, *31*, 04016086. [CrossRef]

34. Yang, B.; Tan, K.H. Robustness of Bolted-Angle Connections against Progressive Collapse: Experimental Tests of Beam-Column Joints and Development of Component-Based Models. *J. Struct. Eng.* **2013**, *139*, 1498–1514. [CrossRef]

35. Jamshidi, A.; Koduru, S.; Driver, R.G. Reliability Analysis of Shear Tab Connections under Progressive Collapse Scenario. In *Structures Congress 2014*; ASCE: Reston, VA, USA, 2014.

36. Qin, C.B.; Chian, S.C.; Yang, X.L. 3D Limit Analysis of Progressive Collapse in Partly Weathered Hoek–Brown Rock Banks. *Int. J. Geomech.* **2017**, *17*, 04017011. [CrossRef]

37. Oosterhof, S.A.; Driver, R.G. Shear connection modelling for column removal analysis. *J. Constr. Steel Res.* **2016**, *117*, 227–242. [CrossRef]

38. Shen, J.; Astaneh-Asl, A. Hysteresis model of bolted-angle connections. *J. Constr. Steel Res.* **2000**, *54*, 317–343. [CrossRef]

39. Stylianidis, P.M.; Nethercot, D.A.; Izzuddin, B.A.; Elghazouli, A.Y. Progressive collapse: Failure criteria used in engineering analysis. In *Structures Congress 2009: Don't Mess with Structural Engineers: Expanding Our Role*; ASCE: Reston, VA, USA, 2009.

40. *Seismic Provisions for Structural Steel Buildings*; American Institute of Steel Construction: Chicago, IL, USA, 2010.

41. Zhong, W.; Meng, B.; Hao, J. Performance of different stiffness connections against progressive collapse. *J. Constr. Steel Res.* **2017**, *135*, 162–175. [CrossRef]

42. Faridmehr, I.; Osman, M.H.; Tahir, M.B.M.; Nejad, A.F.; Hodjati, R. Seismic and progressive collapse assessment of SidePlate moment connection system. *Struct. Eng. Mech.* **2015**, *54*, 35–54. [CrossRef]

43. Wang, W.; Fang, C.; Qin, X.; Chen, Y.; Li, L. Performance of practical beam-to-SHS column connections against progressive collapse. *Eng. Struct.* **2016**, *106*, 332–347. [CrossRef]

44. Yang, B.; Tan, K.H. Experimental tests of different types of bolted steel beam–column joints under a central-column-removal scenario. *Eng. Struct.* **2013**, *54*, 112–130. [CrossRef]

45. Mckay, A.; Gomez, M.; Marchand, K. *Non-Linear Dynamic Alternate Path Analysis for Progressive Collapse: Detailed Procedures Using UFC 4-023-03*; Protection Engineering Consultants: Castle Hills, TX, USA, 2010; (Revised July 2009).

46. Pekelnicky, R.; Engineers, S.D.; Poland, S.C.; Engineers, N.D. Seismic Evaluation and Retrofit Rehabilitation of Existing Buildings. In Proceedings of the SEAOC, Santa Fe, NM, USA, 12–15 September 2012.

47. Daneshvar, H.; Driver, R.G. Behaviour of shear tab connections in column removal scenario. *J. Constr. Steel Res.* **2017**, *138*, 580–593. [CrossRef]

48. Liu, X.C.; Cui, F.Y.; Zhan, X.X.; Yu, C.; Jiang, Z.Q. Seismic performance of bolted connection of H-beam to HSS-column with web end-plate. *J. Constr. Steel Res.* **2019**, *156*, 167–181. [CrossRef]

49. Kermajani, M.; Ghaini, F.M.; Miresmaeili, R.; AghaKouchak, A.; Shadmand, M. Effect of weld metal toughness on fracture behavior under ultra-low cycle fatigue loading (earthquake). *Mater. Sci. Eng. A* **2016**, *668*, 30–37. [CrossRef]

Article

Prediction of Damage Level of Slab-Column Joints under Blast Load

Kwang Mo Lim [1], Do Guen Yoo [1], Bo Yeon Lee [2] and Joo Ha Lee [1,*]

[1]   Department of Civil and Environmental Engineering, the University of Suwon, Gyeonggi-do 18323, Korea; kwangmolim@suwon.ac.kr (K.M.L.); dgyoo411@suwon.ac.kr (D.G.Y.)

[2]   Department of Architecture, the University of Suwon, Gyeonggi-do 18323, Korea; bylee@suwon.ac.kr

*   Correspondence: leejooha@suwon.ac.kr; Tel.: +82-31-229-2159

Received: 23 July 2020; Accepted: 21 August 2020; Published: 23 August 2020

Abstract: The behavior of a slab-column joint subjected to blast loads was studied by numerical analysis using a general-purpose finite element analysis program, LS-DYNA. Under the explosive load, the joint region known as the stress disturbed zone was defined as a region with a scaled distance of 0.1 m/kg$^{1/3}$ or less through comparison with ConWep's empirical formula. Displacement and support rotation according to Trinitrotoluene (TNT) weight and scaled distance were investigated by dividing in and out of the joint region. In addition, fracture volume was newly proposed as an evaluation factor for blast-resistant performance, and it was confirmed that the degree of damage to a member due to blast loads was well represented by the fracture volume. Finally, a prediction equation for the blast-resistant performance of the slab-column joint was proposed, and the reliability and accuracy of the equation were verified through additional numerical analysis.

Keywords: blast loads; slab-column joints; prediction model; damage level

1. Introduction

Recently, a number of cases have been reported in which structures are seriously damaged by explosion loads caused by terrorism or accidents. It is necessary to protect citizens' property and lives through blast-resistant design of structures against such extreme events.

The explosion load is characterized by very high pressure in a very short time. In addition to the peculiarity of the load, the inhomogeneity of the material of concrete causes difficulties in the blast-resistant design of reinforced concrete (RC) structures [1–3]. For this reason, the blast-resistant design of RC structures has generally been overly conservative, mainly by increasing the thickness of the members or by reinforcing bars excessively. Therefore, reliable blast-resistant performance evaluation and prediction methods are required for the rational blast-resistant design of RC structures.

While many studies have been conducted on the blast-resistant behavior of single members, such as columns and beams, sufficient studies have not been conducted in the case of joints composed of two or more single members [4–6]. However, the joint may cause the collapse of the entire structural system upon its failure, so more attention is required in design [7,8]. Therefore, this study focused on the slab-column joints, which is one of the common types of joints.

Currently, support rotation is the only evaluation factor for the blast-resistant performance of RC members in various design standards, including American Society of Civil Engineers(ASCE) and Department of Defense (DoD) [9,10]. Moreover, no specific guide is provided for the blast-resistant design of joints connecting structural members. According to the previous study, it was confirmed that the support rotation alone was not sufficient to evaluate the blast-resistant performance of the joint [11]. In some cases, even though the support rotation was within the limits of the design criterion, severe fractures in the joint region were observed. As such, the need for additional evaluation factors for blast-resistant performance was confirmed, but no specific solution has been proposed so far. Therefore, this study analyzed the behavior of slab-column joints under blast loads from various viewpoints as well as displacement and support rotation. As a result, it was found that effective fracture volume could be used as a new blast-resistant performance evaluation factor. The effective fracture volume is the total volume of the joint concrete destroyed by the blast load, excluding the concrete cover. In addition, a model for predicting the damage level of the slab-column joints according to the amount and location of explosives was proposed using the effective fracture volume. This could be used not only for the blast-resistant design of new structures but also for determining the level of blast-resistant performance of existing structures.

2. Modeling of Slab-Column Joints

2.1. Analysis Variables

According to the scaled distance, Z (Equation (1)), the behavior of the RC joint due to the blast load was investigated through numerical analysis. That is, the weight of the explosive load, W (kg), and the distance between the explosive material and the joint, R (m), were set as analysis variables.

$$Z = R/W^{1/3} \tag{1}$$

As shown in Table 1, the charge weight of the explosion load was determined to describe the terrorist situation in a general facility [12]. Explosives applied to the analysis were limited to a weight of 30 kg or less. It is an explosive material that can be transported by briefcase or bicycle. In addition, 10 kg of explosives is the size of a vest bomb commonly used by terrorists, which can be defined as the minimum weight that can be used in terrorism [13]. Therefore, in this study, 10 kg, 20 kg, and 30 kg Trinitrotoluene (TNT) weights were set as variables for the purpose of simulating a single terrorist situation where no vehicle or heavy equipment is used. As shown in Table 2, when simulating such a small bombing situation, it had a fairly small Z value compared to a free air burst situation with Z of $0.147 \text{ m/kg}^{1/3}$ to $40 \text{ m/kg}^{1/3}$ in general [14]. In this case, it could be seen how the structure behaved when a blast load was applied near the member surface.

Table 1. The typical weight of explosive materials [12].

Bomblet	Loaded Weight (kg)	Material Type
Small briefcase	2~4	
Large briefcase	4~12	Military and commercial explosives
Suitcase	12~22	(such as Trinitrotoluene (TNT))
Bicycle	30	

Table 2. Analysis variables.

TNT Weight (kg)	Standoff Distance (m) *	Scaled Distance (m/kg$^{1/3}$)
10	0~0.19	0~0.153
20	0~0.25	0~0.160
30	0~0.30	0~0.163

Standoff distance *: the vertical distance from the location of the explosive to the surface of the column or slab.

2.2. Modeling of Slab-Column Joints

Numerical analysis was performed on the slab-column joint, which is the most common type of joint in RC structures. Figure 1 shows the structural details of the slab-column connection. Numerical analysis was performed using LS-DYNA, a general-purpose finite element analysis program whose reliability has been verified through many previous studies [15,16]. The upper and lower surfaces of the column were completely constrained, and four sides of the slab were constrained in the horizontal direction. In this study, Mat_072R3 was selected from the concrete material models provided by the analysis program LS-DYNA. This material model reflects the strain-rate effect and has already been found in several works of literature to be suitable for analyzing concrete structures under high strain-rate [17–19]. However, Mat_072R3 was unable to exhibit local damage caused by explosions, such as crater spalls, which are associated with structural failure and erosion [20]. Therefore, to simulate these characteristics, LS-DYNA's 'Add_Erosion' keyword option was applied to the concrete material model. To model the reinforcing bars, LS-DYNA's Mat_024 was applied, which is defined as an elastic-plastic material with arbitrary stress versus strain curve and an arbitrary strain-rate dependency. The fracture of Mat_024 is based on plastic deformation [19]. Table 3 shows the properties of the concrete and reinforcement used in the analysis.

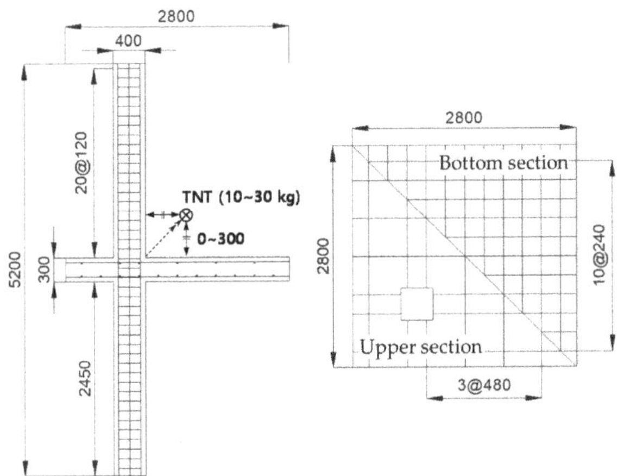

Figure 1. Details of slab-column joints.

Table 3. Material properties for analysis.

Properties	Slab	Column	Reinforcements
Compressive strength (MPa)	30	50	-
Yield strength (MPa)	-	-	475
Tensile strength (MPa)	-	-	751
Density (kg/m^3)	2400	2500	7850
Poisson's ratio	0.18	0.18	0.3

Numerical analysis results may vary depending on the mesh size of the element [21,22]. According to the previous studies, when simulating a structure subjected to an explosive load, a mesh size of 25 to 30 mm led to the analysis results, most similar to the experimental results [23,24]. In this study, before the main analysis was conducted, various mesh sizes were evaluated in terms of accuracy and efficiency of analysis, then it was determined that a mesh size of 20~25 mm was the most reasonable. Therefore, the concrete mesh used a 20 mm cubic, 8-node solid element (C3D8), and the rebar was modeled as a 2-node beam element.

The interaction between concrete and reinforcing bars has a great influence on the behavior of RC structures. In particular, interactions, such as bond-slip, are very difficult to simulate. In RC structures subjected to explosion load, modeling reinforcing bars as solids elements and defining contact conditions can be considered. However, it is not recommended because this method dramatically increases the run time of the analysis and often causes analysis errors. Therefore, a method of tying nodes was recommended to simulate the structure's actual behavior and to provide the simplicity of analysis [18,25]. In this study, the nodes of the reinforcing bar and concrete were connected to each other to provide accurate structural performance.

Blast loads were modeled using the LBE (load blast enhanced) method, where the explosive pressure is applied directly to the element surface. The LBE method has already been verified through many studies [26–29].

Table 4 shows the analysis conditions. Considering the enough converge of kinetic energy, the end time of analysis was set to 2000 ms. Analysis running time was approximately 6 h and 45 min with slight differences for each variable.

Table 4. Analysis conditions.

Analysis Conditions		Value
Number of elements	Solids	392,000 ea (20 × 20 × 20 mm)
	Beam	10,648 ea (20 mm)
Time step		0.1 ms
Analysis of end time		2000 ms
Analysis of running time		6 h 45 m

3. Defining a Joint Region under Blast Load

In the slab-column connection under static load, the section at half the effective depth of the slab from the column surface is considered as the critical section [30]. The joints, which are D-regions (disturbed or discontinued region) with complex strain distributions, may exhibit behavior different from that under a static load in a high strain rate region, such as an explosion load. Therefore, in this study, prior to performing the parametric analysis in Table 2, a preliminary numerical analysis was conducted to define the area of the joint in the slab-column connection under the explosive load.

The U.S. Department of Defense (DoD) suggested shock wave parameters, such as pressure and impulse, according to the scaled distance for single members [10]. According to this, it shows a specific pressure or impulse value for an arbitrary scaled distance regardless of the type of members, such as a column or a slab. Therefore, in this study, through the preliminary numerical analysis of the slab-column connection, the scaled distance, where the analysis results for the slab and the column are similar to each other and at the same time to the value suggested by the DoD, was considered as the point separating the joint region from the single-member region.

Figure 2 shows the pressure and impulse according to the scaled distance. The graphs for slab and column are the results obtained through numerical analysis. Here, the pressure and impulse represent the maximum reflected values that occur in the entire region of the modeled slab and column, according to the location of TNT 30 kg. Figure 2 also includes the values proposed by DoD by using the ConWep model [10]. For both pressure and impulse, the analysis results of the column and the slab differed from each other when the scaled distance was small, but from about 0.1 m/kg$^{1/3}$ or more, the difference was markedly reduced and showed similar values. It should also be noted that from about 0.1 m/kg$^{1/3}$, both the column and the slab showed pressure and impulse similar to those of the ConWep model. As mentioned above, considering that the ConWep values of DoD are the results derived for a single member, the region with a scaled distance of 0.1 m/kg$^{1/3}$ or less from the surface of the slab and column can be considered as the joint region, as shown in Figure 3.

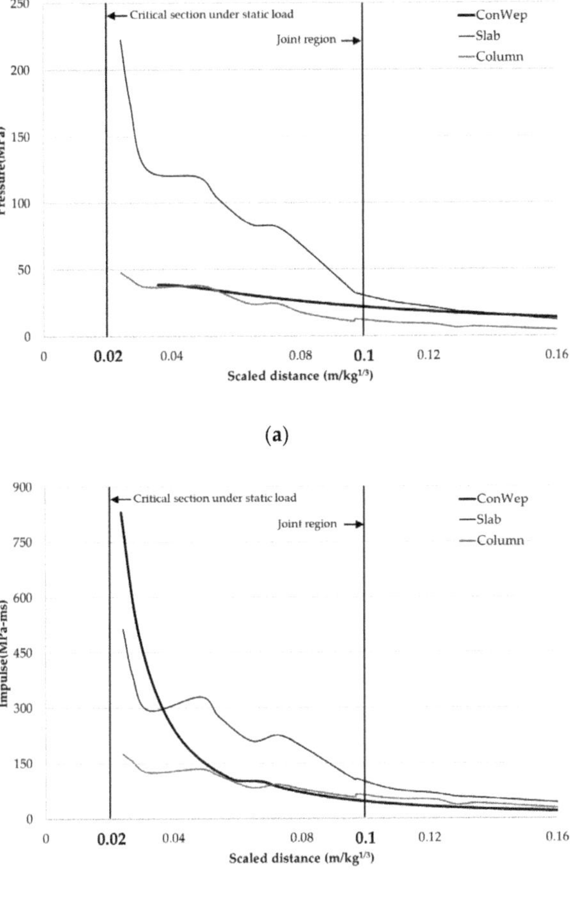

(**a**)

(**b**)

Figure 2. Defining joint region under blast load; (**a**) Pressure; (**b**) Impulse.

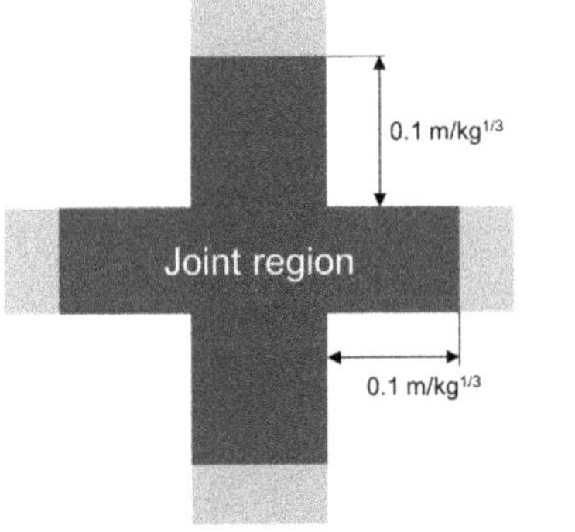

Figure 3. Joint region ($Z = 0.1$ m/kg$^{1/3}$).

4. Numerical Analysis Results and Discussion

4.1. Displacement

Figure 4 shows the results of the maximum displacement of the slab and the column. The displacement for various scaled distances was observed by increasing the standoff distance of each TNT of 10 kg, 20 kg, or 30 kg. In Figure 4, the displacement represents the maximum displacement generated on the central axis of the cross-section of the slab and column. In addition, according to the results of preliminary numerical analysis, it was divided into a joint region and a single member region based on a scaled distance of 0.1 $m/kg^{1/3}$, and the maximum displacement in each region was investigated. Here, the distance from the column or slab surface of TNT 10 kg, 20 kg, and 30 kg corresponding to a scaled distance of 0.1 $m/kg^{1/3}$ was about 215 mm, 271 mm, and 311 mm, respectively. In Figure 4, it was noted that the maximum displacement was not a value at a fixed location but represented the maximum value among displacements that occurred in each region, such as joint and the single-member region, according to different scaled distances.

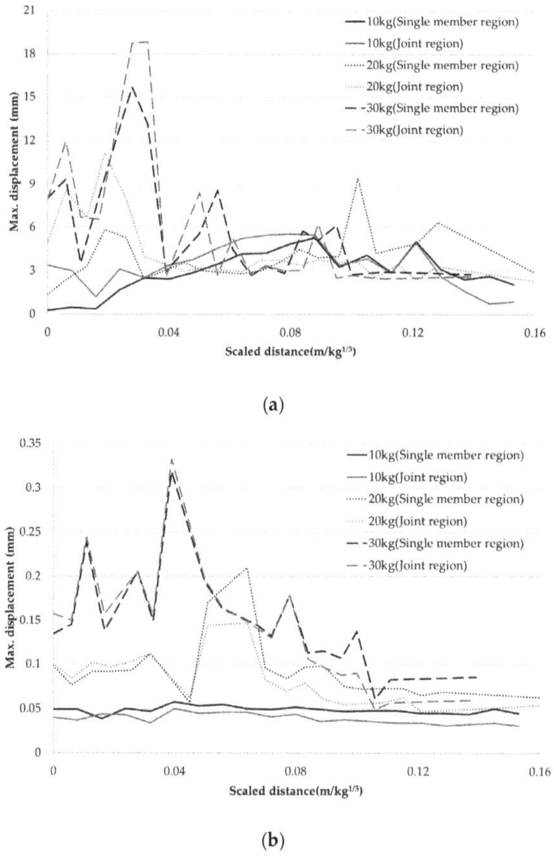

Figure 4. Maximum displacement of slab and column; (**a**) Slab; (**b**) Column.

In all cases, the displacement of the slab was greater than the displacement of the column. This was in line with the results of preliminary numerical analysis, which resulted in greater pressure and impulse in the slab. When the scaled distance was relatively small, the larger the TNT weight, the greater the amount of displacement. However, as the scaled distance became larger, the difference in displacement amount according to the TNT weight was not large. It appeared that the magnitude of shock wave parameters, such as pressure and impulse, transmitted directly to the member decreased significantly as the position of the explosion moved away from the member. When comparing the amount of displacement of the joint region and the single-member region, the slab generally had slightly larger deformation in the joint region, but the difference was not significant. Columns were also observed to have almost the same maximum strain in both regions. However, since there was no clear tendency in the relationship between the scaled distance and the displacement, it was difficult to predict the blast-resistant behavior of the joint based on this relationship.

4.2. Support Rotation

Figure 5 shows the support rotation calculated by the displacement of the slab. According to the criteria, the limit of support rotation to effectively resist the moment is 2 degrees [9,10]. As shown in Figure 5, in all cases, except that the scaled distance between 0.026 m/kg$^{1/3}$ and 0.037 m/kg$^{1/3}$ for TNT 30 kg explosive loads, the support rotation was less than the American Society of Civil Engineers/Structural Engineering Institute (ASCE/SEI) and DoD criteria limit of 2 degrees [9,10]. However, as shown in Figure 6, the top and bottom surfaces of the slab suffered severe fracture damage at all scaled distances for the 30 kg TNT. Similar phenomena were observed in all specimens with 20 kg of TNT. This means that even if the support rotation is smaller than the criteria limit, substantial destruction can occur in the member. In order to properly evaluate the blast-resistant performance of the joint, other evaluation factors besides support rotation are required.

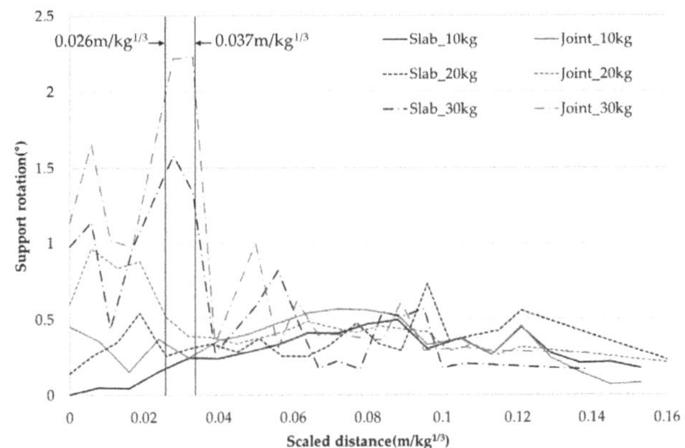

Figure 5. Support rotation of slab.

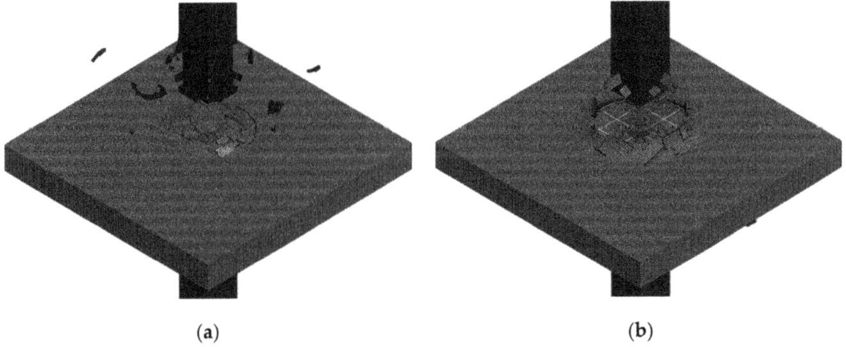

(a)   (b)

Figure 6. Failure shapes (30 kg, Z = 0.067 m/kg$^{1/3}$); (a) Top surface; (b) Bottom surface.

4.3. Fracture Volume

In general, support rotation and displacement are used as criteria for evaluating the blast-resistant performance of RC members, but as shown in the previous analysis, the support rotation and displacement alone were not sufficient to accurately evaluate the performance of the slab-column joint. Moreover, it was difficult to use the support rotation or displacement to predict the blast-resistant performance of the slab-column joint because the support rotation and displacement according to the blast load condition did not have a certain tendency. Therefore, it is necessary to examine additional factors to evaluate the performance of the joint subjected to a blast load.

In this study, fracture volume was analyzed from the analysis results as an additional evaluation factor. Fracture volume was expressed as a percentage of the volume lost due to the explosive load relative to the total volume for the joint region within a scaled distance of 0.1 m/kg$^{1/3}$ in Figure 3. It should be noted that this fracture volume is the effective volume of joint excluding the concrete cover.

Figure 7 shows the effective fracture volume of the joint according to the scaled distance for each TNT weight of 10 kg, 20 kg, and 30 kg. Not surprisingly, as the weight of TNT increased, more damage occurred in the joint region. Interestingly, in all cases, the effective fracture volume decreased almost uniformly as the scaled distance increased.

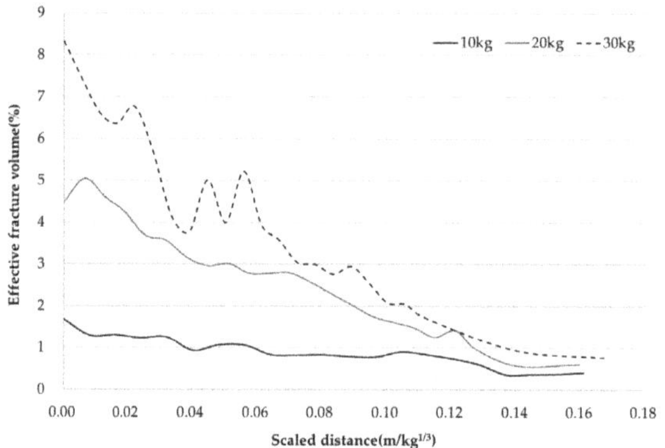

Figure 7. Effective fracture volume.

5. Prediction Model

5.1. The Suggestion of Prediction Model

In this study, the effective fracture volume was used to predict the blast-resistant performance of the slab-column joint. Based on the trends identified in Figure 7, the correlation between the effective fracture volume per unit weight of TNT and the scaled distance was derived. As shown in Figure 8, regardless of the total amount of TNT applied, the effective fracture volume per unit weight of TNT showed almost similar value at any scaled distance and showed a certain tendency to decrease with increasing scaled distance. The trend line equation is shown in Figure 8, and the coefficient of determination (R^2) of the equation for the entire data was 0.870.

As a result, Equation (2) was proposed to predict the damage level of the slab-column joint subjected to blast load through the weight of the explosive material and the standoff distance.

$$\text{Effective fracture volume} = W \times (0.2375 \times 10^{-5.1374Z}) \tag{2}$$

where Effective fracture volume: Effective fracture volume percentage of slab-column joint (%). W: TNT weight (kg). Z: scaled distance (m/kg$^{1/3}$).

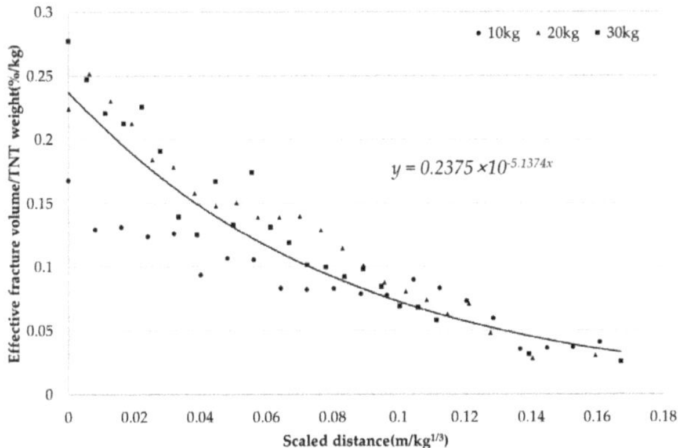

Figure 8. Effective fracture volume per unit weight of TNT according to scaled distance.

5.2. Verification of Prediction Model

In order to verify the prediction model proposed in this study, numerical analysis was additionally performed. As shown in Table 5, numerical analysis results for various concrete strength of slab and column, slab thickness, TNT weights, and scaled distances were compared with predicted values of the proposed equation.

Table 5. Verification analysis variables and results.

Case	Column Strength (MPa)	Slab Strength (MPa)	Slab Thickness (mm)	TNT Weight (kg)	Scaled Distance (m/kg$^{1/3}$)	Effective Fracture Volume (%) Verification Analysis	Prediction Model	Prediction Differences (%)
1				6	0.03	0.73	1.00	0.27
2				12	0.06	1.21	1.40	0.19
3				18	0.09	1.84	1.47	0.37
4	50			24	0.12	1.55	1.38	0.17
5		30	300	24		4.19	4.00	0.19
6				40	0.03	6.39	6.66	0.27
7				50		8.07	8.33	0.26
8	30					1.72		0.32
9	40					1.63		0.23
10	50	20		12	0.06	1.92	1.40	0.52
11	50	40				0.90		0.50
12	50	30	320			1.41		0.01
13	50	30	340			1.43		0.03

Comparing cases 1 to 7 with the same concrete strength and slab thickness as the numerical analysis conditions for deriving the prediction model, the model predictions were in good agreement with the verification numerical analysis results for various TNT weights and scaled distances. It is noteworthy that reliable predictions were shown in all cases of cases 1 to 4 with relatively small effective fracture volume and cases 5 to 7 with relatively large effective fracture volume. Interestingly, although the proposed equation was derived based on a small TNT within 30 kg, the predictions for cases 6 and 7 of TNT 40 kg and 50 kg, respectively, agreed well with the analysis result.

In cases 8 to 13, the effects of variables not included in the prediction equation, such as concrete strength and slab thickness, on the effective fracture volume were examined. As the concrete strength increased, the effective fracture volume decreased. In cases 8, 9, and 2, as the column concrete strength increased to 30 MPa, 40 MPa, and 50 MPa, the effective fracture volume decreased to 1.72%, 1.63%, and 1.21%, respectively. In cases 10, 2, and 11, as the slab concrete strength increased to 20 MPa, 30 MPa, and 40 MPa, the effective fracture volume decreased to 1.92%, 1.21%, and 0.90%, respectively, showing a greater reduction than the column. This was in line with the preliminary analysis results in which the slab showed relatively larger displacement, pressure, and impulse than the column. In cases 12 and 13, which had the same concrete cover thickness and reinforcement details as in case 2, the effect of the slab thickness on the effective fracture volume was investigated. It was observed that the effective fracture volume slightly increased as the slab thickness increased, but the difference in cases 12 and 13 was only 0.02%. It seemed that the effective fracture volume was not significantly affected by column concrete strength and slab thickness. For cases 8 to 13, the proposed equation yielded a constant value of the effective fracture volume due to variables not included in the equation. Although the predictions did not show much difference from the analysis results in the range of variables of concrete strength and slab thickness used in the verification analysis, further research is needed to propose a prediction equation that can take all these variables into account. Figure 9 shows the failure shape for some notable cases.

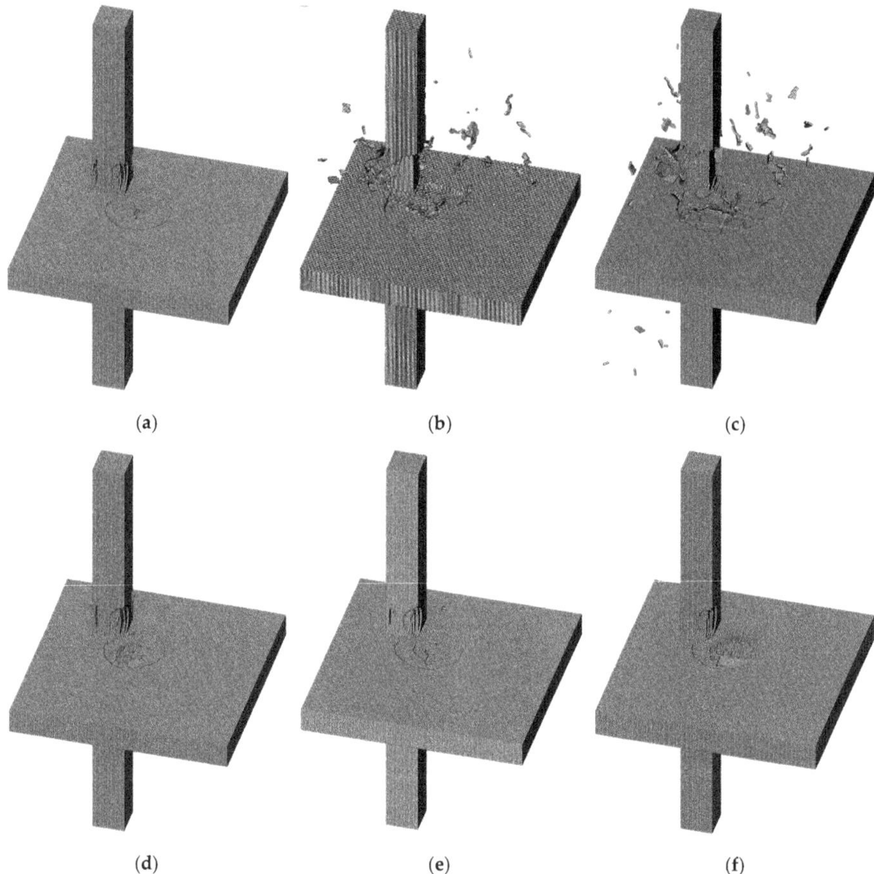

Figure 9. Failure shape for some verification analysis cases; (**a**) Case 3; (**b**) Case 6; (**c**) Case 7; (**d**) Case 8; (**e**) Case 10; (**f**) Case 13.

6. Conclusions

In this study, numerical analysis was performed using LS-DYNA for the behavior of a slab-column joint under an explosive load. The results of this study can be summarized as follows.

(1)   ConWep's empirical values for shock wave parameters of a single member, such as a slab and a column, were compared with the pressure and impulse of a slab-column joint by numerical analysis. As a result, a region with a scaled distance of less than 0.1 $m/kg^{1/3}$ was defined as a joint region.

(2)   The explosion created more pressure and impact on the slab than the column, thereby causing a larger displacement in the slab. In addition, it was observed that the damage of the member decreased sharply as the explosion position moved away from the member.

(3)   Even if the support rotation of the slab after the explosion was less than the limit of 2 degrees, it was observed that serious damage, such as spalling, occurred over a wide range of slab. Therefore, in addition to the support rotation and displacement, which are mainly used to evaluate blast-resistant performance, other evaluation factors are required.

(4)   Effective fracture volume was proposed as an evaluation factor for blast-resistant performance. Effective fracture volume was a good indication of the actual degree of damage to the member depending on the TNT weight and the explosive distance.

(5)   A prediction equation for the damage level of the slab-column joints through the TNT weight and the standoff distance of explosives was proposed. The reliability and accuracy of the proposed equation were verified through additional numerical analysis.

Author Contributions: Conceptualization, performing analysis, and writing—original draft, K.M.L. and J.H.L.; Writing—review and editing, D.G.Y., B.Y.L., and J.H.L.; supervision, J.H.L. All authors have read and agreed to the published version of the manuscript.

Funding: The paper was supported by the research grant of the University of Suwon in 2017.

Acknowledgments: The paper was supported by the research grant of the University of Suwon in 2017.

Conflicts of Interest: The authors declare no conflict of interest.

References

1.   Brun, A.; Batti, A.; Limam, A.; Gravouil, A. Explicit/implicit multi-time step co-computations for blast analyses on a reinforced concrete frame structure. *Finite Elem. Anal. Des.* **2012**, *52*, 41–59. [CrossRef]

2.   Hajek, R.; Fladr, J.; Pachman, J.; Stoller, J.; Foglar, M. An experimental evaluation of the blast resistance of heterogeneous concrete-based composite bridge decks. *Eng. Struct.* **2019**, *179*, 204–210. [CrossRef]

3.   Kumar, V.; Kartik, K.V.; Iqbal, M.A. Experimental and numerical investigation of reinforced concrete slabs under blast loading. *Eng. Struct.* **2020**, *206*, 110–125. [CrossRef]

4.   Abdulsamee, M.H.; Yazan, B.A.T.; Amin, H.A.; George, Z.V. The effect of shape memory alloys on the ductility of exterior reinforced concrete beam-column joints using the damage plasticity model. *Eng. Struct.* **2019**, *200*, 109676.

5.   Cumhur, C.; Ahmet, M.T.; Atakan, M.; Turgay, C.; Guven, K. Experimental behavior and failure of beam-column joints with plain bars, low-strength concrete and different anchorage details. *Eng. Fail. Anal.* **2020**, *109*, 104247.

6.   Yim, H.C.; Krauthammer, T. Load-impulse characterization for steel connections in monolithic reinforced concrete structures. *Int. J. Impact Eng.* **2009**, *36*, 737–745. [CrossRef]

7.   Yoon, Y.S. *Mechanics & Design of Reinforced Concrete*; CIR: Seoul, Korea, 2013.

8.   Park, J.Y. *Modern Protective Structures*; CIR: Seoul, Korea, 2011.

9.   ASCE/SEI. *ASCE/SEI 59-11: Blast Protection of Buildings*; American Society of Civil Engineers, American Society of Civil Engineers/Structural Engineering Institute: Reston, VA, USA, 2011.

10.   Department of Defense. *UFC 3-340-02: Structures to Resist the Effects of Accidental Explosions*; Department of Defense: Arlington, VA, USA, 2008.

11. Lim, K.M.; Shin, H.O.; Kim, D.J.; Yoon, Y.S.; Lee, J.H. Numerical assessment of reinforcing details in beam-column joints on blast resistance. *Int. J. Concr. Struct. Mater.* **2016**, *10*, 87–96. [CrossRef]

12. Yandzio, E.; Gough, M. *Protection of Buildings against Explosions*; The Steel Construction Institute: Ascot, UK, 1999.

13. Homeland Security. *Improvised Explosive Devices*; The National Academies and the Department of Homeland Security: Washington, DC, USA, 2019.

14. Giversen, S. *Blast Testing and Modelling of Composite Structures*; DCAMM Special Report, No. S167; DTU Mechanical Engineering: Kongens Lyngby, Denmark, 2014.

15. Andrew, R.; Nicola, B.; Giuseppe, C.; Stefano, D.M.; Gianluca, I.; Sonia, M.; Elio, S.; Sara, S.; Gabriel, T. Full scale experimental tests and numerical model validation of reinforced concrete slab subjected to direct contact explosion. *Int. J. Impact Eng.* **2019**, *132*, 103309.

16. Azer, M.; Stijn, M.; Bachir, B.; David, L.; John, V. Blast response of retrofitted reinforced concrete hollow core slabs under a close distance explosion. *Eng. Struct.* **2019**, *191*, 447–459.

17. Brannon, R.M.; Leelavanichkul, S. *Survey of Four Damage Models for Concrete*; Sandia National Laboratories: Albuquerque, NM, USA, 2009. SAND2009-5544.

18. Crawford, J.E.; Wu, Y.; Choi, H.J.; Magallanes, J.M.; Lan, S. *Use and Validation of the Release III K&C Concrete Material Model in LS-DYNA*; Karagozian & Case technical report, TR-11-36.5; Karagozian & Case, Inc.: Glendale, CA, USA, 2012.

19. LSTC (Livermore Software Technology Corporation). *LS-DYNA Keyword User's Manual Volume II Material Models*; Livermore Software Technology Corporation: Livermore, CA, USA, 2013.

20. Ling, L. Local Damages and Blast Resistance of RC Slabs Subjected to Contact Detonation. Master's Thesis, Korea University, Seoul, Korea, 2013.

21. Gonzalez, H.A.; Zapatero, J. Influence of minimum element size to determine crack closure stress by the finite element method. *Eng. Fract. Mech.* **2005**, *72*, 337–355. [CrossRef]

22. Krauthammer, T. Mesh, gravity and load effects on finite element simulations of blast loaded reinforced concrete structures. *Comput. Struct.* **1997**, *63*, 1113–1120. [CrossRef]

23. Foglar, M.; Kovar, M. Conclusions for experimental testing of blast resistance of FRC and RC bridge decks. *Int. J. Impact Eng.* **2013**, *59*, 18–28. [CrossRef]

24. Thiagarajan, G.; Kadambi, A.V.; Robert, S.; Johnson, C.F. Experimental and finite element analysis of doubly reinforced concrete slabs subjected to blast loads. *Int. J. Impact Eng.* **2015**, *75*, 162–173. [CrossRef]

25. Shi, Y.; Stewart, M.G. Damage and risk assessment for reinforced concrete wall pannels subjected to explosive blast loading. *Int. J. Impact Eng.* **2015**, *85*, 5–19. [CrossRef]

26. Bae, D.M.; Zakki, A.F. Comparisons of multi material ALE and single material ALE in LS-DYNA for estimation of acceleration response of free-fall lifeboat. *J. Soc. Nav. Archit. Korea* **2011**, *48*, 552–559. [CrossRef]

27. Jang, I.H. Sloshing Response Analysis of LNG Carrier Tank Using Fluid Structure Interaction Analysis Technique of LS-DYNA3D. Master's Thesis, Korea Maritime and Ocean University, Busan, Korea, 2007.

28. Zahra, S.T.; Jeffery, S.V. A comparison between three different blast methods in LS-DYNA: LBE, MM-ALE, Coupling of LBE and MM-ALE. In Proceedings of the 12th International LS-DYNA User Conference, Dearborn, MI, USA, 3–5 June 2012.

29. Huh, Y.C.; Chung, T.Y.; Kim, K.C.; Jung, H.J.; Choi, H.H. A study on the modeling techniques of air blast load using LS-DYNA. In Proceedings of the Korean Society for Noise and Vibration Engineering Conference, Jeju Island, Daejeon, Korea, 26–28 April 2012; pp. 375–376.

30. ACI-ASCE Committee 352. *ACI352.1R-11: Guide for Design of Slab-Column Connections in Monolithic Concrete Structures*; American Concrete Institute: Farmington Hills, MI, USA, 2012.

MDPI

St. Alban-Anlage 66

4052 Basel

Switzerland

Tel. +41 61 683 77 34

Fax +41 61 302 89 18

www.mdpi.com

Applied Sciences Editorial Office

E-mail: applsci@mdpi.com

www.mdpi.com/journal/applsci